Lecture Notes in Computer Scienc

Edited by G. Goos and J. Hartmanis

R. G. Herrtwich (Ed.)

Network and Operating System Support for Digital Audio and Video

Second International Workshop
Heidelberg, Germany, November 18-19, 1991
Proceedings

Springer-Verlag

Berlin Heidelberg New York
London Paris Tokyo
Hong Kong Barcelona
Budapest

Series Editors

Gerhard Goos
Universität Karlsruhe
Postfach 69 80
Vincenz-Priessnitz-Straße 1
W-7500 Karlsruhe, FRG

Juris Hartmanis
Department of Computer Science
Cornell University
5149 Upson Hall
Ithaca, NY 14853, USA

Volume Editor

Ralf Guido Herrtwich
IBM European Networking Center
Tiergartenstr. 8, W-6900 Heidelberg 1, FRG

CR Subject Classification (1991): H.5.1, C.2, D.4, H.4.3

ISBN 3-540-55639-7 Springer-Verlag Berlin Heidelberg New York
ISBN 0-387-55639-7 Springer-Verlag New York Berlin Heidelberg

Typesetting: Camera ready by author/editor
Printing and binding: Druckhaus Beltz, Hemsbach/Bergstr.
45/3140-543210 - Printed on acid-free paper

Preface

On November 18 and 19, 1991, 60 researchers participated in the Second International Workshop on Network and Operating System Support for Digital Audio and Video, held in cooperation with ACM SIGCOMM and SIGOPS at the IBM European Networking Center in Heidelberg, Germany. This workshop was the second in a series started in November 1990 at the International Computer Science Institute (ICSI) and the University of California at Berkeley. It brought together again researchers in networks and operating systems to discuss the needs of multimedia applications and how they may be satisfied.

The trend towards powerful workstations and high-speed networks has enabled applications to communicate and manipulate digital audio and video (continuous media). These new media differ from traditional discrete media such as text and graphics in that they have stringent delay and bandwidth requirements. The mechanisms used to transport ordinary data over networks and many of today's communication protocols are insufficient to communicate continuous media. Special operating system support must also be provided to meet the requirements of both discrete and continuous media in future multimedia applications.

The workshop dealt with virtually all system aspects of multimedia support. Among the topics discussed were

- Communication Systems for Multimedia
- Multimedia Endsystem Architectures
- Operating System Extensions for Multimedia
- UNIX and Multimedia
- Resource Management (Workstation and Network)
- Delay Management
- Performance of Multimedia Systems
- Program Models and Structuring Principles
- Application Program Interfaces for Multimedia
- Synchronization Between Media Streams
- Multimedia Applications

Prospective workshop participants had to submit a position paper. Each program committee member refereed all submissions. From the position papers received, the program committee chose the participants of the workshop and determined 34 papers for presentation.Those authors selected to give presentations were asked to submit full versions of their papers for inclusion in these workshop proceedings. These proceedings, however, contain not only the workshop papers, but also summaries of the workshop sessions compiled by the session chairmen.

Many people have contributed to making this workshop a success. First, I would like to thank all the workshop participants for sharing their ideas with us during the two days in Heidelberg and for taking part in the very lively discussions. All program committee members and session chairmen put a lot of effort in selecting papers, presenting, and summarizing them. Our secretaries assisted me with the local arrangements for the workshop in the best possible way.

Finally, I would like to express my personal gratitude to my colleagues Hermann Schmutz and Ralf Steinmetz who established multimedia systems as a research topic at the IBM European Networking Center well before I joined them. Without their work, the ENC would hardly have been chosen to host the workshop. In this context, I would also like to thank Greg Flurry and Wayne Blackard from IBM Austin for their continued support of our multimedia activities.

Ralf Guido Herrtwich

Program Committee

Ferrari, Domenico; University of California at Berkeley and ICSI, USA
Herrtwich, Ralf Guido; IBM European Networking Center, Heidelberg, Germany
    (Chair)
Hopper, Andy; Olivetti Research Center and University of Cambridge, UK
Katseff, Howard; AT&T Bell Laboratories, Holmdel, USA
Naffah, Najah; Bull, Massy, France
Northcutt, Duane; SUN Microsystems Laboratories, Mountain View, USA
Popescu-Zeletin, Radu; Technical University of Berlin and GMD-FOKUS, Germany
    (Co-Chair)
Rosenberg, Jonathan; Bellcore, Morristown, USA
Shepherd, Doug; Lancaster University, UK
Steinmetz, Ralf; IBM European Networking Center, Heidelberg, Germany
Tokuda, Hideyuki; Carnegie-Mellon University, Pittsburgh, USA
Topolcic, Claudio; Corporation for National Research Initiatives, Reston, USA

Workshop Participants

Almeida, Jose Nuno; INESC, Porto, Portugal
Anderson, David; University of California at Berkeley, USA
Banerjea, Anindo; University of California at Berkeley, USA
Biersack, Ernst; Bellcore, Morristown, USA
Binding, Carl; IBM Research Division, Zurich, Switzerland
Blakowski, Gerold; University of Karlsruhe, Germany
Bulterman, Dick C. A.; CWI, Amsterdam, The Netherlands
Calnan, Roger; SUN Microsystems Laboratories, Mountain View, USA
Cordes, Ralf; Telenorma, Frankfurt, Germany
Cousin, Bernard; ENSERB-LaBRI, Talence, France
Crumley, Paul; Carnegie-Mellon University, Pittsburgh, USA
Delgrossi, Luca; IBM European Networking Center, Heidelberg, Germany
Dupuy, Sylvie; University of Paris 6, France
Dykeman, Doug; IBM Research Division, Zurich, Switzerland
Effelsberg, Wolfgang; University of Mannheim, Germany
Ferrari, Domenico; University of California at Berkeley and ICSI, USA
Field, Brian; University of Pittsburgh, USA
Fritzsche, J. Christian; Johann-Wolfgang-Goethe University of Frankfurt, Germany
García, Francisco; Lancaster University, UK
Gibbs, Simon; University of Geneva, Switzerland
Goldberg, Steve; IBM Personal Systems Division, Palo Alto, USA
Griefer, Allan; IBM Research Division, San Jose, USA
Gusella, Riccardo; Hewlett-Packard Laboratories, Palo Alto, USA
Gutfreund, Yechezkal Shimon; GTE Laboratories, Waltham, USA

Hanko, Jim; SUN Microsystems Laboratories, Mountain View, USA
Hehmann, Dietmar; IBM European Networking Center, Heidelberg, Germany
Henkelmann, Chris; Saint Mary's University, Halifax, Canada
Herrtwich, Ralf Guido; IBM European Networking Center, Heidelberg, Germany
Hinz, Ralf; Daimler-Benz AG, Ulm, Germany
Hoepner, Petra; GMD-FOKUS, Berlin, Germany
Hopper, Andy; Olivetti Research Center and University of Cambridge, UK
Horn, Francois; CNET, Issy les Moulineaux, France
Ingold, Daniel; ETH, Zurich, Switzerland
Jeffay, Kevin; University of North Carolina at Chapel Hill, USA
Kalfa, Winfried; Technical University of Dresden, Germany
Lamparter, Bernd; University of Mannheim, Germany
Leydekkers, Peter; PTT Research, Groningen, The Netherlands
Little, Thomas D. C.; Boston University, USA
Mackert, Lothar; IBM European Networking Center, Heidelberg, Germany
Maresca, Massimo; University of Genoa, Italy
Meißner, Klaus; Philips-Kommunikations-Industrie, Siegen, Germany
Meyer, Thomas; University of Mannheim, Germany
Nagarajan, Ramesh; University of Massachusetts at Amherst, USA
Northcutt, Duane; SUN Microsystems Laboratories, Mountain View, USA
Pink, Stephen; Swedish Institute of Computer Science, Stockholm, Sweden
Popescu-Zeletin, Radu; Technical University of Berlin and GMD-FOKUS, Germany
Rangan, Venkat; University of California at San Diego, USA
Rosenberg, Jonathan; Bellcore, Morristown, USA
Ruston, Lillian; Bellcore, Morristown, USA
Scherer, Manfred; Siemens AG, Munich, Germany
Schmutz, Hermann; IBM European Networking Center, Heidelberg, Germany
Shepherd, Doug; Lancaster University, UK
Steinberg, Daniel; SUN Microsystems Laboratories, Mountain View, USA
Stevens, Scott; Carnegie-Mellon University, Pittsburgh, USA
Stüttgen, Heinrich; IBM European Networking Center, Heidelberg, Germany
Tokuda, Hideyuki; Carnegie-Mellon University, Pittsburgh, USA
Tritsch, Bernhard; Fraunhofer-Gesellschaft, Darmstadt, Germany
Wolf, Lars; IBM European Networking Center, Heidelberg, Germany
Wolfinger, Bernd; University of Hamburg, Germany
Zölzer, Udo; Technical University of Hamburg-Harburg, Germany

Contents

Session I: Real-Time Support

Chair: Ralf Guido Herrtwich, IBM European Networking Center

The first workshop session was dedicated to the issue of real-time support for multimedia applications. Audio and video are commonly referred to as time-dependent continuous media. Their timing dependencies need proper support by the computer system. This calls for the use of classical techniques from real-time computing. Yet, audio and video have slightly different requirements than traditional real-time applications.

During the first talk, Jim Hanko from SUN Microsystems presented a paper co-authored with Eugene Kuerner, Duane Northcutt, and Gerald Wall on "Workstation Support for Time-Critical Applications," introducing the topic. The initial thesis of the talk was that while today's workstations have a great deal of computational power, this power cannot be effectively delivered to support multimedia applications because system resources are not organized and managed in the necessary manner.

Existing workstations provide abundant CPU capacity, but poor I/O support. While in traditional computing the lack of I/O power can be masked by caching, continuous-media applications that do not reuse data cannot benefit from it. Workstations do not dedicate resources to I/O processing as mainframes do. The single CPU has to help out to perform I/O, constituting a performance bottleneck. That the CPU is scheduled based on "fairness" rather than urgency worsens the problem.

The authors point out that existing real-time systems cannot be used to overcome this problem because they rely on deterministic behavior and known load, assumptions not well suited for a workstation environment. Their statement that existing real-time scheduling techniques such as rate-monotonic scheduling are not appropriate for multimedia has, however, generated opposition from the audience.

A new resource management technique, Time-Driven Resource Management (TDRM), is proposed. The technique bases its decisions on the requester's deadline, importance, and expected resource requirements. The technique does not provide sharp guarantees for a certain quality of service, it rather encourages "graceful degradation."

The matter of how a resource management technique should affect system behavior was subject to major discussion. Many reservation schemes proposed (e.g., those from Anderson or Ferrari of the University of California at Berkeley) provide a fixed quality of service and reject new service requests when the provision of the service would endanger previously given guarantees. To a user, this may mean that his request for retrieving a video is turned down. Such a system can be compared to the telephone system: If a busy signal is received, the user will have to wait and dial again.

Adaptive policies provide more flexibility, but cannot guarantee fixed quality levels. It was mainly agreed that an ideal resource management technique should be a combination of both cases. Some fictitious "costs" can be chosen to decide which application can use resources. If conflicts occur, the application (not the system) should decide on how to cope with the reduced bandwidth available.

In the second talk of session, Kevin Jeffay presented joint work with Donald Stone and Donelson Smith at the University of North Carolina at Chapel Hill on "Kernel Support for Live Digital Audio and Video." The main objective of this effort is to provide DVI-based video conferencing across Token Ring. The work is rooted in the design of a new real-time kernel for multimedia applications.

The kernel called YARTOS (yet another real-time operating system) is based on Kevin's real-time producer/consumer paradigm, which allows to reason about message-passing systems based on the rates of task execution. Tasks are deadline-scheduled and all requests for resources (shared software objects) are directed to the kernel to prevent priority inversions. For a given workload, YARTOS tries to guarantee that all tasks complete before their deadlines and that no shared resource is used simultaneously by more than one task.

A schedulability test is made prior to establishing a video conference. Hence, the system follows the traditional approach of fixed quality of service. However, YARTOS provides information about each execution sequence that would lead to a deadline violation. If a user can certify that the sequence found by YARTOS never occurs in practice, then the system will establish the call regardless of the schedulability test.

The system constructed runs on IBM PS/2 machines across Token Ring utilizing the Action Media 750 hardware. It was found that a delay of 6 to 7 frame times occurs for any end-to-end video and audio transmission. This translates into $200-230$ ms delay which was found to be sufficient for the intended application. It is notable that the major portion of the delay occurs in the DVI pipeline of frame processing.

The Token Ring was found to be suitable to transport compressed multimedia data. However, it was found that DVI packets should be sent with a higher priority to minimize ring access time. During the discussion it was argued that such high priorities (or more general: real-time scheduling mechanisms) are more essential to support time-critical applications that a reservation process.

The third and final talk of the real-time session was given by Hideyuki Tokuda of Carnegie-Mellon University. His joint paper with Clifford Mercer entitled "Priority Consistency in Protocol Architectures" continued the discussion of network access – or more general: protocol processing – for continuous-media data. The goal is to prevent priority inversions in the communication subsystem, i.e., situations in which a high-priority activity is delayed by the execution of a lower-priority action. Several techniques for protocol processing are compared and analyzed. This analysis provides users with guidelines on the technique to choose, based on the ratio of protocol processing time to context switch time.

The authors have found that for a protocol processing time to context switch time ratio of 5-to-1 or 10-to-1 preempting protocol processing to avoid priority inversion makes sense. With a ratio of 2-to-1 and 1-to-1 their simulation showed that it is not worth preempting the protocol processing. This means in particular that for these scenarios it may not be worthwhile to generate individual threads for accomplishing protocol processing: The traditional UNIX method of protocol processing at the software interrupt level may suffice and lead to better overall response times.

Concluding the session on real-time support, several workshop participants brought up the issue of whether new system architectures would facilitate the handling of multimedia data in real time. The question was asked whether new switch-based rather than bus-based workstation architectures would not be a better solution to the resource contention problem. It was generally agreed that such new architectures would facilitate the introduction of audio and video in computers, but that there are strong user demands for providing multimedia functions in today's systems.

Workstation Support for Time-Critical Applications

James G. Hanko, Eugene M. Kuerner, J. Duane Northcutt, Gerard A. Wall
Sun Microsystems Laboratories, Inc.

Current workstations have a great deal of computational power. However, this power cannot be effectively delivered to support time-critical applications because the system resources are not organized nor managed in the necessary manner. What is needed is hardware and software platform technology that manages (i.e., acquires, processes, transfers, coordinates, and delivers) these time-critical data streams. This research into properly constructing, organizing, and managing system resources (according to principles of time-driven resource management) will enable workstations to solve many problems currently handled by dedicated-function embedded systems or point-solution add-on devices, while retaining the essential characteristics of the distributed workstation system environment.

Introduction

There are entire classes of problems to which today's workstation cannot be applied. This application area is characterized by the inclusion of information or activities whose value to the user is a function of time. Multimedia audio and video, visualization, virtual reality, transaction processing, and data acquisition are example application classes. Current workstation system architecture lacks support for these classes of applications. When problems such as these are encountered, they tend to be addressed in a specialized, *ad-hoc*, manner. One domain where this is particularly apparent is in the area of multimedia applications.

The workstations of today are not able to effectively deliver their power to time-critical applications. The answer to this problem is not give the users the "bare iron" (as is done with embedded systems and PC's), but rather to change the nature of the workstation to extend support for this new set of requirements. To achieve the desired goals, workstation system architecture must evolve in a number of areas: processor power and organization, memory structure and bandwidth, system interconnection schemes and topology, and I/O capabilities.

Problem Statement

To realize the promise of new application areas, workstations must be able to manipulate time-critical data streams. In the past, the time-critical nature of the data was often ignored or handled by the brute-force application of excess resources. However, new and demanding applications are emerging in which this approach cannot work. One of these application areas is videoconferencing, which deals with multiple, continuous, high-bandwidth data streams that must be mutually synchronized.

Time-critical information represents a fundamentally different form of data than is found in workstations today, and imposes a totally different set of requirements on system hardware and software. There are fundamental limitations in the structure of current workstation hardware and software that prevent them from being effective in dealing with time-critical data. Furthermore, the embedded real-time techniques usually applied to these applications will fail in the workstation environment.

The support of time-critical data does not necessarily conflict with the workstation's traditional requirements, but does impose additional requirements on the system architecture.

Workstation Shortcomings

Current workstations do not provide the proper hardware and software support for applications which require the manipulation of time-critical data streams. The reasons for this are threefold: insufficient resources (both in kind and amount), inappropriate organization of the resources, and inadequate management of the resources.

Workstations have primarily been designed to maximize performance on compute-intensive applications. For example, although today's workstation CPU is faster than traditional mainframe computers, the memory and I/O bandwidth of a workstation is only a small fraction of the mainframe's. This is usually masked by providing the CPU with a large cache and main memory, so that the memory and I/O loading is reduced.

Applications that manipulate continuous, high-bandwidth, time-critical data streams do not have the data reuse property that allows caching schemes to work. Therefore, much greater demands are placed on system resources such as I/O and memory bandwidth, and the effective utilization of such resources becomes increasingly important in supporting time-critical applications.

In addition, some applications need more or different processing capabilities than current workstation processors provide. For example, workstation and mainframe architectures differ in that workstations do not dedicate processing resources to I/O operations as mainframes do (e.g., channel-processors). As a result, any processing associated with I/O must be done by the workstation's main, general-purpose, processor(s). Furthermore, general-purpose processors lack the capacity (in raw operations per second) to perform many functions (such as standard compression of full frame rate video) in real time, thus requiring special processing elements.

Finally, even if the necessary resources exist and are organized properly, the workstation system software is not able to manage the resources properly for this class of application. Resource scheduling decisions are typically based on a "fairness" criteria because of the workstation's time-sharing heritage. Individual resources, such as memory and CPU cycles, are managed in *ad-hoc* ways, with no way to coordinate them to ensure that time-critical activities are completed on time. Because system software also has no concept of time-criticality, all applications run, in effect, open-loop. Therefore, time-critical applications must currently accept the resources as provided to them as the result of seemingly random resource management decisions.

Shortcomings of Embedded Real-Time Systems

PC-class machine designers have addressed the problems of time-critical media applications by turning to an "embedded real-time" systems approach. Within a limited context, this approach appears to meet the applications' needs. However, it is an inappropriate approach for the workstation environment because it requires dedicated resources and deterministic behavior. In addition, a system using this approach must be re-architected when new applications or capabilities are introduced.

Traditional embedded real-time systems statically allocate excess resources for time-critical problems to "guarantee" that resource demands never exceed the supply. This can be seen in PC-class multimedia systems, where the entire CPU is dedicated to a single application.

Furthermore, traditional (strict priority and rate monotonic) real-time scheduling doctrine decrees that you can only use up to approximately 70% of the available processing time and still meet timeliness requirements [Liu 73]. Therefore, the system is artificially limited short of its full potential. This is unacceptable in a workstation environment where arbitrary amounts of application load may be impressed upon the system at the discretion of the user.

The "guarantees" that embedded real-time systems make are predicated on the assumption that all system activities are deterministic or that hard bounds can be placed on all variability. The workstation environment is inherently non-deterministic. For example, the user may activate another time-critical application, or a remote command might arrive via the network. A system architecture that relies on determinism fails in this environment.

Finally, there is no uniform way to handle unanticipated competing resource requirements. This means that the resulting systems are fragile; when separate resource demands are allowed to be made simultaneously, or when new features are added, the entire system must be re-architected. For example, two applications that handle a video and audio stream, respectively, may each have static resource allocations that allow them to run well in isolation. However, when they are run simultaneously by a user trying to have both sound and pictures, the assumptions of determinism and excess resources built into each application are violated. The results are unpredictable. The system must be reanalyzed and tuned before these applications can be used together.

Systems Implications

What is called for here is a general systems solution to the problem of managing time-critical activities. It is not sufficient for the system to simply "get out of the way". Rather, it is the system's responsibility to actively manage its resources in such a fashion as to, wherever possible, permit the applications' time-constraints to met.

The solution is not to cripple systems by reducing them to the level of PC's, nor simply to add mainframe style I/O channels, but rather to extend the architecture of the workstation to fully support these applications.

The Time-Critical Systems research group at Sun Microsystems Laboratories, Inc., is creating hardware and software, platform-level technology that manages (i.e., acquires, processes, transfers, coordinates, and delivers) collections of time-critical data streams.

The approach being followed involves ensuring that the proper resources exist in the system, that they are organized correctly, and that they are managed in a manner that supports the needs of time-critical applications.

Ensure Proper Resources Exist

The group's initial work is centered around a research testbed system that is being constructed in conjunction with David Sarnoff Research Center and Texas Instruments, as a part of a DARPA-sponsored research project — i.e., the High Resolution Video (HRV) Workstation project [HRV 90a, HRV 90b]. Through the development and use of this testbed hardware and software, several basic ideas, including: the implementation of core algorithms, the development of realistic workload generators, and the collection and analysis of system performance statistics are being explored. This involves examining the system architecture of the testbed in depth, and comparing its capabilities to that of today's workstations. The testbed system will also allow other groups (and their applications, within and outside of Sun) to provide us additional insights into the problem area through their use of this system.

The Time-Critical Systems research group is considering the difficulties which arise in the manipulation of time-critical data streams, and devising solutions which provide the necessary capacity to support time- critical applications. The group is also in the process of identifying the bandwidth and processing requirements for the sample applications, which will provide insight into the limitations of (and trade-offs between) processing elements, memory, I/O, and system interconnect topology.

Ensure Resources are Properly Organized

Support for time-critical data streams is a system-level problem which requires a system-level solution. As a part of this work, hardware architectures are being explored which provide support for time-critical resource management at the system-software-level. Also, the mere existence of interesting hardware that enables time-critical resource management is not enough. This research will investigate and provide insights into proper ways of organizing the hardware (i.e., hardware architectural features that lend support to the management of time-critical information).

This effort involves investigating architectures which make use of multiple, concurrent paths to high-bandwidth memories, in order to permit the execution of simultaneous, independent, time-critical activities in future workstations. It is also likely that this exploration into issues of system topology and interconnection may involve the study of different forms of switch-connected machine architectures (e.g., high-speed non-blocking switches, point-to-point links, hybrid bus/crossbar, etc.), as opposed to global-bus-based structures.

To date, this work has suggested that the requirements of multimedia-like, time-critical applications call for globally symmetrical resource accessibility (e.g., as provided by global bus-structured systems). However, the high-bandwidth, stylized information flow patterns that accompany such applications suggests the need for concurrent operation (e.g., as provided by switch-connected architectures [Tennenhouse 89, Hayter 91]).

By introducing more concurrency into the workstation interconnect, highly asynchronous activities can be scheduled in parallel with less (or little) impact on each other. Alternative paths for data movement may greatly increase system scalability and flexibility. Additional data paths will also allow the exploration of insertion into the system of specialized processing elements such as data compression engines.

Ensure Resources are Properly Managed

Time-Driven Resource Management (TDRM) is our basic approach to solving the time-critical systems problems defined above. We are adding timeliness as a system-supported attribute of the data streams that exist within the workstation.

It is our contention that all resource management decisions should be made on the basis of timeliness and importance of the applications requesting the resources. Our work involves the creation of system resource management software that takes timeliness and importance into account when making its decisions concerning processor allocation (i.e., scheduling), memory allocation, synchronization operations, interprocess communication, I/O operations, etc. The classical arbitrary, priority-based resource allocation mechanisms will be compared to the new techniques and the results analyzed to understand the costs and benefits of such an approach.

According to this model, resource management involves making decisions based upon the requesters' deadline, importance, and expected resource requirements. That is, rather than strictly using deadlines, which fail in overload cases or priorities which are too static, a time and value evaluation will be made to choose among all of the requesters for system resources [Jensen 75].

Rather than trying to guarantee that overload or timing exceptions do not occur, it is assumed that overload is a common occurrence in a workstation. Resource overload should be resolved by application and user-defined policies. The system will detect overload and lateness exceptions, make the appropriate evaluations based on timeliness and value, and then follow user-defined policies (or signal the applications impacted by the resource overload) to resolve or react to the overload situation. It is application programs that should choose how to utilize less of an overloaded resource, and the manner of that reduction, rather the having the system choose for all applications [Northcutt 87].

The system will be architected to directly implement and encourage "graceful degradation" strategies. Reserving resources and handing them out blindly until a "brick wall" is hit is unacceptable since it directly detracts from the base nature of a workstation. Instead, best-effort use of the resources should be encouraged, short-term aberrations noted, and the system should gracefully adjust as overload is reached.

Conclusion

We are applying a methodology of effectively providing, organizing, and managing system resources to the problem of integrating multimedia into the distributed workstation programming environment. We believe that this work will lead to fundamental changes in the nature of workstation hardware and software. Furthermore, the techniques being developed here also have the promise being applicable to resource management in other areas (e.g., high-speed networks).

References

[Escobar 91]   J. Escobar, D. Deutsch, and C. Partridge
                 A Multi-Service Flow Synchronization Protocol
                 Technical Report, Bolt Beranek and Newman, March 1991.

[Hayter 91]   M. Hayter and D. McAuley
                 The Desk Area Network
                 Technical Report No. 228, University of Cambridge Computer
                 Laboratory, May 1991.

[HRV 90a]   High Resolution Video Workstation Project
                 *High Resolution Video Workstation: Requirements and Architectural
                 Summary*
                 HRV Project Technical Report #90051, May 1990.

[HRV 90b]   High Resolution Video (HRV) Workstation Project
                 *System Support for Time-Critical Media Applications: Functional
                 Requirements*
                 HRV Project Technical Report #90101, November 1990.

[Jensen 75]   E. D. Jensen
                 Time-Value Functions for BMD Radar Scheduling
                 Technical Report, Honeywell System and Research Center, June 1975.

[Lui 73]   C. L. Liu and J. W. Layland
                 *Scheduling Algorithms fo Multiprogramming in a Hard Real-Time
                 Environment*
                 J. ACM 20, 1, 1973.

[Northcutt 87]   J. D. Northcutt
                 *Mechanisms for Reliable Distributed Real-Time Operating Systems: The
                 Alpha Kernel*
                 Academic Press, Boston, 1987.

[Tennenhouse 89] D. Tennenhouse and I. Leslie
                 A Testbed for Wide Area ATM Research
                 In Proceedings ACM SIGCOMM, September 1989.

Kernel Support for Live Digital Audio and Video*

Kevin Jeffay   Donald L. Stone   F. Donelson Smith

University of North Carolina at Chapel Hill
Department of Computer Science
Chapel Hill, NC, USA   27599-3175
{jeffay,stone,smithfd}@cs.unc.edu

Abstract: We have developed a real-time operating system kernel which has been used to support the transmission and reception of streams of live digital audio and video in real-time as part of a workstation-based conferencing application. An experimental environment consisting of a number of workstations interconnected with a 16 Mbit token ring has been created and used to evaluate quantitatively the performance of the kernel and conferencing application, as well as the quality of the conferences they are capable of supporting. Our early experiences with these systems are described.

Introduction

Recent advances in video compression algorithms — and their realization in silicon — have made it possible to consider introducing streams of digitized audio and video into the processing workload of workstation operating systems. For example, by outfitting workstations with off-the-shelf video cameras, microphones, digital video and audio acquisition and compression hardware, and audio amplifiers, it is possible to construct multimedia applications such as integrated voice/video/text documents and browsers [Hopper 90] as well as communication utilities such as workstation-based video and/or audio conferences [Terry & Swinehart 88, Jeffay & Smith 91].

While the hardware for such systems is readily available, existing operating systems and network communication protocols are inadequate for supporting multimedia applications such as browsing a video document or conferencing. This is due to the real-time processing requirements of digital audio and video, specifically, rigid throughput and latency requirements. For browsing or conferencing in a distributed system, frames of video must be acquired at a remote workstation (either from a camera or from a disk file) and transmitted so as to arrive at the local workstation and be displayed at the (precise) rate of one frame every 33 ms. Problems such as these have stimulated programs of basic research in many of the traditional areas of operating systems such as file systems [Rangan & Vin 91], and scheduling and inter-process communication [Anderson *et al.* 90, Govindan & Anderson 91, Jeffay *et al.* 91].

---

* This work supported in parts by grants from the National Science Foundation (numbers CCR-9110938 and ICI-9015443), and by the Digital Equipment Corporation and the IBM Corporation.

Our interest lies in the development of operating system infrastructure for the processing of *live* digital audio and video, specifically, workstation-based conferencing. Applications requiring live digital audio and video, such as conferencing, are unique in that their real-time throughput and latency requirements are particularly demanding. A conferencing application fundamentally requires that the audio and video data be processed *as it is generated* (*i.e.*, with zero or one buffer). To do otherwise implies that either portions of the conference will not be reproduced (*e.g.*, frames will be dropped) or that artificial latency is imposed between acquisition and display processes. In order for a system to be usable as a conferencing tool, we should minimize, if not avoid altogether, the occurrence of these events. Ideally a workstation-based conferencing system should be indistinguishable from the more traditional analog (*i.e.*, non-computer based) system.

We have constructed an experimental network of workstations capable of processing live digital audio and video and are using this system to experiment with operating system and network support for continuous-time media. In this note we describe some early experiences with a prototype conferencing application constructed on top of an operating system kernel we have built. The kernel provides real-time computation and communication services that enable a programmer to both specify real-time throughout requirements and to assess end-to-end latencies. Moreover, the kernel supports a tasking model for which it is possible to determine *a priori* if sufficient processing resources are currently available to meet an application's requirements.

In designing the conferencing application, our approach has been to view the problem as one of real-time resource allocation and control. We hope to demonstrate by example that much of the technology developed within the real-time systems community is directly applicable to the problem of supporting applications that manipulate digital audio and video. Although we have chosen to focus on live digital audio and video, we believe the infrastructure we have developed is applicable to applications that manipulate both live and recorded media.

The following section briefly describes the operating system kernel used to support real-time digital audio and video conferencing. We follow with a description of the architecture of the conferencing prototype itself and describe our initial experiences using this system. We conclude with a brief discussion of the appropriateness of the performance guarantees provided by the kernel and assess how well the kernel's programming model responds to the real-time processing needs of the conferencing application.

Kernel Overview

Operating system support for our conferencing application is provided by an operating system kernel we have developed called YARTOS (Yet Another Real-Time Operating System) [Jeffay *et al.* 91]. This kernel was developed to experiment with a paradigm of process interaction called the *real-time producer/consumer (RTP/C) paradigm* [Jeffay 89]. The RTP/C paradigm defines a semantics of inter-process communication that provides a framework for reasoning about the real-time behavior of programs. This semantics is realized through an application of some recent results in the theory of deterministic scheduling and resource allocation. We believe YARTOS to be a "general purpose" real-time operating system kernel. In addition to the conferencing application, YARTOS prototypes have been used in a 3-dimensional interactive graphics system used for research in *virtual realities*, and a HiPPI data link controller.

The programming model supported by YARTOS is an extension of Wirth's discipline of real-time programming [Wirth 77]. In essence it is a message passing system with a semantics of inter-process communication that specifies the real-time response that an operating system must provide to a message receiver. This allows us to assert an upper bound on the time to receipt and processing of each message. The exact response time requirement is a function of such factors as the rate with which a process receives messages on a given input channel. Ultimately, these rates are functions of the rates at which data arrives from external sources. These semantics provide a framework both for expressing processor-time-dependent computations and for reasoning about the real-time behavior of programs. The programming model is described in greater detail elsewhere [Jeffay 89].

YARTOS itself supports two basic abstractions: *tasks* and *resources*. A task is an independent thread of control (*i.e.*, a sequential program) that is invoked in response to the occurrence of an event. An event is a stimulus that may be generated by processes external to the system (*e.g.*, an interrupt from a device) or by processes internal to the system (*e.g.*, the arrival of a message). We assume events are generated repeatedly with a (non-zero) lower bound on the duration between consecutive occurrences of the same event. Each invocation of a task must complete execution before a well-defined deadline. The invocation intervals and deadlines for a task are derived from constructs in the higher-level programming model. During the course of execution, a task may require access to some number of resources. A resource is a software object (*e.g.*, an abstract data type) that encapsulates shared data and exports a set of procedures for accessing and manipulating the data. Like a monitor, resources guarantee mutually exclusive access to the data they encapsulate. Resources are accessed indirectly through the kernel. Support for resources is included to ensure *priority inversions* do not occur — a phenomena in which low priority processes exclude high-priority processes from accessing time-critical data, thus causing the high priority processes to miss deadlines [Sha *et al.* 90].

For a given workload (a set of tasks and resources), the goal of YARTOS is to guarantee that (1) all requests of all tasks will complete execution before their deadlines and (2) no shared resource is accessed simultaneously by more than one task. We have developed an optimal (preemptive) algorithm for sequencing such tasks on a single processor [Jeffay 90]. The algorithm is optimal in the sense that it can provide the two guarantees whenever it is possible to do so. Moreover, an efficient algorithm has been developed for determining if a workload can be guaranteed a correct execution. This algorithm forms the basis for a resource reservation protocol that is executed prior to the start of a video conference by workstations participating in the conference. In addition to its academic value, the optimality of the YARTOS resource allocation policies are important for effectively trading-off processing requirements for guaranteed response time. If YARTOS cannot guarantee a correct execution to a process, feedback can be provided on why the guarantee is not possible. A programmer can typically achieve a compromise guarantee by either relaxing the response time constraint or by improving the execution time of one or two specified processes. The optimality properties ensure that the reasons for the lack of a guarantee are fundamental in nature.

Workstation-Based Conferencing

Motivation

Our emphasis on workstation-based conferencing arises from an interest in using computers and communication networks to facilitate collaboration among scientific and technical professionals [Smith *et al.* 90]. From a technological standpoint, the goal is to support multiple, concurrent streams of digital audio and video in a distributed network of computer workstations. These streams may either form disjoint conferences within the network or involve one workstation in multiple simultaneous conferences.

While the requirements for media in conferencing systems could be met using a combination of conventional digital and analog (audio/video) technology (*e.g.*, a workstation with an adapter for analog video such as the Parallax card), by manipulating audio and video in an entirely digital format we leverage *existing* communications infrastructure (*e.g.*, local-area networks) to construct new and powerful tools for collaboration with remote colleagues. Moreover, digital formats admit the possibility of writing software to implement functions that now require specialized hardware in teleconferencing systems (*e.g.*, voice-activated controls to put the current speaker's image in a window; multi-image windows "quad-split" so that up to four participants are simultaneously visible).

Experimental Set-Up

We have built a private network to experiment with live digital audio and video. Currently we have a small number of IBM PS/2 workstations (Intel 80386 processor) interconnected with a 16 Mbit token ring network. We use IBM-Intel ActionMedia 750 adapters (based on Intel's Digital Video Interactive (DVI) technology) for the acquisition, compression, decompression, and display of digital audio and video [Luther 90].[1] The compression component of the ActionMedia system can be programmed to perform any of a number of compression algorithms that trade-off execution time for image size and quality. We use an algorithm capable of supporting real-time (*i.e.*, full-motion — 30 frames per second) compression or decompression of full color images at a resolution of 256×240 but with considerable loss of image quality. Newer high-performance versions of the display and image processors should provide larger and better quality images at 30 frames per second [Harney *et al.* 91].

Each workstation is configured with a set of ActionMedia adapters and is connected to a video camera, microphone, audio amplifier, and video monitor. Two ActionMedia adapters are required to acquire digital audio and video: (1) a capture adapter that digitizes RGB signals from a video camera along with two channels of analog audio inputs, and (2) a delivery adapter that provides (along with many other functions) capabilities for video compression, decompression, display control, and audio signal processing. The capture adapter alone is required for playback of an audio and video stream. With a pair of ActionMedia adapters, a given workstation may be either the originator or receiver of a audio/video stream, but cannot be both concurrently. With the addition of a second delivery adapter, a workstation can transmit and receive video streams simultaneously. In the following we consider only a single unidirectional audio/video stream between two workstations.

In our experiments we have used a frame-independent compression algorithm (*i.e.*, one in which the compression of a frame is not influenced by the compression of previous frames). In our present configuration, the resulting bandwidth requirement for transmitting a stream of compressed audio and video over a local area network is approximately 2 Mbits per second. A compressed audio frame is approximately 0.5 Kbytes and a compressed video frame is, depending on the scene, approximately 6.5-7.5 Kbytes.

In addition to the digital system, we have also constructed a separate analog video conferencing system using an existing in-house CATV system. The output from the workstation camera simultaneously supplies the digital and analog systems with video input thereby providing a convenient mechanism for assessing the (qualitative) latency and image quality of the digital system. A preliminary comparison of the two systems is reported below.

---

[1] ActionMedia are Digital Video Interactive are registered trademarks of the Intel Corporation.

Software Architecture

The conferencing application runs as a user program on top of the YARTOS kernel. There are separate applications for originating and receiving a conference. Figure 1 shows design of the conference origination application. The conference reception application is similar. The origination application is responsible for acquiring, compressing, and transmitting a video stream. As shown in Figure 1, the application consists of 5 tasks (represented by circles) and 2 resources (represented as a collection of shaded circles). In Figure 1 single headed arrows indicate message channels. In YARTOS these provide control flow information. Double headed arrows indicate global data flow. Omitted from Figure 1 are the kernel-level interrupt handlers.

The origination application is controlled by two externally generated signals. The network adapter signals an interrupt for one of two events. The *Transmit Ready* interrupt, TR, is generated sometime after an application initiates a network transmit. This signals that the network adapter is ready for the data. The *Transmit Complete* interrupt, TC, is generated when a packet has been transmitted. The ActionMedia hardware signals an interrupt for one of three events. Two *Vertical Blanking Interrupts*, VBI0 and VBI1, are generated after each half of a frame of video has been scanned. A *Compression Complete* interrupt, CC, is generated after a frame of video has been compressed.

Each video frame goes through a three stage pipeline: digitization, compression of the digitized image, and transmission of the compressed data over the network. At any given time, there are three frames in the pipeline. The pipeline is initiated by a VBI1 signal. A digitize is initiated ("scheduled") by providing an address of a buffer to the capture adapter. Two VBI0s later (the first VBI0 signals that the digitize has begun — the second VBI0 signals that it is complete) the frame has been digitized. After the second VBI0, the compression is initiated. As the frame is compressed, the hardware reads digitized video from one buffer and deposits compressed video data into another. The CC interrupt signals the end of this operation. At this point, the transmission over the network is initiated. Lastly, when the network adapter is ready to transmit, it raises the TR interrupt and network packets are transferred onto the adapter. The results of a stage in the pipeline are communicated to later stages through two buffer pools (two YARTOS resources) that are shared between software tasks (as shown in Figure 1) and hardware tasks (as described in Table 1). One buffer pool holds digitized frames, the other holds compressed frames.

The timing of the execution of all tasks is critical to the functioning of the pipeline. The response time guarantees provided by the YARTOS kernel are used to ensure correct operation of the pipeline. For example, since a digitize always begins at a VBI0, the Schedule Digitize task (invoked by VBI1 messages) must be completed before the next VBI0. VBI interrupts occur at intervals of 16.5 ms, so this task must complete within 16.5 ms. of its invocation. Table 1 summarizes the sequence of hardware and software operations required to acquire, compress, and transmit a frame of video and specifies the response time requirements of the software tasks and the assumed hardware timing characteristics.

Experimental Results

We have run a number of experiments with unidirectional conferences (one origination application and one reception application) to test both the performance of our system and the qualitative nature of the conference. (For unidirectional digital conferences, we have the capability of providing audio and video channels in the reverse direction with our analog system.) The measures of interest are the intra-workstation and intra-network latency for a frame of video and the effect of dropped or missed frames on the perceived quality of the conference.

Table 1: Hardware/Software pipeline for transmitting a video frame.

Operation	HW/SW	Function	Timing Properties
Schedule Digitize	SW	Get a free digitize buffer a and initiate the digitization of the next frame of video.	Must complete within 16.5 ms.
Digitize	HW	Scan a frame of video into a digitizing buffer.	Completed in 33 ms.
Schedule Compress	SW	Get a digitizing buffer and initiate the compression of the most recently digitized video frame.	Must complete within 16.5 ms.
Compression	HW	Compress a frame of video from a compress source buffer into a compress sink buffer.	Completed in 20 ms.
Transmit	SW	Get a compress sink buffer, make it into a network packet, and initiate a network transmit.	Must complete within 20 ms.
Initiate Transmit	HW	Signals that a buffer is available on the token ring adapter.	Completed in less than 1 ms.
Network Driver	SW	Copy data from a compress buffer onto the token ring adapter.	Must complete within 16.5 ms.
Transmission	HW	Physically transmit data.	A function of network load.

For the conference origination application we define the intra-workstation latency for a frame of video as the difference between the time a network packet containing the data for a frame is transmitted and the time the digitization of the frame started. We can demonstrate analytically that the worst case latency is 3.5 frame times (approximately 116 ms.) for this application. (This assumes an otherwise idle network.) The conference reception application is structured similarly and has similar latency. Figure 2 illustrates the worst case interleaving of the software and hardware operations for the origination application. In Figure 2, hardware processes execute throughout the intervals shown for hardware processes. Software tasks execute somewhere between the left and right endpoints of the intervals shown for software tasks.

In practice we estimate the end-to-end latency in a conference to be between 6 to 7 frame times (200 ms. - 230 ms.). This latency in the digital system is easily noticeable when compared side-by-side with the analog system. However, when the system is used for conferencing, specifically, when users are physically separated and there there is no reference standard, we have found the digital system to be adequate. We conjecture that this is because in conferencing there is little physical movement in front of the camera and often only one individual speaks at a time. We further conjecture that without synchronizing the sending camera and the receiving display, it is not possible to achieve a worst case end-to-end latency of less than 5 frame times in the present generation ActionMedia system. (A latency of 5 frame times would require an infinitely fast processor and network.)

Although we are limited in our present experimental configuration to only originating or receiving 30 frames of video per second,[2] we are confident that the latency guarantees provided by YARTOS for a single video stream would remain unchanged as the number of video streams manipulated by a workstation increases — provided that the processor does not

---

[2] We simulate multi-person conferences by displaying video images at a receiving station at a reduced rate (*e.g.* 15 frames per second for displaying two remote conferees) and replaying the audio from only one stream at a time. (Two streams of audio can be played simultaneously by playing one channel from each stream.)

become saturated. If a task with a minimum inter-invocation time of p time units is guaranteed a response time of p time units, then YARTOS is effectively reserving c/p of the processor, where c is the cost of executing the task, for the execution of this task. So long as YARTOS does not over commit the processor, the task will be guaranteed a response time of p time units independent of the number and processing requirements of other tasks in the system.

We have run a number of experiments on our network to determine how the quality of a conference degrades as a function of the delay in the network. In the initial experiments reported here we use do not use any of the priority reservation mechanisms of the token ring. Conference quality is assessed in terms of the number of "frame incidents" observed in an interval. A frame incident occurs at the conference receiver when a previously played frame has to be replayed. This occurs when either a frame is discarded by the origination application (because the network has not transmitted frame n by the time frame $n+1$ is ready for transmission), or when a frame arrives "late" at the receiver.

In assessing the quality of a conference, it should be noted that for our use of the ActionMedia system, faithful reproduction of the audio component is paramount. This is because audio data is acquired in fairly large blocks (33 ms. worth). In the conferencing application, an observer can (easily) detect a single dropped or replayed audio frame whereas several consecutive video frames need to be dropped in order for a user to notice. Since we currently transmit audio and video frames in the same network packet it is not possible to drop/replay one frame without doing the same to the other. Therefore while our emphasis is primarily on audio frames, we will speak only of dropped/replayed frames.

In the current implementation of a unidirectional conference, there is no explicit synchronization between the sending and receiving workstations. (To add synchronization would fundamentally add latency.) The origination workstation transmits frames at an aggregate rate of 1 frame every 33 ms. (with a measured jitter of ±2 ms.). The display tasks on the receiving workstation side are driven off (local) VBI interrupts. On each VBI0, a frame from a queue (typically of length one) of received frames is inserted into a decompression/display pipeline. If the receiving side is ahead of the sending side, *i.e.* a frame arrives late, then the receiver replays the previous frame. Such an event typically occurs at most once (and typically within the first few frames) for a light to moderately loaded network.

Our preliminary observations indicate that a frame incident rate of more than 2-4 incidents per 1000 frames is noticeable (*i.e.*, annoying).[3] Moreover, we observe that as the delay in the network (in our case due to artificially generated traffic) as seen by a workstation originating a conference, approaches 16 milliseconds, the occurrence of frame incidents increases dramatically and hence the audio quality and the quality of the conference itself deteriorates rapidly. This can be explained by noting that some of the latency in the conference is due to the fact that the (hardware) video processes generating VBI interrupts on the originating and receiving machines are not synchronized. If there were no delay in the network then one would expect that on average a frame of audio and video would be queued for 16 ms. (*i.e.*, half a frame time) at the receiving workstation before entering the decompression/display pipeline. Therefore, on average, the receiving workstation should be immune to substantial delays in the network. (For a 16 Mbit token ring, a 10 ms. delay would correspond to a utilization of the network bandwidth of approximately 75% [Bux 89].)

---

[3] These experiments were performed using a CD player as the audio input source on the originating machine. It is not clear how the threshold on frame incidents would differ for voice. We plan to perform additional experiments with live voice input.

On Performance Guarantees

The notion of a response time or other performance guarantee is central to our work and is indeed essential for supporting applications that manipulate live digital audio and video. As one example, if the Schedule Digitize task does not complete execution within 16 ms. of receiving a VBI1 message, then a frame of video and audio necessarily will be lost. Through careful attention to processor scheduling, YARTOS provides the desired guarantees. Moreover, we can demonstrate both analytically and empirically that frames are not dropped by a workstation originating a conference because of the workload in the system. There are, however, two aspects of these performance guarantees that need to be examined in closer detail to determine how well YARTOS supports the real-time requirements of the conferencing application.

The first concerns the usefulness of the guarantees. For a given programming model that includes repetitive real-time processing constraints, it is typically not very difficult to derive sufficient conditions on the operating environment — called *schedulability* conditions — that will ensure that the real-time response properties of tasks will be met. The more interesting question is how accurate these conditions are. That is, if a set of tasks do not satisfy the conditions then does this necessarily imply that a deadline will be missed if the tasks are executed? Moreover, if this is indeed the case then how can the programmer ameliorate this situation? For YARTOS, it is the case that if a set of tasks do not satisfy the schedulability conditions then it is possible to demonstrate a maximal, finite set of orderings of events (*e.g.*, message arrivals) that will necessarily lead to a missed deadline. If the programmer is willing to certify that none of these sequences of events can happen then no deadlines will ever be missed.

The second, and likely more important issue, concerns the fit between the programming model exported by YARTOS and the processing needs (both real-time and non-real-time) of the conferencing application. Although YARTOS can provide meaningful real-time guarantees, these guarantees come at a cost of a fairly restrictive programming model. YARTOS was designed to to support applications whose real-time processing constraints arise from the need to process data at a precise rate. Therefore, each task has a notion of a minimum inter-arrival time of activation messages and a deadline based on this inter-arrival time. This meshes nicely with the processing required to respond appropriately to VBI interrupts — *i.e.*, it is periodic and has a well-defined deadline. It is not well suited, however, to support some of the processing required to respond to a CC interrupt. This is because there are several logical operations that are executed in response to a CC interrupt but not all of these operations have the same deadline. In particular, some of these deadlines are not a function of the inter-arrival time of the CC interrupt. One important function has been omitted from Figure 1: the movement of digitize and compress buffers from the compress source and transmit queues respectively, back to their corresponding free queues. In the case of digitize buffers on the compress source queue, at the occurrence of a CC interrupt, a digitize buffer on the compress source queue may be moved to the free queue.

A careful analysis of Figure 2 reveals that the conference origination application can, in principle, work with three digitize buffers. This can be seen by noting that at the fourth VBI0 in Figure 2 (the start of the digitization of the unshown fourth frame), the hardware compression of frame 1 must have completed and hence the digitize buffer used for frame 1 can be reused for frame 4. If, however, the operation to free the digitize buffer is performed as part of the Transmit task, it is possible that the free operation for the digitize buffer used for frame n may not take place before execution of the Schedule Digitize task for frame $n+3$. This is because under YARTOS the deadline for the Transmit task can be expressed only in terms of the when the next CC message is expected to arrive. In general, this deadline is insufficient for ensuring that a digitize buffer can be freed before the next invocation of the Schedule

Digitize task (*i.e.*, before the next VBI1). Therefore, if the buffer free operation is performed by the Transmit task then the implementation of the digitize buffer resource must provide four buffers.

Abstractly, this is solely an issue of efficient resource utilization since the addition of a fourth digitize buffer would have no impact on the real-time performance of the application. In the current configuration of our system, however, this is a critical issue as digitize and compress buffer resources are implemented in memory on the ActionMedia adapter and there is insufficient space for four digitize buffers.

The root of the problem is that the true deadline of the buffer free operation is not a function of the interarrival time of CC interrupts and hence cannot be directly supported with a YARTOS task. The solution we have adopted is to associate an *eventcount* [Reed & Kanodia 79] with each message port in the system. Whenever a message is sent to a task the eventcount is incremented. The eventcount is then used for producer/consumer synchronization on events as described in [Reed & Kanodia 79]. In the case of the digitize buffer-free operation, this operation can now be performed when it is needed as a side effect of the resource call to remove a free digitize buffer. When this call is made, the kernel is invoked to check the eventcount for the Transmit task. The value of this event count indicates the number of CC interrupts that have occurred and this corresponds to the number of times digitize buffers have been freed. If the CC eventcount differs from the VBI1 eventcount (indicating the number of times free digitize buffers have been removed) by less than the number of digitize buffers in the system there is indeed a free digitize buffer.

In general, the eventcount mechanism must be a kernel provided function because the recognition of the event and the increment of the eventcount must be indivisible in order to preserve the semantics of eventcounts.

Future Directions

In the near term we will be experimenting with alternate conference application designs and measuring their performance. The goal is to characterize the cost of an "acceptable" quality conference in terms of the observed rate of frame incidents, the latencies induced by (a generic model of) the video hardware, the operating system and the network, and to demonstrate how fluctuations in the values of one component affect the others and the conference itself.

With regard to the conferencing application itself, of particular interest is the separation of audio and video into independent network packets. In the ActionMedia system, compressed audio data for a frame is available 33 ms. before the corresponding compressed video data and hence could be transmitted significantly earlier than the video data. Given that the quality of the audio in a conference is the primary indicator of overall conference quality, by transmitting audio as soon as it is available we should be able to tolerate substantial network delays (*e.g.*, delays on the order of 80 ms.). This is, of course, an artifact of the 200 ms. end-to-end latency for video data in our system.

Given that our emphasis on workstation-based conferencing arises from an interest in using such systems to facilitate collaboration among scientific and technical professionals, we are striving to integrate our kernel and conferencing applications into a computing environment that includes UNIX workstations. To this end, we are currently adding ethernet support and porting a TCP/IP implementation to the YARTOS kernel. The token ring network (and hopefully its FDDI successor) will remain primarily a private network for experiments with real-time communications protocols. In addition we are working on porting an X server to YARTOS. In essence we hope to construct a real-time multimedia workstation that provides a window onto existing computing environments while providing new, real-time communications and continuous media services.

Summary and Conclusions

The YARTOS programming model provides response time guarantees to tasks based on their minimum inter-invocation time. While this basic mechanism has not supported the conferencing application seamlessly, it has been sufficient to construct the system. In particular, the accuracy of the YARTOS schedulability analysis has been most useful as it has allowed us to concentrate on issues of logical correctness while ignoring efficiency considerations. (For example, we have constructed, and received the desired performance guarantees, for conferencing applications that utilize close to 80% of the CPU.)

Concerning our experiments, we have empirically determined that when the delay in the network exceeds 16 ms. then the perceived quality of the conference falls off sharply (more than 4 dropped/replayed frames per 1000). Therefore, for our system we posit that if P(*network delay* < 16 ms.) > .996, no reservation or priority mechanisms are required to ensure good fidelity conferences on our (token ring) network.

References

Anderson, D.P., Tzou, S.-Y., Wahbe, R., Govindan, R., Andrews, M., 1990. *Support for Continuous Media in the DASH System*, Proc. Tenth Intl. Conf. on Distributed Computing Systems, Paris, France, May 1990, pp. 54-61.

Bux, W., 1989. *Token-Ring Local-Area Networks and Their Performance*, Proc. of the IEEE, Vol. 77, No. 2, (August), pp. 238-256.

Govindan, R., Anderson, D.P., 1991. *Scheduling and IPC Mechanisms for Continuous Media*, Proc. ACM Symp. on Operating Systems Principles, ACM Operating Systems Review, Vol. 25, No. 5, (October), pp. 68-80.

Harney, K., Keith, M., Lavelle, G., Ryan, L.D., Stark, D.J., 1991. *The i750 Video Processor: A Total Multimedia Solution*, Comm. of the ACM, Vol. 34, No. 4 (April), pp. 64-79.

Hopper, A., 1990. *Pandora — An Experimental System For Multimedia Application*, ACM Operating Systems Review, Vol. 24, No. 2, (April), pp. 19-34.

Jeffay, K., 1989. *The Real-Time Producer/Consumer Paradigm: Towards Verifiable Real-Time Computations*, Ph.D. Thesis, University of Washington, Department of Computer Science, Technical Report #89-09-15.

Jeffay, K., 1990. *Scheduling Sporadic Tasks With Shared Resources in Hard-Real-Time Systems,* University of North Carolina at Chapel Hill, Department of Computer Science, Technical Report TR90-038, August 1990. (In submission.)

Jeffay, K., Smith, F.D., 1991. *System Design for Workstation-Based Conferencing With Digital Audio and Video*, Proc. IEEE Conference on Communication Software: Communications for Distributed Applications and Systems, Chapel Hill, NC, April 1991, pp.169-178.

Jeffay, K., Stone, D., Poirier, D., 1991. *YARTOS: Kernel support for efficient, predictable real-time systems*, to appear: Proc. IFAC Workshop on Real-Time Programming, Pergamon Press.

Lederberg, J., Uncapher, K., (eds.), 1989. *Towards a National Collaboratory: Report of an Invitational Workshop at the Rockefeller University*, March, 1989. Distributed by the National Science Foundation.

Luther, A.C., 1990. "Digital Video in the PC Environment," McGraw-Hill, Second Ed.

Rangan, P.V., Vin, H.M., 1991. *Designing File Systems for Digital Video and Audio*, Proc. ACM Symp. on Operating Systems Principles, ACM Operating Systems Review, Vol. 25, No. 5, (October), pp. 81-94.

Reed, D.P., Kanodia, R.K., 1979. *Synchronization with eventcounts and sequencers*, Comm. of the ACM, Vol. 22, No. 2, (February), pp. 115-123.

Sha, L., Rajkumar, R., Lehoczky, J.P., 1990. *Priority Inheritance Protocols: An Approach to Real-Time Synchronization*, IEEE Trans. on Computers, Vol. 39, No. 9, (September), pp. 1175-1185.

Smith, J.B., Smith, F.D., Calingaert, P., Hayes, J.R., Holland, D., Jeffay, K., Lansman, L., 1990. *UNC Collaboratory Project: Overview*, University of North Carolina at Chapel Hill, Department of Computer Science, Technical Report TR90-042.

Terry, D.B., Swinehart, D.C., 1988. *Managing Stored Voice in the Etherphone System*, ACM Trans. on Computer Systems, Vol. 6, No. 1, (February), pp. 3-27.

Wirth, N., 1977. *Toward a discipline of real-time programming*, Comm. of the ACM, Vol. 20, No. 8 (August), 577-583.

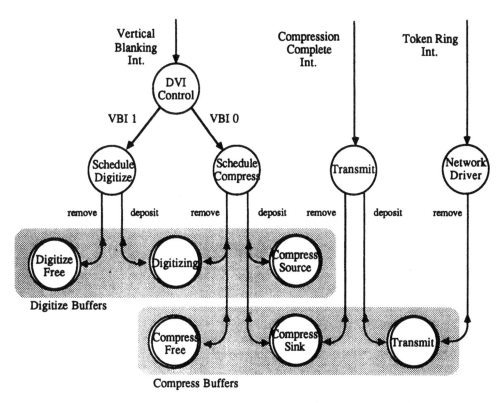

Figure 1: Conference origination application.

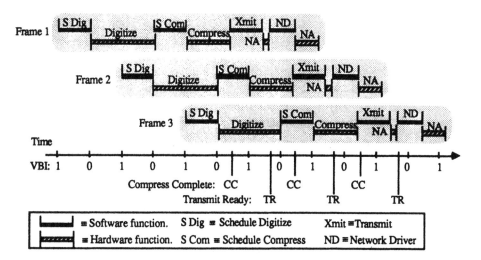

Figure 2: Execution of the conference origination application for the processing of three frames of video.

Priority Consistency in Protocol Architectures

Clifford W. Mercer and Hideyuki Tokuda
School of Computer Science
Carnegie Mellon University
Pittsburgh, Pennsylvania 15213
{cwm,hxt}@cs.cmu.edu

Abstract

The protocol processing software of a multimedia operating system must provide fast response time and predictable delays for time-constrained data streams like digital audio and video streams. This paper describes several different techniques for scheduling the protocol processing of messages. These techniques are analyzed and their (simulated) performance compared using various metrics. One of these metrics is the *priority inversion factor* which provides a way of quantifying priority inversion in the system. Protocol processing time and context switch time are given as parameters in the simulations, and we present guidelines for choosing between the message scheduling techniques based on the ratio of protocol processing time to context switch time for a given system.

1 Introduction

A continuous media operating system must be highly preemptable so that *priority inversion* (the situation where a high priority activity is delayed by the execution of a lower priority activity) can be minimized. This provides good response time for high priority activities and better adherence to the priority structure defined by the system designer. Commercial real-time operating systems guarantee fast response time by structuring the kernel in a way that is highly preemptable. For example, if the scheduler is required to search a list of ready processes in order to make a scheduling decision, interrupts must be disabled to preserve the invariants of the list. But if the list is long, the scheduler may delay pending interrupts for a time. To avoid this problem, the search may be broken up into pieces to allow pending interrupts to be serviced at appropriate points during the course of the search. Designing a high degree of preemptability into a simple uniprocessor system is relatively easy, but the problem becomes more difficult in

This research was supported in part by the U.S. Naval Ocean Systems Center under contract number N66001-87-C-0155, by the Office of Naval Research under contract number N00014-84-K-0734, by the Federal Systems Division of IBM Corporation under University Agreement YA-278067, and by the SONY Corporation. The views and conclusions contained in this document are those of the authors and should not be interpreted as representing official policies, either expressed or implied, of NOSC, ONR, IBM, SONY, or the U.S. Government.

the context of a distributed environment where network devices must be handled and protocols must be executed.

In this position paper, we describe our on-going study of various structures for protocol processing software where we concentrate on the preemptability properties and efficiency of the techniques. We first describe the problem with current techniques used in network operating systems, and then we identify several techniques which vary in degree of preemptability and in efficiency. We introduce a metric for quantifying the priority inversion in a system, and we use this to compare the various techniques.

2 Protocol Processing Techniques

The techniques we consider in our study range from a method where protocol processing is done at a non-preemptable software interrupt level (a priority level higher than any user process but lower than the hardware interrupt level) to a method where prioritized, preemptable threads perform the protocol processing. We consider several techniques between these two extremes and attempt to identify the degree of preemptability that is required to service what we think might be a typical continuous media workload. We start with the software interrupt technique (SOFTINT) used in BSD4.3 in which priority inversion arises as a result of the FIFO queueing of messages as well as the non-preemptable nature of the software interrupt. Then we consider a single-threaded technique (T1F) where the thread executes at the highest priority in the system and the messages are again queued in FIFO order. This gives us the same basic characteristics except that there is some overhead in going to a threaded approach. The single-threaded technique with priority queueing allows us to evaluate the performance of the priority message queueing with a non-preemptable service which is easier to implement. This technique gives rise to priority inversion from the non-preemptability of the service, but the priority inversion due to queueing is eliminated. The problem with these approaches is that if the protocol thread is running at the highest priority in the system, low priority messages will be serviced at the expense of high priority local activity. So we consider a single-threaded technique where the priority of the thread is inherited (based on some mapping of message priority to thread priority) from the priority of the message (T1PI); i.e. the priority of the thread is low when servicing a low priority message. Of course, if a high priority message arrives while the low priority message is being serviced, that high priority message may have to wait for any medium priority threads which preempted the protocol thread (at low priority). This is another case of priority inversion. To solve this problem, we use priority "bumping" to increase the priority of the protocol thread when a high priority message is enqueued (T1PIB).

Finally, we consider two additional techniques for multi-threaded protocol engines which may have any number of threads; aside from the differences in these techniques, the number of threads has important implications for the performance and predictability as well. The two techniques differ in their method of assigning priorities to threads and their method of matching incoming messages to threads. The first technique assigns a fixed priority to each thread and queues incoming messages based on a match between the message priority and the thread priority. In the second technique, the priority changes dynamically depending on the priority of the message being processed. This technique is an extension of the single-threaded priority inheritance technique mentioned above, and the important issue is how to allocate threads to incoming messages.

In our previous work [2, 9], we determined that the prioritized multi-threaded protocol processing model gives better schedulability (by means of a higher degree of preemptability) than the software interrupt technique employed in the 4.3BSD system. And furthermore, this increase in schedulability costs only about 10% in additional overhead. The effects can be seen in actual applications [8] as well. However, this earlier work did not shed any light on the relative importance of associating priorities with messages vs. using preemptable workers for the protocol processing. The more recent work [4] gives a better indication of the performance implications of these techniques for a variety of different types of computation including distributed continuous media streams, local continuous media computations, background traffic on the network, and low priority activities on the local host.

We evaluated the techniques by simulating a task set where tasks send audio streams across the network. The tasks are:

- 1 16-bit audio task at high priority (20 ms period),

- 1 8-bit audio task at medium priority (40 ms period),

- 1 high priority local computation (20 ms period) and

- n low priority tasks (100 ms period).

The value n is a background load parameter, a larger n indicates more background load. Figure 1 shows the task set. The 16-bit audio stream is generated by τ_1 and received by τ_2. The 8-bit audio stream is sent from τ_3 to τ_4. There is a local computation, τ_5, and the background messages go from τ_6 to τ_7. These two tasks are replicated n times, and this replication determines the size of the background load. It is also important to note that the periodic background activity is bursty; measuring the performance of the techniques under strenuous conditions is our objective. The scheduling of the activities on the machine labeled PCPU4 is our main interest.

An important consideration in comparing these techniques is the ratio of protocol processing time to context switch overhead. Intuitively, a large ratio would indicate that preemption should be used to reduce the effects of priority inversion, and a small ratio would mean that preemption would be detrimental to efficient completion of the protocol processing. Our convention for describing the ratio is to use the one-way protocol processing time and the thread context switch time. In our simulation studies, we found that for a protocol processing time to context switch time ratio of 5-to-1 or 10-to-1, preempting protocol processing to avoid priority inversion makes sense. With a ratio of 2-to-1 or 1-to-1, however, the simulation showed that it is not worth preempting the protocol processing.

The table shows the protocol processing time and context switch time for two operating systems we are concerned with. The numbers are based on RPC protocol performance where we measure the round-trip response time. From this, we conclude that for systems such as ARTS [7] where the ratio is about 20-to-1 (using a one-way protocol processing time of about 4500 μs), the multi-threaded model is appropriate. Systems such as the x-kernel [1], with a ratio of about 50-to-1 could also benefit from this approach. For protocol processing engines implemented in hardware [5], however, the ratio may be more like 1-to-1, and so non-preemptable processing is appropriate in this case.

Operating System	Time	
ARTS		
context switch time	260	μs
protocol processing time (RPC)	9000	μs
x-kernel		
context switch time	38	μs
protocol processing time (RPC)	4000	μs

Table 1: Operating System Timing Measurements

3 Measuring the "Degree of Preemptability"

In evaluating the performance of the techniques in our simulation studies, we use several traditional metrics, and we introduced a new measure as well. We measure the mean and variance for the response times of the simulated activities, and we count the missed deadlines for the time-constrained, periodic continuous media activities.

We have introduced [4] a new metric which we call the *priority inversion factor* to measure the degree of preemptability in a system. Until this work, there was no way to describe the level of preemptability in a system except by specifying bounds on interrupt response time which is a very limited way to measure this effect. Interrupt response time, for example, does not give any indication of priority inversion in lower priority activities where hardware interrupts are not involved.

This priority inversion factor is a utilization-based method of measuring the preemptability. Utilization is the fraction of time during some interval that the processor was actually in use. During the time that the processor is not idle, it will be running some activity which is hopefully the highest priority activity available in the system. It is possible, however, that the running activity is not the highest priority (ready) activity, i.e. the system is suffering from priority inversion. To find the priority inversion factor, we consider only the time during which the processor is not idle, and we define the priority inversion factor to be the fraction of this time that the *wrong* activity is running. That is, the priority inversion factor is the time during which some priority inversion occurs divided by the time during which the processor is active. The product of the priority inversion factor and the total utilization is called the *priority inverted utilization*.

In order to illustrate this measure, we consider two systems which we simulated: one uses a single-threaded protocol processing engine with FIFO queueing and the other uses a multi-threaded engine with priority queueing. Furthermore, the multi-threaded engine has one thread for each message priority level in the system, and context switch time is taken to be zero, just to see what happens when preempting is free. We put the periodic task set described above on each system and measured the utilization and priority inverted utilization. Figure 2 shows the measured utilizations from the single-threaded case (T1F). The total utilization is given by the solid line; the priority inverted utilization is given by the broken line. In this case the priority inverted utilization is quite large compared with total utilization.

In Figure 3, we show the utilization for the W4P case. The total utilization is again given by the solid line, but the priority inverted utilization is zero. This is because each message priority has its own thread and the context switch time (a non-preemptable critical region in the system) is zero.

The idea of quantifying the predictability of a task set comes from real-time scheduling theory. The *schedulable bound* [3] provides a numerical measure of the schedulability of a task set. If a task set contains independent, periodic tasks with fixed computation times and if the utilization of a task set is less that the schedulable bound, then no deadlines will be missed; if the utilization is greater than the schedulable bound, no guarantees can be made. Furthermore, if these tasks are allowed to contain shared critical regions, the schedulable bound is effectively reduced, admitting fewer task sets [6]. And the more priority inversion is associated with this resource sharing, the more the schedulable bound is reduced. Thus, priority inversion adversely affects the schedulable bound, and quantifying the priority_inversion for comparison among competing techniques is therefore a reasonable way to evaluate the predictability of the system.

In our simulation study, the priority inversion factor observed for the T1PI, T2P, and T4P techniques ranged from .20 for the protocol processing time to context switch time ratio of 20-1 to .06 for a ratio of 1-1. For the T1F and T1P techniques, the priority inversion factor increases linearly (with the size of the background spike) for a ratio of 20-1, but the factor is constant at about .06 for the case where the ratio is 1-1.

4 Conclusion

The simulation results we have obtained [4] indicate that the method for structuring the protocol processing software should be chosen based on the ratio of protocol processing time to context switch time. This choice affects the predictability of continuous media streams on the network as well as the performance of local computations on each host in the system. And the timing characteristics of current operating systems indicate that these systems can benefit from the use of multi-threaded protocol engines. The priority inversion factor provides a way to measure the schedulability of a particular technique and allows us to evaluate the relative performance of the techniques.

References

[1] N. C. Hutchinson and L. L. Peterson. The *x*-Kernel: An Architecture for Implementing Network Protocols. *IEEE Transactions on Software Engineering*, January 1991.

[2] Y. Ishikawa, H. Tokuda, and C. W. Mercer. Priority Inversion in Network Protocol Module. *Proceedings of 1989 National Conference of the Japan Society for Software Science and Technology*, October 1989.

[3] C. L. Liu and J. W. Layland. Scheduling Algorithms for Multiprogramming in a Hard Real Time Environment. *JACM*, 20(1):46–61, 1973.

[4] C. W. Mercer and H. Tokuda. An Evaluation of Priority Consistency in Protocol Architectures. In *Proceedings of the IEEE Conference on Local Area Networks*, October 1991.

[5] Protocol Engines, Inc., Santa Barbara, CA. *XTP Protocol Definition, Revision 3.5*, September 1990. PEI 90-120.

[6] L. Sha, R. Rajkumar, and J. P. Lehoczky. Priority Inheritance Protocols: An Approach to Real-Time Synchronization. *IEEE Transactions on Computers*, 39(9), September 1990.

[7] H. Tokuda and C. W. Mercer. ARTS: A Distributed Real-Time Kernel. *ACM Operating Systems Review*, 23(3), July 1989.

[8] H. Tokuda, C. W. Mercer, and S. E. Breach. The Impact of Priority Inversion on Continuous Media Applications. In *Proceedings of the International Workshop on Network Operating System Support for Digital Audio and Video*, November 1990. Available in International Computer Science Institute Techical Report TR-90-062.

[9] H. Tokuda, C. W. Mercer, Y. Ishikawa, and T. E. Marchok. Priority Inversions in Real-Time Communication. In *Proceedings of 10th IEEE Real-Time Systems Symposium*, December 1989.

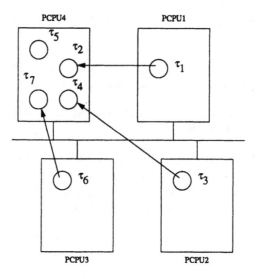

Figure 1: Simulated Task Set

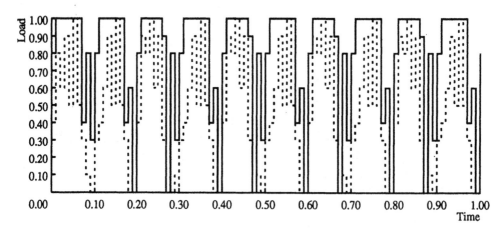

Figure 2: Utilization (T1F, 2000-0)

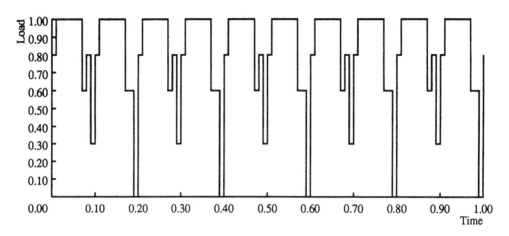

Figure 3: Utilization (T4P, 2000-0)

Session II: Architectures

Chair: Radu-Pospescu Zeletin, Technical University of Berlin and GMD-FOKUS

Session II was dedicated to architectures of multimedia systems. Each of the session papers introduced a certain multimedia architecture. During the talks, however, each speaker dealt with on one special aspect of the architecture presented. Dietmar Hehmann concentrated on architectural constraints of a multimedia transport system. Frankie Garcia focused on where to put multimedia synchronization. Lillian Ruston elaborated on the differences of multimedia architectures from telecommunications companies on one side and data processing companies on the other side.

The first presentation introduced multimedia work being done at the host site of the workshop, the IBM European Networking Center (ENC) in Heidelberg. In "Implementing HeiTS: Architecture and Implementation Strategy of the Heidelberg High-Speed Transport System" Dietmar Hehmann, Ralf Guido Herrtwich, Werner Schulz, Thomas Schütt, and Ralf Steinmetz describe the ENC's multimedia transport system. Beyond this, they discuss the system architecture in which the transport system is embedded. Their system, HeiTS, is aimed at a heterogeneous environment comprising several computers (RS/6000's and PS/2's), different operating systems (AIX and OS/2), and a variety of underlying networks (most notably, 16 Mbps Token Ring, FDDI, and ATM-based Broadband ISDN).

HeiTS is embedded in a special real-time environment within the system software. According to the authors' model, all system entities handling multimedia data (such as HeiTS) become so-called stream handlers. Stream-handler code is executed by real-time threads which are either deadline or rate-scheduled. Management functions permit to connect stream handlers and to control the flow of data through them.

The system relies on the support of underlying buffer and resource management subsystems. The buffer management subsystem avoids the needless copying of data within the system. It extends the traditional Mbuf approach to facilitate direct adapter-to-adapter data transfer. The resource management subsystem allocates resources for stream handlers to guarantee a certain throughput, delay, and reliability.

In closing, Dietmar Hehmann discussed several design and implementation decisions for the actual multimedia transport protocol. The system uses upcalls to keep control over the pacing of incoming data. No multiplexing is used above the data link layer which facilitates directing data to the right stream handler. Segmentation is avoided in the system. Flow control is purely rate-based.

During the discussion, Dietmar elaborated on the network protocol used in the system, ST-2 (described in a later talk by Stephen Pink). In continuation of the discussion in the first session, the reservation model of ST-2 was criticized. The ENC wants to use pessimistic and optimistic reservation in their future implementations.

Francisco Garcia presented a paper on "Protocol Support for Distributed Multimedia Applications" co-authored by Geoff Coulson, David Hutchison, and Doug Shepherd. He described the distributed systems platform being developed at Lancaster University to handle multiple media across a number of multimedia workstations. The main focus of his talk was multimedia synchronization.

The authors use the ANSA architecture for realizing their distributed system support. This architecture provides trading and binding services and basic facilities of threads and their synchronization. In the Lancaster system, three categories of basic services are developed: network services, multimedia devices, and storage services. An additional service is provided by the synchronization manager. It implements both event-driven synchronization (as it results, e.g., from an interrupting device) and continuous synchronization of two and more multimedia streams. For continuous synchronization, an orchestrator process is inserted at the source and/or sink ot the streams. Its policy for synchronization is user-defined.

The transport system developed for multimedia transfer is designed for running on transputer hardware. Rate control of the network (the key element of the protocol) and orchestration are tightly coupled. While the authors proposal is aimed at audio and video transfer, they point out that similar rated-based protocols can be used to achieve reliable data transfer with selective retransmission. They see the main application of their protocol in future ATM networks.

This final paper of the session by Lillian Ruston of Bellcore and Gordon Blair, Geoff Coulson, and Nigel Davies of Lancaster University continued (at least to some extent) the discussion about the multimedia architecture developed in Doug Shepherd's group in Lancaster. In "Integrating Computing and Telecommunications: A Tale of Two Architectures" Lillian Ruston presented multimedia as an add-on to two kinds of systems: a telecommunications network and a typical distributed computing environment.

Bellcore's Touring Machine architecture is heavily influenced by the traditional partitioning in the telecommunications industry. The system distinguishes between the telecommunications infrastructure and the customer's equipment. The infrastructure provides session management, resource allocation, and directory services. It is accessible through a well-defined API through which the policy of infrastructure usage is specified by the application.

The ANSA-based Lancaster design mentioned above does not make such a distinction. Whereas the Touring Machine builds an object hierarchy with restricted access to some objects, in the Lancaster model all objects have equal status. While the first approach allows for a more "controlled" and "secure" system architecture through one special API, the open computing model from Lancaster lets the programmer select the appropriate object level for his need; all objects are accessible. A combination of two models can be achieved by integrating a layering facility into the Lancaster model.

The discussion following the talk took up the theme of telecommunications vs. computer industry and their role in multimedia support. It was agreed that with ATM integration into future computers, the traditional network boundary will more and more vanish. The telecommunications industry wants to make up for it by providing higher-level information services ("teleservices") as the new API to the multimedia network. It was believed that such a development is only possible as a joint effort with the computer industry.

Implementing HeiTS:
Architecture and Implementation Strategy
of the Heidelberg High-Speed Transport System

Dietmar Hehmann
Ralf Guido Herrtwich
Werner Schulz
Thomas Schütt
Ralf Steinmetz

IBM European Networking Center
Tiergartenstr. 8
D-6900 Heidelberg 1

Abstract: The **Heidelberg High-Speed Transport System** (HeiTS) is a new-generation end-to-end communication system currently under development at the IBM European Networking Center (ENC) in Heidelberg. HeiTS is aimed at a heterogeneous environment comprising several computers with different operating systems and a variety of underlying local, metropolitan, and wide-area networks. It incorporates both end-system and gateway communication functions.

1.0 Introduction

In new high-speed networks, an integration of different traffic types can be observed: Whereas, e.g., Ethernet was designed for data traffic only, FDDI supports data-type traffic (asynchronous) as well as voice and video traffic (synchronous), albeit in a rudimentary form. Such an integration in the physical network should lead to an integrated communication system as a whole. HeiTS is designed to support fast data traffic and multimedia communication, in particular the transfer of digital audio and video. The HeiTS prototype is conceived to be a generic basis for high-speed data transport applications such as CAD file transfer, and multimedia communication applications such as video distribution.

For many years, networks were the bottleneck of data transmission. Processing equipment in the end-systems and gateways was faster than the transmission lines so that bandwidth usage had to be optimized over processing. With the upcoming high-speed networks, this paradigm is changing: Shortage of resources is now in the nodes. To achieve high performance, both architecture and implementation of HeiTS are oriented to optimize processing over bandwidth use.

HeiTS is primarily directed to two platforms within IBM's Small Systems line: the PS/2 under OS/2 and the RISC System/6000 under AIX. This imposes the burden to interface HeiTS with two disjoint operating systems, but the common Microchannel bus architecture of both machines offers the opportunity to use identical communication network adapters on both systems. The primary networks to be operated by HeiTS are Token Ring, FDDI, and Broadband-ISDN. For B-ISDN, a first version will attach to an STM network, later versions will interface with an ATM network.

This paper discusses our implementation decisions for HeiTS. Section 2 introduces the overall multimedia system architecture into which HeiTS will be integrated. Section 3 elaborates on the support functions that serve as building blocks for the construction of HeiTS. Section 4, finally, discusses the actual communication functions that HeiTS provides. For the objectives of the system please refer to an earlier paper [5].

2.0 Overall Multimedia System Architecture

HeiTS is just one module within a general platform for multimedia applications. It, therefore, has to fit into the overall architecture of such a platform. This section discusses the external constraints on the implementation of HeiTS which result from the need for integrating HeiTS with a surrounding system.

2.1 Stream Handlers

Multimedia data usually enters the computer through an input device and leaves it through an output device (where storage can serve as an I/O device in both cases). It is less common that the data is generated by the computer itself, i.e., calculated or interpreted by the CPU. Nevertheless, this case occurs in simulation and control applications; a multimedia architecture needs to take it into account, too.

An entity generating or consuming a raw stream of continuous multimedia data is commonly called a *stream handler*. The term "raw" shall indicate that no stream-handler-specific data is contained in the stream. For example, in a transport system (which would be a typical stream handler), the data must not contain any protocol headers or trailers; they have no meaning to other stream handlers, e.g., those responsible for video output. We make no restriction on the encoding of a raw stream. In particular, a stream can be compressed or uncompressed. Any stream should be typed to prevent it from being directed to the wrong stream handler.

Stream handler functions are often distributed among the main CPU and on-board processors. In case of video decompression adapters most of the stream handler software executes on the on-board DSP of the adapter. Additional software completes the stream handler to access the board from an application. Whenever the adapter does not deliver raw data, but some additional information, the stream handler software that executes on the main CPU becomes more complex. Multimedia file or transport systems are examples of such stream handlers. In applications where multimedia data is generated or consumed by the computer, a stream handler may run on the CPU only.

In a modern system which follows the microkernel approach, one will arrive at the following three-level code structure for a stream handler implementation:

- *Hardware portions* of stream handlers are executed on an adapter board. In many cases, one will not be able to change these stream handler portions. However, some modern hardware stream handlers are microprogrammable.
- *Device drivers* constitute the stream handler portion executed inside the operating system kernel to interface to the hardware adapter.

- *Software portions* of stream handlers execute in user space, on top of the operating system kernel. From a software engineering viewpoint, the majority of the stream handler code should belong to this portion. Stream handlers which do not require special hardware support (e.g., simple filter functions) can be implemented in software only.

This layered stream handler structure is continued within each stream handler portion. In particular, the stream handler implementation of a transport system will contain the traditional communication layers.

2.2 Threads

Stream handlers need to obey the inherent real-time requirements of audiovisual data: They have to deliver their output before a certain deadline to make it available in time for its presentation or consumption. In addition, they may have to reduce jitter between the delivery of adjacent output items or synchronize the presentation of their data items with those of other stream handlers.

The following implementation structure makes it easy to take these timing criteria into account: All stream handler software portions are encoded as functions of a single task. This task assumes the role of an *audio/video server* which − somewhat similar to the X server − provides a common input/output environment for time-critical multimedia data. The functions of the AV server are executed by *threads* which escort a single piece of multimedia data from input to output. They wait to obtain the data from the device driver, then execute the layered functions of the input stream handler in an upcall fashion [2], execute the functions of the output stream handler in a downcall fashion and finally submit the data to the output device through the corresponding device driver functions.

Threads, much better than messages themselves, can take the timing requirements of multimedia data into account. Each thread can be scheduled according to the urgency of the data item it handles using real-time scheduling techniques. It also can be synchronized with the execution of other threads through well-known process synchronization functions. An appropriate synchronization point is the switch from the upcall to the downcall segment. A thread can also be paced at this point by delaying its execution for some time.

2.3 Stream Management

A multimedia application runs on top of the AV server outside of the real-time environment. Multimedia data usually does not pass through the application; the application merely manages the flow of data in the AV server. To manage the data flow, we distinguish between *device control operations* that determine the content of a multimedia stream and *stream control* operations that determine its direction.

Device control operations depend on the individual I/O device. Devices may be grouped into classes and their device control operations may be derived generically as suggested in [10]. For storage devices, typical control operations include *fast forward*, *reverse* and *seek*. Other operations are *zoom* for cameras and *volume* for speakers.

Stream control operations are the same for all stream handlers. They include

- *open/close* (applied to individual stream handlers, yielding *stream handles*),
- *connect/disconnect* (applied to pairs of stream handles, yielding *stream identifications*), and
- *start/stop* (applied to stream identifications).

The *open* function is a hybrid between a device control and a stream control operation. It usually requires information specific to the stream handler that also determines stream content. Except for stream handlers that cannot be multiplexed (such as those for microphones and speakers), opening the stream handler is not enough: In a file system, the file to be accessed needs to be known. In a transport system, the address of the communication partner is required. In a video display, the area where to display the data on the screen is needed. This information is provided by handler-specific parameters of the *open* call.

The *connect* function generates a thread to escort data from the input to the output stream handler. When the connection is established, system resources are allocated to ensure that the thread can perform its function according to the application's requirements (on time, with a certain reliability, etc.). In distributed applications, such resource reservation has to be made from end to end, including the network.

In addition to the above functions, an application can also specify that a connection (= thread) shall be synchronized. In this case, the above-mentioned synchronization mechanism is enabled and threads are potentially delayed.

Figure 1 shows the overall architecture of HeiTS.

The devices drivers (DD), the transport system (TS) and the audio/video handler (AV) are examples of stream handlers. Data flows from a network adapter through the corresponding device driver to the transport system, where the communication functions are provided (see 4.0), passed over to the audio/video handler and written to the adapter using the AV-device driver. The buffer management system (BMS) enables the handling of the data between the different stream handlers without copied them (see 3.1). The resource management system (RMS) allocates and manages the resources (see 3.2). The stream management system (SMS) provides the interface to the application.

3.0 Support Functions

In protocol implementations, the protocol machine contributes only a small fraction to the overall processing time. Most computation power goes into support functions such as data administration, communication between modules (e.g., processes), etc. This is even more true for light-weight protocols with their streamlined protocol machines.

In HeiTS our goal is to handle data in real-time. First of all, this means that some delay bounds for the data handling can be guaranteed. As delay bounds for audiovisual data are tight, this automatically translates into fast data handling. To handle data efficiently, a sophisticated *buffer management* is needed. To schedule the resources for real-time data handling, we need a *resource management* which reserves the resources in HeiTS and guarantees the commitments made.

3.1 Buffer Management

Conceptually, data always flows from one stream handler into another. However, if a significant portion of the stream handler is realized in software and makes use of the CPU, this flow of the data should not imply costly data copying, in particular copying from kernel to user space and back.

The buffer management system (BMS) enables the transfer of the data "below" the stream handlers to achieve higher performance. In this case, special device capabilities such as direct adapter-to-adapter transfer can be utilized and the BMS hides differences between buffers on different adapters and in main memory.

The BMS not only avoids copying while data is flowing between stream handlers, it also provides features needed for efficient protocol implementations such as chaining of buffer fragments (headers from different layers, data, possibly trailers) and locking (e.g., to keep buffers for retransmissions).

A BMS buffer consists of one or more blocks of memory (called fragments) which are linked together as shown in figure 2. The information describing the buffer is contained in a buffer descriptor, so no buffer management information has to be stored in the fragments.

A fragment consists of three parts:

- The *data area* is filled with information.
- The *empty area* is free and can be used to store information (e.g., to add a protocol header).
- The *dirty area* cannot be used (it can contain adapter specific information).

The pointers to the different areas are stored in the buffer descriptor. Space in a fragment is allocated from the back to the front, so each layer can add its header. If there is no empty space left in the fragment, a new fragment is allocated and is linked to the current buffer. So it is not necessary to copy the buffer.

A side effect of this scheme is, that segmenting and recombining of data units is possible without copying the data. To segment a data unit into, e.g., two pieces, the BMS allocates a second buffer descriptor which points to the same fragment(s) and only changes the pointers to the different areas accordingly.

Normally the last stream handler handling the buffer gives the buffer back to the BMS after the data is copied to the external device and the buffer is freed. Under some circumstances a stream handler may want to keep the data for later use, e.g., the transport layer may keep a buffer for retransmissions if it has to provide a reliable service to its user. In this case the respective stream handler can tell the BMS to *lock* the buffer. When a locked buffer is returned to the BMS, it is not actually freed until the lock is removed.

3.2 Resource Management

The resource management system (RMS) allocates resources for connections to guarantee a certain *throughput*, *delay* and *reliability* [6]. The workload model used in HeiTS is the *Linear Bounded Arrival Process (LBAP)*, which was introduced by Cruz [4] and employed, for example, in SRP [1]. Whenever a new connection is established, the RMS makes sure that this connection does not violate performance guarantees already promised to other connections. There are two types of connections: best effort and guaranteed.

The RMS consists of submodules for each resource (e.g., local resources like CPU processing capacity and network resources like bandwidth) to perform schedulability testing, reservation, and resource scheduling. The RMS also reserves buffer space. The buffers needed for a connection are allocated statically from a buffer pool. The amount of buffer space to be reserved is based on the throughput and the burst size.

Let us discuss the reservation and scheduling of the CPU as an example for the resource management techniques employed. For any new connection, a schedulability test is performed: Based on the maximum message rate of a connection and the processing time needed per message it is calculated whether the acceptance of this new connection could violate guarantees for other connections. If this is the case the new connection is rejected. The values calculated for accepted connections are stored in a local database. This information is used by the RMS for further schedulability tests and by the scheduler for scheduling the thread processing the message.

For CPU scheduling we are currently using three priority classes:

1. critical threads,
2. critical threads that have used up their processing time, as specified by their workload specification, but require further processing (this is based on our optimistic assumption, that this message can make up for the lost time in later resources and is especially useful for best effort connections), and
3. threads that are no stream handlers (the normal system threads).

Currently we investigate to use a fourth class for workahead messages, but it is not clear if the potentially better CPU utilization is worth the scheduling effort (which consumes CPU capacity itself).

The scheduling within the different classes is currently based on a modified rate-monotonic scheme, where the priority of a thread is based on the message rate for the respective connection. For incoming messages the urgency is calculated and depending on the outcome the message is passed to the thread handling the connection directly or it is hold back by the scheduler.

In the future we plan to also use deadline scheduling where the priority of a thread is calculated based on the urgency of an arriving message.

4.0 Communication Functions

Using the support environment introduced in the previous chapters, HeiTS realizes the function of the lower 4 layers of the OSI reference model as a stream handler providing endsystem-to-endsystem transfer of multimedia data items.

4.1 Design Decisions

Based on the application requirements either a reliable or an unreliable communication path between a sender and a single or multiple recipients can be established by HeiTS. QOS parameters are used to specify such requirements. Formally, the services provided are based on the ISO transport service standard document [7]; however suitable enhancements had to be defined to cover typical multimedia requirements. Additionally, the use of certain ISO defined optional facilities had to be restricted to ensure isochronity requirements could be met.

Specific design decisions have been made in the following areas:

- *Calling conventions:* Downcalls are used for outbound communication (i.e., requests and responses in the OSI terminology), upcalls for incoming indications and confirmations.

  This architectural decision ensures in particular that incoming data can not only be offered for processing to the application − as is standard practice in today's data-oriented communication systems −, but that any required processing can directly be initiated by HeiTS at the correct time. Side effects of this design decision include reduced elasticity buffer requirements and that immediate connection-specific processing can be done in the user-provided indication and confirmation routines.

  Entry points into these user provided functions are passed to the transport subsystem at the latest possible occasion.

- *Multicast:* Multicast is supported by the network layer where the topology of the network is known. Two forms of multicast are distinguished: In traditional "sender-initiated" multicast, the sender enumerates its communication partners. In "receiver-initiated" multicast, a receiver may join an existing communication (probably without even informing the sender about its presence).

- *Multiplexing:* Multiplexing is not supported for time-critical traffic above the data link layer, i.e., a data link connection will always be mapped onto a single transport connection. It is necessary to support multiplexing in the data link layer since some networks support only one physical connection. Its exclusion for network and transport layer allows for easier identification of the receiving process for incoming data.

- *Splitting:* Splitting is not supported, i.e., a single upper-layer connection does not use more than one lower-layer connection. In particular, one transport connection never sends data over two or more network adaptors. Splitting was once used to let a fast processor output data to several slow networks, increasing the overall throughput of a connection. In an environment, where networks become faster than processors, splitting becomes obsolete.

- *Segmentation:* Segmentation should be avoided, but is supported by HeiTS. Segmentation to and reassembly of very small data units (such as ATM cells), however, shall be accomplished in hardware − HeiTS is not optimized for this function.

- *Flow control:* Flow control consists of end-to-end flow control to prevent the receiver from being flooded with data and access control to prevent the network from being overloaded. For time critical unreliable traffic, HeiTS applies a rate-based control scheme for connections and enforces the rate of connections through leaky bucket algorithms. For conventional reliable data communication the standard techniques are used.

4.2 Implementation Structure

Internally the communication subsystem is structured into 3 layers:

*   *Transport Sublayer:* The transport subsystem provides reliable and unreliable end-to-end communication services enhanced by provisions for multimedia data transfer. The use of these provisions results in an isochronous data delivery whenever a respective quality of service was negotiated. As a starting point, an extended ISO transport service is considered. A modified ISO transport protocol class 1 [8] has been implemented. Other protocols (e.g., XTP) are under consideration.

*   *Network Sublayer:* The focus of the network layer work is in the areas of multicast support and LAN/WAN internetworking. Two different protocols are currently being implemented for experimentation: As a result of previous ISO work, a modified X.25 version is used for unicast experiments in gateway scenarios. The Internet protocol ST-II [12] is used to experiment with multicast communication over guaranteed-performance channels.

*   *Connectivity Subsystem:* The connectivity subsystem provides a data link service interface to HeiTS. On most of the already available adapters for high-speed networks a data link protocol is implemented. The connectivity subsystem hides the different interfaces of the drivers for the various network adapters. Network adaptors currently under consideration are 4 and 16 Mbit Token Ring, FDDI, and B-ISDN.

A stream handler interface is built on top of the HeiTS stack.

4.3 Sample Session

Let us discuss a "typical" example scenario from the transport system interface perspective. Assume the head of a small company wants to give his Monday morning speech (a monologue, of course) to his employees sitting in their offices with their multimedia workstations switched on and ready to listen to their boss. When the chief is ready to begin, we will see the events at the transport service interface as shown in figure 3.

In this example the symbol " > " is used to identify upcalls. "@" stands for "address of".

The example illustrates some of the key choices made for the HeiTS design: First, the concept of specialized service access points is used to distinguish multimedia and regular data traffic. Second, a set of QOS parameters is used to specify the application requirements. In this case typical values for the distribution of compressed video are given. In particular, error indication but no correction is specified. This enables the output stream handler to substitute, e.g., a corrupted video frame by either a previous full frame or a zero delta frame which will prevent the error from being visible. However, strong isochrony for the individual pictures and voice sample is required. Third, only addresses are passed at the procedure interfaces whenever data need to be handed over an interface.

5.0 Summary

HeiTS is designed to handle high-speed data applications as well as multimedia applications within IBM's Small System line (PS/2 under OS/2 and the RISC System/6000 under AIX). The main emphasis in this paper was put on the multimedia aspect. In order to meet the real-time requirements of audiovisual data streams HeiTS uses threads to handle such streams. These threads can be scheduled dependent on their real-time requirements. In order to allow this kind of scheduling the Resource Management System has been implemented in HeiTS. It allows best effort and guaranteed connections, and it supplies the scheduler with the necessary information for real-time scheduling.

Another aspect within HeiTS is to minimize the overhead of data handling. For this reason, a Buffer Management System was defined that allows efficient data handling. This includes segmenting and recombining of data units, chaining and locking of buffers, and other features. With this buffer management all unnecessary data movements can be avoided.

With all these supporting functions defined HeiTS is an implementation of the lower four layers of the OSI Reference Model. It allows multicast on the network layer, multiplexing up to the data link layer, segmentation, and ent-to-end flow control.

Currently a modified ISO transport class 1 has been implemented. But other protocols like XTP are under consideration for future implementations. There are also still some open issues in the resource management, e.g., the different threads are scheduled based on a rate-monotonic scheme, where the thread's priority is based on the message rate for the connection. This will be replaced by deadline scheduling, where the priority of a thread is adjusted to the urgency of an arriving message, and is not based on a QOS input parameter.

References

[1] *D. P. Anderson, R. G. Herrtwich, C. Schaefer:* **SRP: A Resource Reservation Protocol for Guaranteed-Performance in the Internet.** ICSI, Berkeley, TR-90-006, Feb. 1990.

[2] *D. D. Clark:* **The Structuring of Systems Using Upcalls.** 10th ACM SIGOPS Symposium on Operating System Priciples, Orcas Island, Washington, Dec. 1985, pp. 171-180.

[3] *D. D. Clark, D. D. Tennenhouse:* **Architectural Considerations for a New Generation of Protocols** SIGCOMM '90 Symposium "Communications Architectures and Protocols", Philadelphia, Pennsyslvania, Sep. 1990, pp. 200-208.

[4] *R. L. Cruz:* **A Calculus for Network Delay, Part I: Network Elements in Isolation.** IEEE Transactions on Information Theory, Vol. 37, No. 1, January 1991.

[5] *D. Hehmann, R.G. Herrtwich, R. Steinmetz:* **Creating HeiTS: Objectives of the Heidelberg High-Speed Transport System.** GI-Jahrestagung, Darmstadt, Oct. 1991.

[6] *R. G. Herrtwich, R Nagarajan, C. Vogt:* **Guaranteed-Performance Multimedia Communication Using ST-II Over Token Ring.** Submitted for publication.

[7]   *International Standards Organisation:* **International Standard 8072, Information Processing Systems − Open Systems Interconnection − Transport Service Definition.** 1986.

[8]   *International Standards Organisation:* **International Standard 8073, Information Processing Systems − Open Systems Interconnection − Connection-Oriented Transport Protocol specification,** 1986.

[9]   *N. Luttenberger, R. v. Stieglitz:* **Performance Evaluation of a Communication Subsystem Prototype for Broadband ISDN.** IEEE Workshop on the Future Trends of Distributed Computing Systems, Cairo, Sep. 1990.

[10] *R. Steinmetz, R. Heite, J. Rückert, B. Schöner;* **Compound Multimedia Objects − Integration into Network and Operating Systems.** International Workshop on Network and Operating System Support for Digital Audio and Video, International Computer Science Institute (ICSI), Berkeley, Nov. 1990.

[11] *D. L. Tennenhouse:* **Layered Multiplexing Considered Harmful.** In: H. Rudin, R. Williamson (Eds.): Protocols for High-Speed Networks, Elsevier (North-Holland), 1989.

[12] *C. Topolcic (Ed.):* **Experimental Internet Stream Protocol, Version 2 (ST-II).** Internet Request for Comment 1190, Oct. 1990.

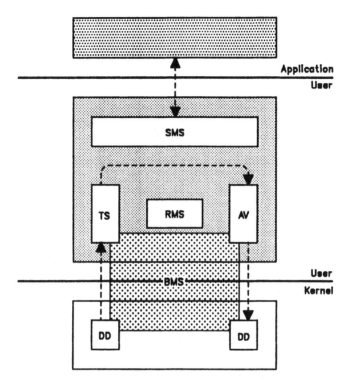

Figure 1. HeiTS: Architecture

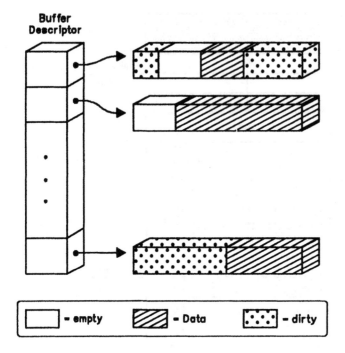

Figure 2. BMS: Buffers and Fragments

```
Chief                          Employees

                               ts_open_sap (MMSAP, ..., @ts_conn_ind, @ts_disc_ind, ...)
ts_open_sap (MMSAP, ...)
ts_conn_req (MMSAP, Employeelist,
            {error_indication, 1.4 Mbps, 25 SDUs per second, 250 msec constant delay},
            @ts_conn_conf, ...)
                               >ts_conn_ind (...)
                               ts_conn_rsp (..., @ts_data_ind, ...)
>ts_conn_conf (...)
ts_data_req (..., @first_picture_data)
                               >ts_data_ind (...)
  .                              .
  .                              .
  .                              .
ts_disc_req (...)
                               >ts_disc_ind (...)
ts_close_sap (MMSAP)
```

Figure 3. Unidirectional Live Distribution of Compressed TV

PROTOCOL SUPPORT FOR DISTRIBUTED MULTIMEDIA APPLICATIONS

Geoff Coulson, Francisco García, David Hutchison and Doug Shepherd

Computing Department
Engineering Building
Lancaster University
Lancaster LA1 4YR
UK

E.mail: mpg@comp.lancs.ac.uk

ABSTRACT

In this paper we describe ongoing work in protocol support for distributed multimedia applications. This work concerns the provision of suitable transport mechanisms to convey multimedia information (text, and digital voice and video) between multimedia workstations in a distributed system. There are two parts to the Lancaster multimedia work. First, we have developed an abstract model for multimedia communications that is based on the use of streams; and second, we have built an experimental system on which to test the implementation of protocols based on this model. This paper reports on both aspects of the Lancaster work, describing the results so far and identifying areas to be investigated further.

1. INTRODUCTION

The data communication world is rapidly evolving towards very high speed, and multiservice, capabilities. At the same time, the equipment that requires communications, such as supercomputers, scientific and graphical workstations, etc., is increasing the demand by producing or consuming data at higher and higher bit rates. These bit rates are gradually growing by a factor of ten every five years or less. The new applications envisaged as being required by users, such as voice and video on multi-function workstations, also demand the use of high speed networks that can provide the required bandwidth for multimedia communications. Fibre and advanced microelectronics technology have made the emergence of high speed networks possible.

Some of the major high speed networks which form part of recent and ongoing research and development, and which in some cases are already commercially available, are: the Broadband Integrated Services Digital Network (B-ISDN) [de Prycker 89], the Fibre Distributed Data Interface (FDDI) [Ross 90], the Distributed Queue Dual Bus (DQDB) [Cooper 91], and the Cambridge Fast Ring [Hopper 88]. It should be emphasised that the capability of these network technologies, together with associated communications protocols and distributed systems support functions, to carry multimedia information is the subject of current investigation. Perhaps the greatest interest lies in the development of the B-ISDN as the basis of the telecommunications networks of the future. During the past several years the B-ISDN has become a major research initiative. A vital consideration, which has so far received little attention, is how to map B-ISDN onto the ISO OSI communications protocol stack .

With the recent developments in powerful workstations, digital audio and video peripheral equipment, and optical mass storage, as well as multiservice high-speed networks providing broadband communications, there is a new user and applications pull in the area of multimedia information handling and exchange, involving text, graphics, audio and moving image. Multimedia applications are evident in areas as diverse as office systems, factory automation, health care, music, advertising, and multiple media information bases. In general, a key requirement is the integration of the various media in both processing and communications. At Lancaster University, a distributed systems platform is being developed that will provide a platform for handling the multiple media across a number of multimedia workstations [Blair 90].

Multimedia applications require the design and implementation of new communication protocols. However, it is vitally important that the requirements of future users and applications are investigated in detail before designing new protocols. These requirements should be studied together with the opportunities presented by the underlying high-speed networks. The necessity for this work is clear when one considers the current international standard communication protocols in OSI which reflect an older generation of communications infrastructure and traditional applications such as file transfer and remote login.

Within the framework of multimedia applications it is expected that the various media will be handled by a range of different styles of communication including datagrams and virtual circuits (VCs), both unicast and multicast. If audio and video are transmitted from a source via different VCs and are supposed to be in 'lip-sync', there will be a requirement for what we call *continuous synchronisation*. Also, the playout of organised multimedia documents and some interactive applications will require that after a sequence of video has been played out, or during the playout at a specific point, some required event should take place. This we call *event driven synchronisation*. For both types of synchronisation mechanisms we believe that support should be provided by the communications protocols as well as by the distributed systems platform.

The main objectives in developing new communications subsystems to support distributed multimedia applications are to support the efficient transmission of the multiple media over new high-speed multiservice networks and to meet the needs of the applications user in regard to Quality of Service (QoS). We believe that rate-based transport protocols are the correct starting point, and this paper describes their adoption and development towards meeting these objectives, in conjunction with the distributed systems platform that is being developed in parallel. Figure 1 summarises the architecture shared by the several multimedia projects at Lancaster University. The figure illustrates the role of rate-based protocols and distributed systems platform within the architecture, and also indicates their relation to the layers of the OSI protocol stack.

An experimental system has been built at Lancaster, using a transputer-based multiprocessor front-end unit to handle the multimedia peripherals and to implement the communications protocols and distributed systems platform [Ball 90, Scott 91]. The work described in this paper is based on that hardware platform, which is known as the *Multimedia Network Interface* (MNI).

2. REQUIREMENTS OF ISOCHRONOUS DATA

The nature of isochronous data such as video and audio impose certain timing constraints which must not be violated by future communication protocols if such real-time data is to be supported over multiservice high speed networks. With real-time data of this nature, first of all a sampling delay is incurred in the coding and digitisation of the analog signal. Secondly there is a transmission delay which depends on the characteristics of the underlying communications sub-system. This transmission delay is end-to-end as it represents the delay incurred from the point at which the sample is generated to the point at which it is presented. Thus the delay is affected by the actual network transmission, packetisation, the buffering required at the source, sink, and intermediate nodes (routers, gateways), and finally depacketisation before the signal is passed up for presentation.

A major requirement of audio and specially video data is high throughput. Studies with transport protocols such as TCP and OSI TP4 have shown that when they are used with high-speed multiservice networks, they present a serious obstacle in achieving high throughput [Clark 87]. There are two major problems which have been outlined with these protocols which restrict throughput; these are: the use of windows as the flow control mechanisms, and the difficulty in handling timers which are employed by the error control mechanisms. Thus new protocols which take the above mentioned characteristics into account, and which use lighter mechanisms to deal with flow and error control are required for the transmission of this type of isochronous data.

The timing constraints imposed on future communication sub-systems by audio and video data are end-to-end delay and delay jitter. Limits on the tolerable packet error rates and/or bit error rates may be employed by the connection management functions of the transport system (or the application user) to adjust the connection if the QoS falls below a specified threshold. The throughput requirements give indications of the bandwidth and buffering requirements to support the respective media.

3. DESIGN AND IMPLEMENTATION

This section outlines the design and implementation of the Lancaster multimedia transport infrastructure. The requirements of this infrastructure are twofold: first, to obtain the high throughput necessary to support audio and video data over the forthcoming multiservice high speed networks; and second, to provide a suitable distributed systems platform for use by applications.

3.1 Distributed System Support

The distributed system support is based closely on the ANSA architecture [ANSA 89]. More specifically, the implementation relies heavily on the facilities offered by *ANSAware*. ANSAware is a suite of software which provides a partial implementation of the ANSA architecture. In particular, it implements the concepts of *trading* for services according to abstract data type descriptions, and *binding* to those services by means of a remote procedure call protocol. It also provides facilities such as lightweight threads and basic primitives (eventcounts and sequencers) for inter-thread synchronisation.

ANSAware runs on a number of environments including UNIX, VMS and MS-DOS, and thus provides the necessary platform to construct distributed applications in a heterogeneous environment. For the purposes of our experimentation, ANSAware was also ported to the transputer environment in order to have a common platform across the entire configuration. It is therefore possible to access ANSA objects in the distributed environment whether they reside on the host workstation, on the MNI unit or on any of the other nodes in the distributed environment.

3.1.1 Base Objects

The distributed systems platform is comprised of a set of base objects (services) which can be subdivided into three general categories as follows: storage services, multimedia devices and network services as shown in Figure 2.

The various components highlighted in the Figure are described in more detail below:-

- *Network Services*

  The main task of the network services layer is to provide a range of protocols (of varying qualities of service) to handle multimedia communications. This required the provision of *stream* services through the ANSA architecture (this is in addition to the standard RPC protocols already supported by ANSA). Streams represent high speed, one way connections between an information source and sink. They are therefore abstractions over multimedia protocols and map on to various protocols provided by a protocol stack. This protocol stack is implemented on one of the transputers in the multimedia network interface and is described in section 3.2. However, as streams are ANSA services, they appear to applications simply as abstract date type interfaces with operations such as *connect* and *disconnect*, together with operations to monitor and control the underlying transport QoS. The network services layer also provides a number of compression services which can be used to reduce the bandwidth requirements of multimedia data. Finally, an interface is provided for synchronisation support that is realised by the transport protocol and orchestrator to be described in section 3.2.

- *Multimedia Devices*

  Each device provides an interface consisting of two parts: a device dependent part and a device independent part. The device dependent part offers a number of operations specific to that device. For example, the camera device has operations to focus, tilt, etc. The device independent part has a number of operations to create communication points (referred to as endpoints) and to send and receive information. At present, a small number of devices have been incorporated into the system including a camera service and an interactive videodisc player (as a source of still and moving image). A windowing system supporting video window capabilities is also provided at this level. To achieve this, the X-server component of X-Windows was ported on to the transputer environment and then modified to support video windows.

- *Storage Services*

  The approach taken at Lancaster is to provide a number of specialised storage services for each media type. This allows each storage service to be tailored to the characteristics of a particular media type, which may involve specialised hardware support or a particular approach to storing and retrieving information. Each of the units of information is an object in its own right and provides the abstraction of a *chain*. Chains are a generalisation of the voice ropes adopted in the Etherphone project for storing audio information [Terry 88]. A chain consists of a number of individual links, e.g. single picture frames, connected together into a sequence. It is perfectly valid to have a degenerate chain consisting of one link.

An important service not described in the above sections is the *synchronisation manager*. This management interface is offered to applications as a service to which all application synchronisation requirements may be delegated. Requirements for event driven synchronisation are expressed as an interpreted script which specifies actions to be taken on the basis of events generated by base services such as devices or streams. Continuous synchronisation is specified according to parameters such as packet ratio between the related streams, the tightness of synchronisation required, and permissible actions to take to regain lost synchronisation (such as packet discard, increased throughput etc.). The synchronisation manager makes use of *orchestrator* processes (described in section 3.2.1) to implement continuous synchronisation. It instantiates instances of the orchestrator as required at the source and/or sink of the synchronised streams, and passes parameters to the orchestrator(s) according to the high level user supplied requirements.

3.2 Outline of Transport Protocol Design

The protocol design is based on techniques employed in the implementation of high-speed communication protocols on parallel processor architectures [Zitterbart 89] capable of supporting the high bandwidths of the up and coming high-speed networks. The transport protocol is therefore subdivided horizontally into a send part and a receive part. These work concurrently and communicate with each other for the exchange of synchronisation information. This design can be extended by implementing the two separate sides on different processors and applying the same approach to the whole of the communications sub-system as discussed in Zitterbart. Results employing these techniques for implementing conventional transport protocols such as TP4 and TCP have shown a noticeable increase in throughput performance [Giarrizzo 89].

Because of the multi-processor architecture of the Lancaster MNI unit [Scott 91], the transport protocol and services are implemented on two transputers. One of the transputers acts as the master through which the required protocols from the stack are selected. This master transputer also implements the connection state tables and the orchestrator(s) functionality. The services provide by the slave transputer are initiated and managed by the master transputer via the use of a transputer link connecting the two transputers together. As mentioned in the design goals above, in each transputer, we have implemented separate send and receive shared memory pools. These pools are available to the individual transport services and the required protocols which are implemented as lightweight process threads. In this prototype configuration, both transputers are required to perform transmission and reception of data, thus individually they can function concurrently but the implemented services between them run under simulated concurrency as context switching takes place. However for our experimentation this is suitable and it provides a platform from which we can learn of the required state synchronisation, protocol profile selection and orchestration, if such a system was to be built using dedicated hardware solutions.

To implement the required concept of streams, simplex VCs are employed to provide the required stream QoS. If a full duplex connection is required, this can be built out of two simplex VCs where information travels in opposite directions. For full duplex VCs, control data required for protocol operation is piggybacked onto the transferred data. In the case where a simplex VC is established for the transmission of real-time data, the transport service establishes what we call a signalling VC in the opposite direction to provide all the required control data. These services are provided transparently to the distributed systems platform. An outline of the type of relaxed rate-based flow control protocol that we are experimenting with is illustrated in Figure 3. By relaxed we mean that no retransmissions are performed in the case of transferring real-time data such as audio and video.

The application process (AP) user is OSI terminology, and represents the device producing the audio/video data or the point at which this data is played out. The *writer process* is concerned with the gathering of samples produced by a particular device into the required buffer size to be used by the protocol profile, that is the burst size. When the buffer is full, the *send process* is responsible for fragmentation into packet sizes to be transmitted across the network and supplying the relevant headers to each packet. Also, it controls the rate at which each burst size is transferred to the network. The same principal is employed when data is received. The *receive process* strips the header information and fills a buffer while the *reader process* empties a filled buffer passing the information up to the point at which it is to be played out.

Like the contemporary rate-based flow control protocols [Cheriton 86, Chesson 89, Clark 87] flow of information is controlled by rate. At connection set up, the transmitting client and the receiving client negotiate the rate at which the receiving client can operate. The rate parameter represents the maximum time period required by the receiving client to consume and process the negotiated burst size. To the sending client, this rate parameter represents the maximum time it has to transmit a burst before transmitting another burst. Thus the transmitting client blasts the packets forming a particular burst as fast as the underlying network allows. This transmitting client then waits for the rate timer to expire before transmitting the next burst, by which time the receiving client should have had enough time to consume and process the previous burst. This way the protocol lends itself for a continuous high throughput provided there is sufficient buffer space at both ends of the connection for more than one burst.

3.2.1 Additional Functionality for Continuous Synchronisation

In order to provide the required functionality to support continuous synchronisation the connection management services have to be extended as discussed below. Continuous synchronisation is achieved by having a process monitoring the various streams to be synchronised. The appropriate location of this process within the OSI Reference Model is, we consider, within the region covered by the rate based flow control protocols and the distributed systems platform (see Figure 1). In fact, the upper layer architecture has control over it, and the communications sub-system provides all the signalling required.

Once the associations between the various simplex VCs to be synchronised have been made and any renegotiation required has also been performed a process, called the *orchestrator*, ensures that the streams are maintained in continuous synchronisation. In its simplest form, each individual transmission side of the simplex VCs transmits a buffer to the receiving sides at opposite ends of the connections and waits until they are signalled via signalling simplex VCs that they may send another buffer. In a perfect environment there would be no network congestion, no packet or bit losses, and the rates and burst sizes would have been negotiated to sustain the required inter-stream timing. Under these circumstances, once the transmission of data had been initiated for the streams at the same time, they would remain synchronised for the duration of the session. Of course, in practice these ideal conditions do not hold, and it is also likely that the streams may take different paths through the underlying network: there is thus a need for the orchestrator to intervene in order to maintain continuous synchronisation as required by the application (see Figure 4).

The error control mechanisms will also inform the orchestrator of any degradation of QoS which may take place on any of the streams. The orchestrator will then block the transmission of the next burst of data from each stream by not sending the required control command to allow the next burst to be transmitted. For the stream or streams which are suffering from degradation of QoS, the orchestrator will relay a command informing the source of data of the relevant action to take. This could be increasing or decreasing the burst size to smooth out jitter, slow down the burst rate etc, the same action of course being performed at the sink for the particular stream. The orchestrator will then be signalled when the relevant changes have taken place, and will resume the transmission of data from each stream via a time stamp. Note that the above explained how the orchestrator functions when the streams merge at a common sink. If the streams extend from a common source, the orchestrator will be signalled by the signalling simplex VCs when the individual destinations are ready to receive the next burst or if degradation of QoS has taken place.

3.2.2 Evaluation

In this type of protocol, elastic buffers are employed, in which increasing or decreasing the burst size as required allows delay jitter to be smoothed out. Also if the receiver or network becomes congested, the rate at which the burst is transmitted may be modified thus preventing fewer packets of information to be dropped at the congested receiver. A second advantage is that the synchronisation between streams is performed at the buffer level and not at the packet level, thus there is no need for synchronisation markers to be inserted in the user's data stream [Shepherd 90]. When blocking takes place, if the streams do fall out of synchronisation, it takes place for each stream at the next logical unit of time. The third advantage is that the AP user has full control of the streams which he requires to have synchronised in parallel, and these may be disconnected individually.

As no retransmission schemes are employed, packet losses may have a serious effect when the media are played out. For audio at 64 kbit/s if a packet is lost an audible click may occur. With video however, the effect may be more pronounced depending on the video coding employed (fixed bit rate or variable bit rate coding) [Verbiest 88]. A single packet loss may corrupt a large part of the frame. However as the number of frames to be transmitted lie in the order of 16 or more per second, these effects may be tolerable and in any case the rate based control protocols will be adjusting themselves to provide the required QoS if packet losses exceed a certain threshold.

3.3 Link Level Support for Rate Control Protocols

Future multimedia transport systems should be general enough so that they can be positioned above any of the evolving multiservice high-speed networks that can support the media in question. For this to become a reality they will require support from the underlying logical link control protocols which provide the direct access to the underlying network. In this section we aim to look at the support required by rate-control transport connections over such networks.

For fast packet networks such as FDDI where each station requesting synchronous bandwidth is assigned a time value during which they can transmit their synchronous data while holding the token, a scheme like the one suggested by Kanakia [Kalmanek 90] may be employed to support rate-based protocol profiles.

The main aim of such a scheme is that for each rate controlled connection established, a connection identifier and a burst size parameter may be passed to the logical link control protocol. This protocol can then implement a simple round-robin queue discipline with the connection identifiers. Thus when a particular connection is serviced its quota of packets are removed from the connections buffers and transmitted, that is the full negotiated burst is transmitted. The serviced connection identifier is then placed at the back of the queue and the next connection is then serviced. The protocol knows the size of the burst for each connection and can therefore implement some form of counter mechanism to ensure that the full burst is transmitted.

It is suggested that a connection should only be serviced if a full burst for that connection is ready in the allocated buffers, otherwise the connection identifier is placed at the back of the queue and another is serviced. This is necessary because otherwise we might interrupt the rate of transmission of other connections waiting to be serviced. The connection which is not ready may be waiting for acknowledgement from a receiver or some other event might have taken place to slow down or stop the flow of information. The protocol will of course have to be notified when a full burst for a particular connection is ready. This could be done by passing the connection identifier and burst size parameter each time a burst is ready for transmission to the underlying logical link protocol which could implement some form of flag notification scheme to determine which connections within the service queue are ready for servicing. The burst size parameter needs to be passed each time because this may change during the lifetime of the connection as described in the previous sections. Connection management services will of course have to ensure the addition and removal of connection identifiers from the queue on connection establishment or release.

If we look at a network such as DQDB which transmits a continuous sequence of 125 µs frames and each frame is further formatted into DQDB slots which each has an access control field, a slot header, and a payload (data) field a different mechanism to that for packet switched networks is required to support rate controlled connections at the link level. In this type of network isochronous traffic is carried by reserving octets in slots, thus reserving a single octet in every frame provides you with a channel bandwidth of 64 kbit/s. For this type of network, the scheme outlined above may be extended as suggested in [Kalmanek 90] to provide the required support.

In the first case the connection data queues are only serviced each time a new frame arrives at the particular node. The logical link control protocol then services each queue, again in a round robin manner, but for the number of allocated slots reserved for that connection. It can then update a pointer value which indicates from where in the queue for a particular connection it should start servicing when the next frame arrives. Notice that for this scheme the burst size is not relevant, what will affect the number of slots reserved for a particular connection is changes (if any) in rate of transmission. This scheme is also suitable for ATM type networks when working under an isochronous mode.

4. CONCLUSION

This paper has described an object-based communications support platform for distributed multimedia applications over high speed networks. This platform is compatible with current proposals for Open Distributed Processing (ODP) standardisation. The stream-based model we have developed appears to be generally promising and has been validated in a number of prototype applications using the transputer-based experimental system.

Also, the paper has presented an outline of a rate based flow control protocol suitable for experimentation with synchronisation between multimedia streams. The protocol is intended for use with audio and video data; similar protocols which employ selective retransmission mechanisms may be employed to handle other data which require error free end-to-end transmission. One of the most promising prospects for this type of protocol is that it is a natural candidate to employ over ATM networks where it is expected that the flow control and error recovery mechanisms will be performed at the end-points [de Prycker 89].

ACKNOWLEDGEMENTS

This paper describes the results of work carried out within two distributed multimedia projects at Lancaster University: the Multimedia Network Interface (MNI) project is supported jointly by the UK Science and Engineering Research Council and by British Telecom Laboratories; and the OSI 95 project is a collaborative ESPRIT project funded by the European Commission.

REFERENCES

[ANSA 89]
ANSA Reference Manual, Release 01.01, APM Ltd., Poseidon House, Castle Park, Cambridge CB3 0RD, UK, July 1989.

[Ball 90]
F. Ball, D. Hutchison, A. Scott, and D. Shepherd, "A Multimedia Network Interface", **3rd IEEE COMSOC International Multimedia Workshop (Multimedia'90)**, Bordeaux, France, November 1990.

[Blair 90]
G.S. Blair, G. Coulson, P. Dark and N. Williams, "Engineering Support for Multimedia Applications in Open Distributed Processing", **3rd IEEE COMSOC International Multimedia Workshop (Multimedia'90)**, Bordeaux, France, November 1990.

[Cheriton 86]
D.R. Cheriton, "VMTP: A Transport Protocol for the Next Generation of Communication Systems", **SIGCOMM'86,** ACM, 1986, pp. 406-415.

[Chesson 89]
G. Chesson, "XTP/PE Design Considerations", in **Protocols for High-Speed Networks,** H. Rudin and R. Williamson (eds.), Elsevier Science Publishers B.V. (North-Holland), IFIP, 1989.

[Clark 87]
D.D. Clark, M.L. Lambert and L.Zhang, "NETBLT: A High Throughput Transport Protocol", **ACM Computer Communication Review,** Vol. 17, No. 5, Special Issue, August 1987.

[Cooper 91]
C.S. Cooper, "High-speed Networks: The Emergence of Technologies for Multiservice Support", **Computer Communications,** Vol. 14 No. 1, January/February, 1991, pp. 27-43.

[Giarrizzo 89]
D. Giarrizzo et al, "High-Speed Parallel Protocol Implementations", **Protocols for High-Speed Networks,** H. Rudin and R. Williamson (eds), Elsevier Science Publishers B.V. (North-Holland), IFIP, 1989.

[Hopper 88]
A. Hopper and R.M. Needham., "The Cambridge Fast Ring Networking System.", **IEEE Transactions on Computers,** Vol. 37, No. 10, 1988.

[Kalmanek 90]
C.R. Kalmanek, H. Kanakia and S. Keshav, "Rate Controlled Servers for Very High-Speed Networks", **GLOBECOM 90,** San Diego, 1990.

[de Prycker 89]
M. de Prycker, "Evolution from ISDN to BISDN: a Logical Step Towards ATM", **Computer Communications,** Vol. 12, No. 3, June 1989.

[Ross 90]
F.E. Ross, "FDDI - A LAN Among MANs", **ACM Computer Communication Review,** Vol. 20, No. 3, July 1990, pp. 16-31.

[Scott 91]
A. Scott, F. Ball, D. Hutchison and P. Lougher, "Communications Support for Multimedia Workstations", **Proceedings of 3rd IEE Conference on Telecommunications (ICT'91),** Edinburgh, Scotland, March 1991.

[Shepherd 90]
D. Shepherd and M. Salmony, "Extending OSI to Support Synchronization Required by Multimedia Applications", **Computer Communications,** Vol. 13, No. 7, September 1990.

[Terry 88]
D.B. Terry and D.C. Swinehart. "Managing Stored Voice in the Etherphone System", **ACM Transactions on Computer Systems**, Vol. 6 No. 1, February 1988.

[Verbiest 88]
W. Verbiest, L. Pinnoo and B. Voeten, "The Impact of the ATM Concept on Video Coding", **IEEE Journal on Selected Areas in Communications**, Vol. 6, No. 9, December 1988.

[Zitterbart 89]
M. Zitterbart, "High-Speed Protocol Implementations Based on a Multiprocessor Architecture", **Protocols for High-Speed Networks**, H. Rudin and R. Williamson (eds), Elsevier Science Publishers B.V. (North-Holland), IFIP, 1989.

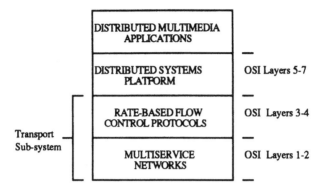

Figure 1: Lancaster multimedia architecture

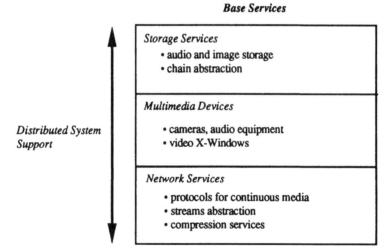

Figure 2: Base services platform

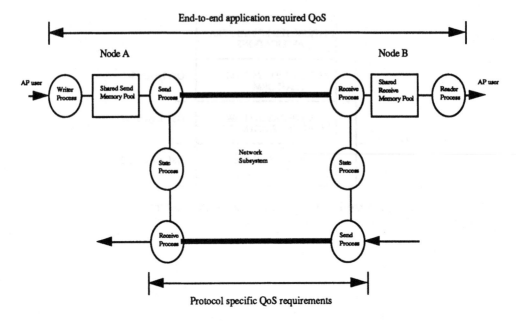

Figure 3: Rate-based flow control protocol profile

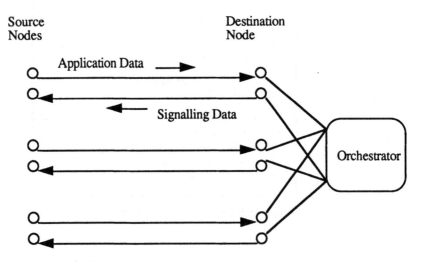

Figure 4: Orchestrator functioning at the sink

INTEGRATING COMPUTING AND TELECOMMUNICATIONS: A TALE OF TWO ARCHITECTURES

Lillian Ruston,[1]
Gordon Blair, Geoff Coulson, Nigel Davies[2]

ABSTRACT

The emerging importance of multimedia applications has led to many ideas about how best to support multimedia in a distributed environment. Lancaster University proposes one approach, i.e. start with models emerging from the Open Distributed Processing (ODP) standardization process, and extend them to incorporate multimedia services. Bellcore's Touring Machine proposes a different approach: let developers of multimedia applications specify an interface, and then design a network that meets the specification. Each approach has relevance, and an ideal environment must incorporate both, but questions remain about how the two perspectives fit together. We have contrasted and compared the Lancaster and Touring Machine approaches in an attempt to integrate their individual perspectives into one design. This paper documents our results.

1. INTRODUCTION

The last few years have seen the emergence of "multimedia computing," a research area focused on implementing multimedia applications on computing equipment. Multimedia computing arises from awareness of the range of applications created by combining information sources such as voice, graphics, and video, and from technological advances that make implementing such combinations possible [Davies,91]. Although most multimedia computing research to date centers on designing and implementing isolated multimedia systems consisting of single applications on stand-alone hardware, a greater potential for multimedia computing lies in implementing multiple multimedia applications on top of some distributed network support. Examples of such implementations include office information systems, medical applications, multimedia conferencing, and distance learning.

Implementing multimedia applications in a distributed environment demands a network architecture that seamlessly integrates the technological achievements afforded by telecommunications research with those afforded by computer science, but research aimed at realizing such an architecture remains at an early stage of development. Telecommunications and computer science have remained isolated from each other, making joint development projects historically difficult to

1. Bellcore, MRE 2A-247, 445 South Street, Morristown, NJ, USA, 07962, 201-829-4958, lillian@thumper.bellcore.com
2. Distributed Multimedia Research Group, Department of Computing, Lancaster University, Bailrigg, Lancaster, LA1 4YR, UK, +44 (0)524 65201, mpg@comp.lancs.ac.uk

accomplish. Telecommunications researchers generally devise methods for extending existing telecommunications architectures to support multimedia services [Bussey,88] [Timms,89], whereas distributed computing researchers devise methods for extending existing distributed systems [McMordie,91] [Nicolaou,91] [Hazard,91]. Integrating the two areas is uncommon.

This paper presents our view about how to integrate telecommunications and distributed computing multimedia architectures. For clarity, we illustrate our ideas by comparing and contrasting two specific projects: Bellcore's Touring Machine project and Lancaster University's extended ANSA architecture. The former exemplifies contemporary telecommunications architecture design; the latter, derived from Open Distributed Processing standards, exemplifies distributed computing.

2. THE TWO PROJECTS

2.1 An Overview of Touring Machine

2.1.1 Background on Touring Machine Bellcore's Touring Machine project [Bates,90] emphasizes an experimental approach to traditional telecommunications software management problems. Researchers involved with the project collectively maintain a multimedia infrastructure, called Touring Machine, which they also use to test out individual research proposals. Touring Machine has been available since fall 1990, and several network services have been implemented on it.

Touring Machine is heavily influenced by the perspective of a public telecommunications network provider. In particular, the architecture reflects the following assumptions:

1. The overall architecture separates into two domains: a telecommunications infrastructure, supplied and maintained by the telecommunications industry, and customer station equipment, supplied and maintained by individual telecommunications customers. Telecommunications applications, provided either by external vendors or individual telecommunications customers and executing on customer station equipment, interact with the infrastructure to establish communications among customers. The infrastructure provides session management, resource allocation, and directory services, among other functions. Interaction between applications and the infrastructure is via an application programming interface (API) that defines the capabilities offered to applications by the infrastructure.

2. The telecommunications infrastructure provides mechanisms for managing resources and other conflicts, but it does not impose policy. Policy determination derives solely from application specification.

2.1.2 Touring Machine Design

2.1.2.1 The Applications Layer Touring Machine does not specify or control objects that reside at a customer station. By necessity, however, Touring Machine assumes the existence of two customer objects: users, and clients. A *user* is the responsible entity on whose behalf service is authorized, and usually implies a human being. A *client* is application software that executes at a particular station for a particular user. Separating users from clients lets a user have multiple clients at a station acting on its behalf. Thus, a user might have one client handle electronic mail and another client handle incoming video calls. Touring Machine provides a directory service through a name server that allows registration of users and clients; clients use this service to locate other clients to set up communications.

Touring Machine postulates the existence of a special client, a *station manager*, at each station. The station manager provides the policy for arbitrating conflicts that occur among client requests at a Touring Machine station. For example, an incoming call to a station where the required local station resources are not available may be accepted by that station but not given station resources – i.e., the policy involves placing such incoming calls "on hold." The interaction between a station manager and clients is not dictated by Touring Machine. A developer may implement different station managers to provide different policies to clients, and a developer may choose not to implement any station manager at all. In the absence of a station manager, Touring Machine provides a default policy for managing local resources.

2.1.2.2 The Control Layer There are two objects in the Touring Machine control layer: station objects and session objects. A *station object* represents the point of contact between clients registered with that station object and the rest of Touring Machine, for it forwards client requests to other Touring Machine objects. Furthermore, the station object provides mechanisms that map and unmap *endpoints*, logical transport terminators, to *ports*, physical transport terminators that attach devices such as microphones, speakers, and cameras. Clients registered with a particular station object may make map and unmap requests to that station object, and the request succeeds if permitted by the station manager policy for arbitrating access to station resources

A *session object* manages the communications interactions among multiple clients. It is the intermediary between clients wishing to communicate and internal Touring Machine resources required for their communication. The clients associated with a session negotiate by passing messages among themselves using session object mechanisms until they agree on a session management policy and on a session transport topology, specified as a set of connectors. A *connector* abstracts a multiway media-specific transport connection among endpoints and hides details of physical resources from applications. Once the clients reach agreement, each client asks its station object to allocate local resources by mapping endpoints to physical ports, and the session object asks transport layer objects to allocate Touring Machine resources that realize the connectors. As with the station object, Touring Machine imposes a default policy by which the session object grants requests to change the session transport topology, but the API gives clients the option to impose a different policy if they choose. For example, the policy may allow all session clients to make requests, or the policy may only accept requests from the initiating client or the billed client.

2.1.2.3 The Transport Layer

The transport layer provides access to physical communication facilities. The transport layer contains two classes of objects: objects existing inside Touring Machine and managed by the session object according to client-defined policy, and objects existing outside Touring Machine and managed by the station object according to station manager policy. Examples of the former include bridges and trunks; examples of the latter include microphones and speakers. We say no more about transport layer objects here, for their description is not necessary in the present discussion.

2.1.2.4 Connection Manager
Touring Machine objects send messages to each other using the Connection Manager, an asynchronous message-passing communication interface layered over available system communications facilities. The Connection Manager provides location independence, simplified message semantics, and a uniform object structuring function.

2.1.3 Example Application Using Touring Machine
As an example, we show how a multi-party audio-visual conference occurs within Touring Machine. In this application, a group of users communicate using both audio and visual connections, and audio and visual information is multicast to all users.

Figure 1 illustrates the configuration of objects required to support the application. As seen in the figure, each user requires at least four ports: an audio input port, an audio output port, a video input port and a video output port.

Establishing the conference involves several phases:

- *Phase 1: Registering with Touring Machine*

  Clients wishing to use Touring Machine first register with a station object. During registration, the station object authenticates and authorizes the client, publishes client information in the name server for use by other registered clients, and determines those local resources available to the client through subsequent client requests.

- *Phase 2: Establishing the session*

  One client asks its station object to create a session object, and then asks the session object to contact other users and add them to the session. The session object uses the name server to determine the clients responsible for handling conference applications for each user requested, and then contacts the clients.

- *Phase 3: Establishing conference transport*

  The clients in the session negotiate with each other to create a session transport topology for the conference. The session object ensures that all necessary clients agree to the topology. Each client asks its station object to allocate local resources by mapping endpoints to physical ports, and the session object asks transport layer objects to allocate Touring Machine resources that realize the connectors.

- *Phase 4: During the conference*

  Any client in the session may ask the session object to alter the established session topology. As before, the session object again ensures that all necessary clients agree to the change before making change requests of transport layer objects, based on the policy agreed to by the participants when the session was created. Eventually, one client asks the session object to end the session, tear down connections, and deallocate resources.

2.2 An Overview of the Lancaster Architecture

2.2.1 Background on the Lancaster Architecture

The Lancaster architecture [Coulson,90] is based closely on the computational model supported by Open Distributed Processing (ODP) standards in general and the Alvey sponsored Advanced Networked Systems Architecture project (ANSA) [ANSA,89] in particular. In this model, services are represented as *objects*, and an object supports one or more *interfaces*. An application consists of several interacting objects.

An object becomes available to the system by *exporting* one or more of its interfaces to a *trader* object. In this sense, the trader acts as a database of services. Similarly, objects wishing to access a service *import* an interface from the trader. The export and import services do not imply a particular implementation; each designer may choose its own methods for supporting these services.

Operations defined on an interface are *invoked* using an underlying *remote procedure call* protocol. This hides the location of objects from the calling object.

The collection of interfaces to an object represent the *type* of the object, the external behavior of the object. Each object type is implemented through one or more templates that describe the internal behavior of the object in terms of data structures and algorithms. In addition, new objects can be created from templates by *factory* objects, i.e. objects that respond to a *new* operation by creating a new instance of a type using a particular template.

2.2.2 The Lancaster Design

To support multimedia computing, we need to extend the basic architecture to give it the capability to support continuous media such as audio, video and animation [Anderson,90]. The approach taken at Lancaster involves introducing additional object types that abstract over the details of multimedia manipulation. We describe these extensions below.

The first extension, *streams*, abstract over the details of multimedia protocols. Streams represent unidirectional m:n connections between a set of media sources and a set of media sinks. To model the variety found in multimedia systems, an object can request a stream with a particular *quality of service*, where the quality of service of a stream defines values for properties of the stream such as throughput, latency, jitter, and error characteristics. The Lancaster design also supports temporal synchronization across two or more streams [Blair,91b], e.g. to achieve lip sync between a voice and a video stream.

The second extension involves introducing objects that abstract over the details of multimedia devices such as cameras, microphones and video windows. A multimedia device object presents two interfaces to the system: a *device-dependent* interface and a *device-independent* interface. The device-dependent interface varies from device to device, but the device-independent interface remains the same for all devices. For example, a camera supports operations such as focus and tilt, and all devices have operations such as start or stop.

The device-independent interface, called a chain, derives from a generalization of the rope concept used in the Etherphone project [Terry,88]. A chain abstracts over a continuous media source or sink and supports operations to control the production and consumption of continuous media data. Each chain can create one or more *endpoints* when requested, and, like streams, the requestor can request creation of an endpoint with a particular *quality of service*. An endpoint acts as a connection point to a stream. Note that a device may support multiple chains and endpoints, thus modeling simultaneous connections to a single device.

Related objects can be collected into a *group*. A group represents a particular type of object, one with two specific interfaces. The first interface, a group invocation interface, allows a requestor to invoke operations on all group members in a single action. The second interface, a *management interface*, includes operations to join or leave the group, and to return the current membership of the group. The management interface also allows a requestor to set policies that constrain group membership or order invocation requests, among others. Groups play an important role in the Lancaster architecture, as becomes evident in the example below.

The Lancaster architecture includes several other extensions to the ANSA architecture that do not pertain to the present discussion, so we do not discuss them here. We refer readers to the references [Blair,91a] and [Coulson,90] for more complete descriptions of the Lancaster architecture.

2.2.3 An Example Application using the Lancaster Architecture We now present an example application using the Lancaster architecture. To compare the Lancaster architecture with Touring Machine directly, we present the same application as we did before: an audio-visual conference.

Figure 2 shows the objects required to support an audio-visual conference. The diagram depicts two connections: one from a group of microphone endpoints through an audio stream to a group of speaker endpoints; and one from a group of cameras through a video stream to a group of video windows. The diagram also shows the device-dependent interface, the endpoint interface, and the chain interface for each device, and the group invocation interface and management interface for each group.

Establishing the audio-visual conference involves the following:

* *Phase 1: Registering devices*

  Each site in the conference assembles the basic configuration of devices required for participation, i.e. a camera, a microphone, a speaker and a video window. The first three objects are persistent objects that already exist on each machine. A new video window is created for the conference by invoking a video window factory. After creating devices, we then create a chain/endpoint pair with appropriate quality of service properties for each device.

- *Phase 2: Creating device groups*

  The next stage involves creating group objects from a factory. These group objects represent the microphone group, the speaker group, the camera group, and the video window group. Next, we join all conference devices to the appropriate device groups

- *Phase 3: Establishing transport connections*

  Establishing connections involves creating streams and joining them to the groups. For our conference, we need a video stream and an audio stream, and so we invoke creation operations on a stream factory, quoting the required quality of service properties for each operation. Following this creation, we connect the microphone group and the speaker group to the audio stream, and we connect the camera group and the video window group to the video stream. This completes the configuration shown in Figure 2.

- *Phase 4: During the conference*

  The conference starts by invoking start operations on all camera and microphone devices through the group invocation interface of the camera and microphone groups. Synchronization within a group is handled by the group's internals; synchronization across different groups has to be handled by some external entity. We do not discuss this latter synchronization here.

  New members can join the conference by creating appropriate devices and placing these objects in the appropriate groups. Moreover, any member may leave the conference at any time by exiting from the appropriate groups.

3. A COMPARISON OF THE TWO PROJECTS

The two projects have several design similarities. For example, both architectures adopt an object-oriented approach, with objects encapsulating services and operations invoked by a message passing protocol. Both architectures provide access to multimedia sources or sinks, and both establish multimedia connections between sources and sinks. The Touring Machine connector resembles the Lancaster stream, the Touring Machine session object and station object resemble Lancaster group objects, and the Touring Machine name server resembles the Lancaster trader. Finally, both projects have communications mechanisms that hide the physical locations of objects.

Nevertheless, the two projects have two major distinctions, one concerned with the *scope* of objects; the other, with *granularity*. This section discusses these distinctions, explains their origin and their influence on the different designs, and then concludes with a discussion on the benefits of an integrated architecture.

3.1 Scope of Object-Oriented Architecture

The first distinction involves the scope of objects. Derived from a telecommunications perspective, Touring Machine separates the telecommunications infrastructure from customer stations, builds a hierarchy of objects, and then denotes some objects as internal, privileged, and restricts access to them accordingly. In contrast, the Lancaster architecture makes no distinction between the telecommunications infrastructure and a customer's station but rather supports an object-oriented architecture that spans all levels of the system. In the Lancaster architecture, all objects have equal status.

The distinction between the telecommunications infrastructure and customer stations, as reflected in Touring Machine, originates from the conflicting concerns of infrastructure providers and network service developers. The infrastructure aims to provide mechanisms that support application policies, but requirements for application flexibility imply that the infrastructure not impose any particular policy on its own. Additionally, the infrastructure is often supplied by one agency per administrative domain, but applications are often created by multiple vendors. Thus, the infrastructure developer frequently controls all internal telecommunications resources, but does not control those resources application developers must supply.

These two concerns lead to a view of a protected, controlled, "inside" network that exists to support unprotected, uncontrolled, "outside" applications. Such a view has a strong influence on architecture design in general, and Touring Machine in particular. Specifically, Touring Machine classifies objects depending on whether they reside at customer stations or inside the telecommunications infrastructure, and uses this classification to restrict resource allocation requests. For example, a Touring Machine client lies at a customer station, and only controls resources at that station. On the other hand, a Touring Machine session object lies within the telecommunications infrastructure, and controls all resources within the infrastructure.

Contrast that view with the approach taken by the Lancaster project. Open Distributed Processing is concerned with standards both between end systems and within end systems. Therefore, there is no distinction between the "inside" and "outside" of a system in ODP. This removes the distinction between the two levels of a system and allows a single approach to span the entire development. The Lancaster architecture reflects this perspective; the object-oriented architecture spans all levels of a system, covering the transport layer, the control layer, and the application domain. The scope of objects is therefore from transport components such as multimedia devices, through control services such as streams, to application concepts such as hypermedia documents.

The Lancaster architecture aims to provide transparent access to all objects, regardless of whether they exist as part of the application layer, the control layer, or the transport layer. Transparent access means that any object with proper authority may invoke operations offered by another object, making security an important concern in Open Distributed Processing. In particular. security management must be provided to restrict the available set of objects, and to restrict the operations defined on these objects. Nevertheless, the amount of generality afforded by the Lancaster architecture does not ease the task of application development, for application developers must individually coordinate the set of operations involved in each communications sessions, or develop a specific object to provide such function.

3.2 Granularity of Objects

We now turn to the second distinction between Touring Machine and the Lancaster architecture, namely the granularity of objects supported in the respective architectures. Again, this distinction stems from the individual backgrounds of the two projects.

Since the Touring Machine focuses on providing an infrastructure for applications, every object in the Touring Machine known to application developers encapsulates only those functions an application developer need consider. Consequently, the Touring Machine contains a small fixed set of objects, specified at a high level of abstraction. Each object has a single, well-defined interface. The Touring Machine API assumes the functions offered by its objects suffice to give application developers all the flexibility they need, yet simultaneously prevents application developers from invoking restricted operations such as infrastructure resource allocation.

In contrast, the Lancaster architecture models all aspects of the system as objects. Therefore, an object in the Lancaster architecture encapsulates a single behavior, regardless of its level of abstraction. Objects represent entities as diverse as communication channels, cameras, or multimedia documents. A hierarchy of objects is therefore created at various levels of abstraction, and, as a consequence, objects range from strings to large segments of video. The architecture intends to provide access to objects at various levels of abstraction, giving application developers the ability to select the most appropriate level for their need. For example, an application writer may choose to access low level services for efficiency (e.g. transport protocols) or higher level services for convenience (e.g. application level protocols). This flexibility is often important in multimedia applications. As mentioned above, a single object in the Lancaster architecture can support multiple interfaces corresponding to the different aspects of the behavior of an object.

3.3 Towards an Integrated Architecture

Based on our experiences with Touring Machine and the Lancaster project, we see some advantages in integrating telecommunications and computing functions into one platform, much like the way a modern operating system integrates privileged and user instructions into one environment. These advantages are discussed below, first from a telecommunications perspective, and second from an Open Distributed Processing perspective:

- *from a telecommunications perspective*

  As discussed above, telecommunications architectures such as Touring Machine use an object-oriented service interface to categorize interactions that cross the boundary between the telecommunications network and customer stations. They do not, however, categorize the interactions that lie above or below the API. Interactions above the API lie in the domain of customer stations applications, which lie outside their control, and services below the API lie in the domain of the infrastructure, which they control completely. Interactions that cross the API must be well-defined, regulated, and restricted, hence the interest in an object-oriented architecture.

We believe extending the object-oriented characteristics of the architecture both above and below the defined API boundary has advantages. Using an object-oriented architecture above the API boundary allows applications to be developed openly. More importantly, additional services can be made available to the network within the same architectural framework. Similarly, using an object-oriented architecture below the API allows object-oriented techniques to be used in the engineering of telecommunications software. This can lead to several well known benefits of object-oriented computing, for example greater re-usability of components and support for evolution and maintenance. Through this approach, it might also be possible to provide some access to lower levels of the system, although such access might lead to a security danger if not well-controlled. At present, Touring Machine has an object-oriented architecture below the API boundary, and it does provide access to lower levels for debugging purposes, but the API, as defined, does not demand it. The current Touring Machine implementation does not address security to a great extent; when it does, the issue of access to lower levels needs to be studied in more detail.

- *from an ODP perspective*

An ODP environment provides a general object-oriented architecture for the development of distributed applications. This can be seen in the Lancaster architecture that deals with objects at various levels of abstraction. It is useful to impose a structure on such a general framework, for a structure frequently eases the task of application development. The interface between the networking infrastructure and clients of such network services gives one such structure. Furthermore, it is useful to consider the level of abstraction offered by services. For example, facilities offered by stations and sessions in Touring Machine are designed to support the rapid development of distributed multimedia applications. It is therefore important to consider such abstractions in designing ODP systems.

4. CONCLUDING REMARKS

This paper presents a comparison between two architectures, Bellcore's Touring Machine architecture and Lancaster University's extended ANSA architecture. The former is heavily influenced by existing telecommunication architectures while the latter is based on current work in the Open Distributed Processing community. Given recent developments in multimedia applications and network services, we believe it important to create a strategy for combining the two styles of architecture. Our belief in this importance motivated this study.

The distinctions between the two architectures derive from their differing influences. Touring Machine has a large grained object model with a clear distinction retained between infrastructure services and customer stations. The Lancaster architecture has a more finely grained and dynamic object model with no distinctions between the different layers in the architecture.

The paper concludes by describing the advantages that result when we combine the two perspectives into one unifying object model. This is a large undertaking, requiring effort from both communities.

5. REFERENCES

[Anderson,90] D.P. Anderson, S.Y. Tzou, R. Wahbe, R. Govindan and M. Andrews, "Support for Continuous Media in the DASH System," Proc. of the 10th International Conference on Distributed Computing Systems, Paris, May 1990.

[ANSA,89] ANSA. "ANSA Reference Manual, Release 01.00," APM Cambridge Ltd., March 1989.

[Bates,90] Bates, P., and Segal, M., "Touring Machine: A Video Telecommunications Software Testbed," First International Workshop on Network and Operating System Support for Digital Audio and Video, Nov. 1990.

[Blair,91a] Blair, G.S., Coulson, G., Davies, N., Williams, N., "Incorporating Multimedia in Distributed Open Systems," Proc. EUUG Spring '91 Conference on Distributed Open Systems in Perspective, Tromso, Norway, May 1991.

[Blair,91b] Blair, G.S., Coulson, G., Davies, N., "Introducing Synchronization in Distributed Multimedia Systems," Internal Report, Computing Department, Lancaster University, Bailrigg, Lancaster, U.K., April 1991.

[Bussey,88] Bussey, H.E., "Integrating Services in Broadband Networks - the Challenge and the Future," Proc. National Communications Forum, Oct. 1988.

[Coulson,90] Coulson, G., Blair G.S., Davies N., and Macartney A., "Extensions to ANSA for Multimedia Computing," Internal Report MPG-90-11, Computing Department, Lancaster University, Bailrigg, Lancaster LA1 4YR, UK., October 1990.

[Davies,91] Davies N.A., Nicol J.R., "A Technological Perspective on Multimedia Computing," Computer Communications, Vol. 14, No. 5, June 1991.

[Hazard,91] Hazard, L., F. Horn, and J.B. Stefani. "Notes on Architectural Support for Distributed Multimedia Applications," CNET/RC.W01.LHFH.001, Centre National d'Etudes des Telecommunications, Paris, France, March 91.

[McMordie,91] McMordie, W. Shane, "Distributed Platform Requirements of New Telecommunications Services," Proc. Telecommunications Information Networking Architecture (TINA) Workshop, March 25-27, 1991.

[Nicolaou,91] C. Nicolaou, "A Distributed Architecture for Multimedia Communication Systems," PhD Thesis, University of Cambridge, Computer Laboratory, April 1991.

[Terry,88] Terry, D.B., and D.C. Swinehart. "Managing Stored Voice in the Etherphone System." ACM Transactions on Computer Systems Vol. 6 No. 1, February 1988.

[Timms,89] Timms, S., "Broadband Communications: The Commercial Impact," Proc. IEEE INFOCOM '89, April, 1989.

68

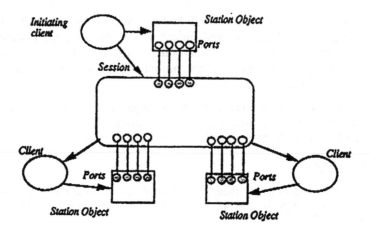

Figure 1. Audio-Visual Conference in the Touring Machine

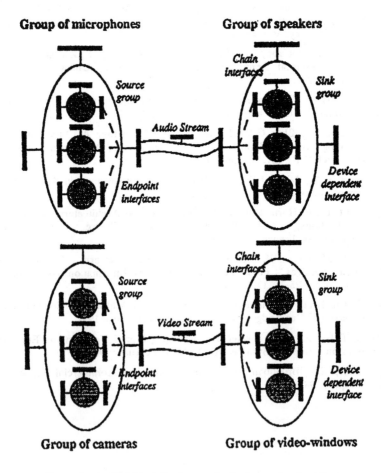

Figure 2. Audio-Visual Conference in the Lancaster Architecture

Session III: Synchronization and Communication I

Chair: Doug Shepherd, Lancaster University

Session III consisted of three presentations, the first one covering the topic of jitter control for packet-switching networks and the other looking at the problems of synchronization. The first of the papers on synchronization was concerned with providing end-to-end synchronization in packet-switched networks and the second on the synchronization of multimedia streams in open distributed environments. In both cases the papers are based on simulation and modeling and indicate that further investigations are required before they can produce definite recommendations as to how synchronization in general should be dealt with.

The first paper, "Design and Implementation of a Delay Jitter Control Scheme for Packet-Switching Networks," by Dominico Ferrari of the Tenet Group at the University of California and ICSI, Berkeley, sets out the ideal situation for a continuous-media connection − i.e., a "perfect delay line" where all packets have equal delay even with fast forward, slow motion, etc. He notes that it is possible to bound jitter to within small limits if the assumptions are made that an upper bound on delay can be guaranteed, sufficient buffering is available at or near the destination, and the source and sink clocks are kept in synchrony. The algorithm is simply to timestamp each packet at the source and to hold back the packet at the sink until the "ideal" time is reached (must be greater or equal to upper bound delay guarantee).

The paper goes on to describe a way in which this sort of scheme can be extended to be implemented in each gateway rather than just at the sink of the connection. The assumed environment for the scheme is then spelled out. It assumes:

1. Store-and-forward network made up of nodes (gateways) and links.
2. The jitter is caused by the nodes − the links produce fixed delays.
3. The existence of a resource allocation protocol which is run at connect time. A message propagates from the source to the sink. At each node, adequate resources are allocated such that a client-supplied minimum delay can be achieved. If the target is achieved at the sink, the message returns along the same path and relaxes any resources that turned out to be over-committed. The result of this protocol is a connection with dedicated resources which means that delays are bounded within known limits.
4. The network can be seen as having a hierarchical structure so that sub-nets are seen as single nodes in a top-level chain of nodes. The assumption is made that the sub-nets have bounded delays which are achieved by any suitable means. These means may be ad-hoc or achieved by recursive application of the protocol in point 3.

The jitter control algorithm is a simple extension of the above "smooth at sink" scheme. At each node there is a canonical delay time between it and the preceding node. Each node simply holds back packets that arrive earlier than their canonical arrival time (i.e., timestamp at previous node plus canonical delay time), and only schedules them for forwarding when the canonical arrival time is reached. (This is why synchronized clocks are needed; it works on absolute times rather than deltas.) In the case of hierarchical networks, the algorithm is recursively applied so that the delay in one top-level node is spread over the multiple lower-level nodes. Again, this may be done by recursively applying the same scheme or by smoothing the jitter produced by the sub-net at the following top-level node if the subnet does not implement the algorithm in its nodes.

Then the paper goes into a discussion. It points out that the algorithm must be implemented at the network layer. This is a potential disadvantage but is not so bad if we already have the resource reservation protocol (above) running on each node. The advantage is that the algorithm does not need increasing buffer space as the distance from the source to the sink grows. The buffering is spread evenly and less buffering overall is required than for the solution with only a sink buffer. Also applications are freed from worrying about jitter. The end-to-end argument is not violated if it is assumed that no additional jitter control needs to be done by applications.

Then a number of scenarios are painted which illustrate uses of the scheme. It is shown that continuous synchronization of related streams is rendered trivial when a "perfect delay line" is possible. It is only necessary to start the streams off together. This applies for both periodic and "jerky" streams, and even where simultaneous arrivals at different destinations are required.

The second paper on "End-to-End Synchronization in Packet-Switched Networks" by Jose Nuno Almeida from INESC in Portugal (joint work with J. Cabral and A. Alves) reports on current simulation results obtained within ESPRIT Project 2054 which is looking at the end-to-end synchronization problem within packet-switched networks (ATM, DQDB) for communication services which require a precise time relation between source and sink (e.g., constant and variable bit rate video).

The authors first describe the hardware architecture employed for the simulation experiments and the I-Series Recommendations for data formats classifying services based on timing relations between source and sink, bit rate, and connection mode. Where possible (as these formats are still being developed) they try to adopt these formats within their own simulation experiments. These I-Series Recommendations are for the B-ISDN ATM Adaptation Layer (AAL).

They describe four methods which may be employed in end-to-end synchronization, these include: slipping (plesiochronous clocks), adaptive clock (FIFO level), time stamping, and methods based on synchronized transmission links. Their initial simulation results are based on the adaptive clock method (FIFO level), where the authors report on how they have constructed their simulation experiments for this particular method, and the results obtained.

They conclude that the adaptive clock method with a hybrid frequency control algorithm seems to be best suited for synchronization mechanisms for constant bit rate services. Work is proceeding to carry out further simulation experiments.

In the final paper of the session on "Synchronization of Multimedia Data Streams in Open Distributed Environments" by Peter Leydekkers and Bertjan Teunissen from PTT Research Netherlands the authors present the requirements for a distributed multimedia conferencing application over wide are networks (WANs), in particular

B-ISDN using ATM as the switching fabric. It provides a brief analysis of continuous-media requirements (voice, video, and real-time data) and examines network characteristics which are required to support both this type of data and the conferencing application. In the examination of these characteristics it suggests whether B-ISDN is suitable or not. It also provides a list of extra QOS parameters which need to be supported by the communication protocols.

The authors continue to give a very brief outline of the synchronization problem experienced when objects are located in geographically distinct places and data from these objects is transmitted via separate channels to one or more users. Each user then requires to compose a multimedia compound object from the information types of the others.

An analysis of three distinct stages where the required synchronization problem may be solved and how is then presented. Synchronization may be solved at the source, network, or receiver. For each of these, the authors briefly look at methods, advantages and disadvantages concluding that the synchronization problem is best dealt with at the receiver. However, two schemes may be employed at the receiver, these are: the synchronization marker concept and the synchronization channel concept.

The synchronization marker concept requires large buffers, modifies the data stream, and only provides basic synchronization services always assuming a simultaneous presentation of information. The authors conclude that the synchronization channel concept is more complex, but allows data to be presented sequentially, simultaneously, or in any other combination.

Finally the authors analyze where such services may be fitted within the OSI Reference Model. They suggest that the current application standard being proposed to deal with multiple associations should be extended to enclose the functionality which may be offered by employing the synchronization marker concept. However, further research is required.

It was clear from the presentations that the problem of jitter was well understood and that provided the networks involved satisfied certain conditions then jitter could be dealt with in an efficient and elegant manner. Work from Berkeley (both in Ferrari's group and by David Anderson and Ralf Guido Herrtwich in connection with the development of the Session Reservation Protocol in the DASH Project) has provided a set of solutions for this problem.

However, the same cannot be said for the problem of synchronization. No way of even specifying what the general problem of synchronization is was presented — instead solutions for specific problems were given. It is clear that considerable research has to be carried out in this area and the associated one of quality of service before any definitive results or recommendations can be produced.

DESIGN AND APPLICATIONS OF A DELAY JITTER CONTROL SCHEME FOR PACKET-SWITCHING INTERNETWORKS

Domenico Ferrari
The Tenet Group
University of California
and International Computer Science Institute
Berkeley, California, U.S.A.

Abstract

Delay jitter is the variation of the delays with which packets traveling on a network connection reach their destination. For good quality of reception, continuous-media (video, audio, image) streams require that jitter be kept below a sufficiently small upper bound. This paper proposes a distributed mechanism for controlling delay jitter in a packet-switching network. The mechanism can be applied to an internetwork that satisfies the conditions detailed in the paper, and can coexist with other schemes (including the absence of any scheme) for jitter control within the same network, the same node, and even the same real-time channel. The mechanism makes the distribution of buffer space requirements more uniform over a channel's route, and reduces by a non-negligible amount the total buffer space needed by a channel. The paper argues that, if these advantages are sufficient to justify the higher costs of the distributed jitter control mechanism with respect to a non-distributed one, it would be useful to offer to the network's users a jitter control service based on the mechanism proposed here.

1. Introduction

Digital audio, video, and image-sequence transmission, both of the interactive and of the non-interactive kind, need a high degree of regularity in the delivery of successive stream-frames[1] to the destination. Ideally,[2] the end-to-end connections that carry those continuous-media streams should delay each element (e.g., a frame) of a stream by the same amount of time; in other terms, each connection should behave as a perfect delay line. This should normally be the case even when the elements of a stream are not transmitted at a constant rate (for example, in image browsing applications, where the user needs to be able to do pausing, fast and slow forwarding, fast and slow rewinding, and so on).

This research was supported by the National Science Foundation and the Defense Advanced Research Projects Agency (DARPA) under Cooperative Agreement NCR-8919038 with the Corporation for National Research Initiatives, by AT&T Bell Laboratories, Hitachi, Ltd., Hitachi America, Ltd., the University of California under a MICRO grant, and the International Computer Science Institute. The views and conclusions contained in this document are those of the author, and should not be interpreted as representing official policies, either expressed or implied, of the U.S. Government or any of the sponsoring organizations.

[1] We call *stream-frames* the data elements (video frames, audio samples, image frames) that constitute a continuous-media stream.

[2] We ignore here the case of stored streams that are shipped in batches and are to be played at the destination upon reception, as well as all the other cases in which input patterns are not to be faithfully reproduced at the output of the network. These either must be regarded as file transfers (i.e., not as continuous-

An ideal constant-delay network could be characterized also as a *zero-jitter* one, where jitter (more precisely, *delay jitter*) is any variation of stream-frame delays. A more realistic requirement for continuous-media transmission is that jitter have a small upper bound. Given an upper bound for delays, which is necessary for interactive continuous media and often desirable even in non-interactive cases, assigning a jitter bound is equivalent to specifying also a lower delay bound fairly close to the upper bound. A jitter bound alone will not guarantee the faithful reproduction of the input pattern at the output. To achieve this result, such a bound will have to be accompanied by a delay bound guarantee.

The computers and the (packet-switching) networks that connect users involved in continuous-media communications may introduce large amounts of jitter, due to the fluctuations of their loads. End-to-end delay bounds can be guaranteed by suitable design of the various components of a connection, including the host computers at the two ends; for instance, a scheme for establishing real-time connections with deterministic or statistical delay bounds, as well as throuput and loss probability bounds, in a packet-switching network is described in [FeVe90]. Is it possible to reduce (below a small upper bound) end-to-end delay jitter as well?

The answer is affirmative, as long as (*i*) a bound on delay can be guaranteed, (*ii*) a sufficiently large amount of buffer space is available at the destination or close to it, and (*iii*) the clocks of the source and the destination are kept in synchrony with each other [Part91]. This jitter control scheme is very simple: the source timestamps each stream-frame, and the destination, knowing the timestamp and the ideal constant delay (which must be greater than or equal to the delay bound guaranteed by the network), buffers each stream-frame until the time it has to be delivered to the receiving user.

The scheme can be implemented at the application level or at a lower level (e.g., at the transport or even the network layer in the communication protocol hierarchy); it is easy to see that the existence of a jitter control mechanism at a lower level, while not automatically providing low jitter to the application, greatly facilitates the achievement of this objective. Essentially, a bounded-jitter packet stream is not a necessary, but may be regarded as a sufficient, condition for obtaining a bounded-jitter sequence of stream-frames, as long as stream-frame reassembly, higher-level protocol processing, and stream-frame delivery to the application do not introduce appreciable amounts of jitter.

An alternative approach to jitter control has been presented in [VeZF91]. While the approach described above (see for example [Part91]) operates primarily at the destination, the one proposed in [VeZF91] is distributed; thus, it can only be implemented by the network layer, not by the upper protocol layers or by the application. In the kinds of networks in which it can be used, the distributed approach has been found to require a substantially smaller total amount of buffer space, and to yield a more uniform buffer allocation along the path of a bounded-jitter connection. In this paper, we introduce a distributed scheme that can be applied to much more complicated networks (in fact, to inter-networks), and argue that, if a network already implements distributed jitter control for

media transmissions) from a storage device to a remote one, or require a receiver that will time their presentation following a given (e.g., periodic) time law. Some form of jitter control may be needed even in these cases.

its internal purposes, it could, at almost no extra cost, offer a jitter control service to its users; indeed, if the receiving systems can be designed to inject only negligible amounts of jitter into the incoming streams, this service would provide the end-to-end jitter bounds needed by continuous-media streams.

A more detailed discussion of the two approaches (distributed and non-distributed) to jitter control can be found in Section 4. That discussion is preceded by a description of the network environment in Section 2, and by a presentation of our scheme in Section 3.

The rest of this section summarizes the approach to real-time communication presented in [FeVe90], which, like the jitter control scheme in [VeZF91], applies to what we will in the sequel call a *simple network*. Such a (packet-switching) network is characterized by hosts and switches interconnected by physical links in an arbitrary topology. The switches operate according to the store-and-forward principle, the links introduce only propagation delays, and the output queues of a switch (one per output link) are managed by a multiple-class version of the Earliest Due Date (EDD) [LiLa73] discipline.

A path from source to destination in such a network may be depicted as in Figure 1. Nodes 2, 3,... N-1 represent the switches traversed by the path. Node 1 and node N implement the protocol layers below the transport layer on the host systems. For these systems to offer end-to-end delay guarantees, they must be scheduled by a policy (such as Multi-Class EDD) that gives higher (and preemptive, when necessary) priority to the real-time activities of the upper protocol layers, as well as to the output queues of the "resident network node" reserved for real-time packets.

The basic abstraction the scheme in [FeVe90] is capable of implementing is the *real-time channel*. A real-time channel is a simplex connection between a sender and a receiver[3] characterized by performance guarantees. The performance guarantees a client may request include a delay bound D for each packet; on a *deterministic channel*, packets are guaranteed to arrive at the destination with a delay not larger than D.

Besides specifying performance requirements such as D, a client will have to provide a simple description of the maximum traffic on the channel to be established. One of the parameters this description consists of in the approach we are summarizing here is x_{min}, the minimum time between successive packets at the entrance to the channel.[4] This will also automatically specify the client's maximum bandwidth requirements.

The creation of a real-time channel may be requested by a client running on the host system that should become the channel's source. An establishment message carrying the performance requirements and the traffic specifications given by the client leaves the source and, while building the connection, tests the ability of each switch to handle the new channel without violating the bounds promised to the other clients; it also collects information, to be used by the destination, on the loading of each switch. If the message reaches the destination without having been rejected, the destination runs the final tests; if also these tests are successful, the destination subdivides the client-given bounds into per-switch bounds, and sends them to the switches with the channel acceptance message.

---

[3] The notion of real-time channel can be extended to the case of multiple receivers. For the sake of simplicity, we will not consider such an extension in this paper.

[4] The client will actually provide the value of the minimum time between successive stream-frames; this will be translated into the minimum interpacket time by the channel establishment protocol.

Thus, switch n receives, for example, the value of its local delay bound $d_{i,n}$ for the new channel i, and will use it to compute the local deadline of each packet transmitted on that channel.

2. The Network

In this section, we describe the environment for which our jitter control scheme has been designed; in particular, the class of networks (to be more precise, internetworks) that can guarantee delay jitter bounds, as well as delay bounds, using the approach to be presented in Section 3. The scheme described in [FeVe90] for simple networks can be extended to a class of internetworks [Ferr91b], which is the one we consider here. The rest of this section gives a brief description of the characteristics of that class.

We consider an internetwork in which it is possible to create real-time channels. We assume that the internetwork offers channels with deterministic (i.e., absolute) delay bounds, and we wish to devise a way to obtain guaranteed jitter bounds for the same channels when the clients need such guarantees.

The path of a channel from the sending host to the receiving host will generally traverse a number of different networks, interconnected by gateways.[5] We consider a hierarchical model of the internetwork in which the level 1 nodes are the gateways. The level 1 path of a channel is a path consisting of gateways interconnected by *logical* links; each one of these links may represent a path through the network of which the two nodes at its extremes are gateways. This path may be very complicated; for example, the network in question might in turn be an internetwork, whose gateways constitute a level 2 network, and so on recursively.

In this hierarchical model, a path through a network may be abstracted by transforming it into a simple logical link at the immediately higher level (note that in this paper higher levels are labeled with lower numbers). This process of abstraction ends with the level 0 path, which is a single logical link connecting the sender and the receiver. This is the logical link whose delay and delay jitter are to be bounded. What are the minimum necessary properties of the underlying internetwork that allow the bounds selected by the client to be guaranteed, if possible?[6] What must we assume about the various networks constituting the hierarchy?

If all the constituent networks were capable of guaranteeing delay and jitter bounds, the internetwork could provide similar guarantees at the end-to-end level. However, such a restrictive condition is by no means necessary. All we need (see [Ferr91b]) is the assumption that the delay on each level 1 logical link be bounded or at least boundable (with a sufficiently small bound), and that those bounds be either known *a priori* or obtainable by a recursive application (at the various levels in the hierarchy) of an extension of the scheme described in [FeVe90].

We also assume that level 1 nodes are store-and-forward, and schedule packets for transmission on each of their output links in such a way that, if the client's request for the channel is accepted, it is possible to guarantee the end-to-end delay bound specified by

---

[5] We use this term in the generic sense of "a computer system interconnecting two distinct networks at the network or higher layer;" thus our "gateways" may be routers as well as gateways proper.

[6] If it is not possible to provide the requested guarantees, the client's request will be rejected.

the client.[7] Many scheduling algorithms satisfy the latter assumption (see [Ferr91b]); for example, round-robin and its extensions and modifications (Fair Queueing [DeKS89], Hierarchical Round-Robin [KaKK90]), deadline-based disciplines (Earliest Due Date or EDD, Multi-Class EDD [FeVe90]), FCFS, and others.

If these assumptions, which are summarized in Figure 2, are satisfied, the internetwork can provide guaranteed delay bounds [Ferr91b], and we assume that the internetworks in the class we are considering indeed do provide them.

For convenience, we break the delay bound $dl_{i,l}$ of a level 1 logical link l for channel i into the sum of two terms, a fixed delay $df_{i,l}$ and a maximum variable delay $dv_{i,l}$, where $df_{i,l}$ is the delay encountered by the smallest possible packet travelling on the logical link under zero-load conditions, and $dv_{i,l}$ is an upper bound to the variable components of packet delays along the same link. Thus, $df_{i,l}$ includes propagation, switching, and transmission delays for the minimum-size packet, while $dv_{i,l}$ primarily consists of queueing delays and the additional transmission delays that a maximum-size packet would incur. The actual delay of a packet on that channel along that level 1 logical link will be between $df_{i,l}$ and $dl_{i,l}$.

As in a simple network, each level 1 node traversed by channel i has a delay bound $d_{i,n}$ assigned to it by the destination at channel establishment time. This bound is the maximum residence time of a packet in node n. The minimum residence time is the packet's processing time in the node (which includes also its transmission time on the output link); we call $dp_{i,n}$ the smallest of these times, corresponding to the idle-node residence time of the smallest packet the channel will carry; by $dq_{i,n}$ we denote the maximum queueing time in the node, which can only be reached by the smallest packet, since for all packets the sum of processing time and queueing time will be at most $d_{i,n}$. Thus, we have $d_{i,n} = dp_{i,n} + dq_{i,n}$.

3. Jitter Control with Perfectly Synchronized Clocks

In an internetwork of the class defined in Section 2, let us assume that the clocks of all level 1 nodes are in perfect synchrony with each other. That is, simultaneous readings of these clocks are always identical. A reasonably good approximation to this ideal situation can be achieved by having the gateways run a clock synchronization protocol (e.g., NTP [Mill89]), or listen to the times broadcast by a satellite and adjust their clocks accordingly. In the level 1 path of a real-time channel, all components of the end-to-end delay bound are known after the establishment of the channel. Each logical link l of channel i has a total delay bound equal to $df_{i,l} + dv_{i,l}$, and each node n has a local bound $d_{i,n}$ assigned to it by the destination (see Section 1 and [FeVe90]). Thus, it is easy to calculate the maximum delay between a node and the next, and to compensate for the jitter due to the variable parts of the delay, which may make the actual delay shorter (sometimes much shorter) than the maximum.

The mechanism is very simple: it requires timestamping each packet's header by the source host, reading the clock at a packet's arrival at each level 1 node, and keeping the

---

[7] Note that no assumptions are made about gateways or routers at lower levels in the hierarchy. The only constraint on them is that they allow level 1 delays to be boundable.

packet until the *canonical* (i.e., zero jitter) arrival time at that node before giving it to the node's scheduler. Each node is informed, when a real-time channel is created, about the maximum delay that separates it from the source for the packets on that channel; this is the *canonical delay* $D^*_{i,n}$ for channel i at node n. Assuming that the clock is read in each node immediately at packet arrival time, the maximum delay with respect to the canonical arrival time at the previous node (node $n-1$) equals $df_{i,l} + dv_{i,l} + d_{i,n-1}$.

Given the time of departure from the source, which is carried by the header, the node calculates the canonical arrival time and compares it with the actual one (which is guaranteed not to exceed the canonical time). The packet will be kept in storage in the node for a time interval equal to the difference between the canonical and the actual arrival times. When that interval expires, the packet becomes *schedulable*; the jitter introduced by the scheduler in node n will be corrected exactly by the same mechanism in node $n+1$. The only jitter that cannot be corrected in this way is therefore the one introduced by the last node on the path. This will have to be corrected by the receiving host, which will need to know the maximum delay assigned to the last node and keep the packet until that bound is reached before giving it to the application. If this is done, a zero-jitter (perfect delay line) transmission is obtained; if it is not done, the only end-to-end jitter is that generated by the last node, since those produced by all previous nodes and (logical) links are perfectly compensated for by the simple mechanism we have just illustrated.

What is the maximum value of jitter when no jitter control is used, and how much is that value reduced by the above scheme? To answer these questions, we refer to the channel whose path is shown in Figure 1.[8] We therefore omit the channel index i in the symbols we will use, and assign to each link the same number as the node immediately downstream.

Without jitter control, the link-node pair numbered n contributes to the total maximum jitter the quantity

$$j_n = dl_n + d_n - dp_n - df_n = dv_n + dq_n, \tag{1}$$

Equation (1) can be explained by observing that the maximum delay through the link-node pair is $dl_n + d_n$, whereas the minimum delay is $dp_n + df_n$.

Thus, the maximum jitter for the packets on the channel is

$$J_{\max} = \sum_{n=1}^{N} j_n = \sum_{n=1}^{N} (dv_n + dq_n). \tag{2}$$

From the above discussion, we have that the maximum jitter permitted by the jitter control mechanism described in this section is[9]

$$J'_{\max} = d_N - dp_N = dq_N. \tag{3}$$

---

[8] The nodes in the figure are level 1 nodes, which run an internetwork protocol. Node 1 represents the network layer and lower layers on the source computer, and shares that computer with source S, which represents the application and the protocols at and above the transport layer.

[9] We use primed symbols to refer to cases in which our distributed jitter control mechanism is present.

Note that the maximum jitter at the input to node n is

$$j'_{in,n} = dv_n + dq_{n-1},$$ (4)

while the maximum jitter at the same point when there is no distributed jitter control is

$$j_{in,n} = dv_n + \sum_{m=1}^{n-1} j_m.$$ (5)

The value of $j'_{in,n}$ equals the maximum amount of time a packet may have to wait in the node before becoming schedulable. Thus, the maximum residence time of a packet in node n with distributed jitter control is

$$r'_n = dv_n + dq_{n-1} + d_n,$$ (6)

and only

$$r_n = d_n$$ (7)

without distributed jitter control; in the last node, however, if jitter is to be absorbed by buffering (a network-layer implementation of the non-distributed approach to jitter control described in Section 1), the maximum residence time must equal the total maximum jitter instead of d_N as dictated by (7):

$$r_N = \sum_{n=1}^{N} j_n.$$ (8)

To calculate the number of packet-sized buffers needed by each channel in node n to avoid packet losses due to buffer overflows, we observe that the worst case corresponds to a stream of packets entering the channel at node 1 with interpacket distances equal to x_{min}. The latest time a packet from the stream can leave node n equals the sum of its departure time from the source, the canonical delay at node n, and the local delay bound in the same node. Similarly, the earliest time a packet can arrive at node n is given by its departure time from the source plus the canonical delay at node n minus the maximum jitter at the input to the same node. If we number the packets in the stream 1, 2, ..., and we assume that node n has allocated B_n buffers to the channel, the packet numbered 1 will have to have left the node before the arrival of the packet numbered $B_n + 1$ even in the worst case. This will be the case in which packet 1 departs at the latest possible time and packet $B_n + 1$ arrives at the earliest possible time. Thus,

$$d_n \leq B_n x_{min} - j_{in,n},$$ (9)

and

$$d_n > (B_n - 1) x_{min} - j_{in,n}.$$ (10)

Combining (9) and (10), we obtain[10]

$$B_n = \left\lceil \frac{d_n + j_{in,n}}{x_{min}} \right\rceil.$$ (11)

---

[10] To be more precise, we should add the maximum transmission time on the output link to the numerator of the fraction in equations (11) and (12); however, this time is often negligible with respect to the terms that appear in that numerator, and we will therefore omit it.

If jitter is controlled with the distributed method described above, our treatment is still valid, as long as we replace $j_{in,n}$ with $j'_{in,n}$. Thus,

$$B'_n = \left\lceil \frac{d_n + j'_{in,n}}{x_{min}} \right\rceil = \left\lceil \frac{r'_n}{x_{min}} \right\rceil. \tag{12}$$

Since $j'_{in,n} \le j_{in,n}$, we usually have $B'_n \le B_n$. In fact, while $j_{in,n}$ (hence also B_n) grows monotonically with the number of nodes traversed along a channel's route (see equation (5)), $j'_{in,n}$ does not (see equation (4)). These simple worst-case calculations show what we believe to be the main advantages of distributed jitter control when no packet loss can be tolerated, i.e., the smaller amount of buffer space to be allocated, and the much more uniform distribution of that space along channel routes. Because of the better balance among per-node space requirements provided by our distributed jitter control scheme, this greater uniformity may be expected to increase the maximum number of real-time channels a given network can support.

It is interesting to observe that the value of B_N is given by equation (11) even when jitter is to be eliminated by buffering in the last node (in this case, however, buffer space utilization in that node will be much higher). This rather counterintuitive result can be explained by noting that, even when the last node is expected to absorb jitter, a maximally delayed packet will not be made to wait in the last node's buffer, and that a minimally delayed packet will be delayed in the last node only until it reaches the maximum delay. The size of the last node's buffer will have to be B_N (if no packet losses can be tolerated) whether or not that buffer is used to absorb jitter. If jitter absorption is done by a separate buffer at the transport or higher layer or in the user's application, that buffer must have the same size as the one in the last node. Thus, unless the two buffers can be implemented in the same physical memory space (which is at least in principle possible, since the last node and the "host" coexist within the same computer system), jitter absorption at the network layer requires half the space needed by any solution that does it at a higher layer.

4. Discussion

Good-quality reception of digital video streams, audio streams, and image sequences over computer networks normally requires bounded delay jitter as well as bounded delay connections. Acceptable jitter bounds can be obtained, in a system with delay guarantees, sufficient buffer space at the destination, and sufficiently accurate synchronization between the source and the destination clocks, by a simple scheme that absorbs the jitter by buffering data at the destination. The scheme can be implemented at one of several different levels; a network-layer implementation may even buffer data in the last switch or the last gateway of a connection rather than in the destination host.[11]

In [VeZF91] and in this paper, two distributed jitter control schemes (which, being distributed, can only be implemented at the network layer) have been introduced. The one described in [VeZF91] only applies to simple networks, but, unlike the one based on buffering at the destination, does not need to make any assumptions about clock

---

[11] This will provide end-to-end controlled jitter only if jitter is kept under control also between the place where data are buffered and the destination.

synchronization. The scheme presented in Section 3 requires clocks to be synchronized (at least loosely [Ferr91a]), but can be used in hierarchical internetworks.

The main advantage of the distributed approach is that it does not require increasing amounts of buffer space as the distance from the source grows; it spreads buffer space requirements much more evenly over the nodes on a channel's path. This will in general concern network designers and managers more than clients, though some users may actually find it inconvenient to have to allocate large buffers in their hosts for each of the several continuous-media streams those hosts may have to receive simultaneously.

The main disadvantage of a distributed scheme is that it complicates the network switches and switching operations (which in high-speed networks should be as simple as possible); however, this problem is not so crucial if the scheme is only implemented in an internetwork's gateways.

If the advantage mentioned above is deemed useful enough, and if the disadvantage just discussed seems tolerable enough, that designers decide to implement jitter control inside the internetwork anyway, the maximum jitter perceived by the clients will decrease very substantially, and it should be much easier and less expensive to reduce the end-to-end jitter below the bounds desired by the clients by buffering data at the destination; furthermore, a jitter control service based on the distributed scheme could be offered to the internetwork's clients, and the clients might be able to rely on it for most of their jitter bound requirements, without any need for further end-to-end jitter reductions. This latter proposal might be regarded as being in conflict with the end-to-end principle, but we are actually assuming that controlling jitter in the network would come so close to achieving end-to-end jitter control that, at least for most applications, no additional jitter control mechanism would be necessary.

If we now compare an application-level jitter control scheme (which can only be a non-distributed one) with a system-level one (which may be distributed or non-distributed), we see that with the former, as already mentioned above, the bound is really an end-to-end one; on the other hand, the latter allows the application programmer to avoid worrying about timestamping, reading the clocks, allocating buffers for jitter control, recovering from synchronization losses, and so on, simply by using the service offered by the system. Of course, this service must be able to provide clients with what they need, and whether or not a generally useful jitter control service can be implemented can only be determined by experimentation.

This advantage of a system-level solution is actually even more important than it looks, since a jitter control service can be used to solve many stream synchronization problems, which would add substantial complexity to the programming of multiple-stream applications (see for example [Ravi91], [ShSa90]). Such applications can be categorized according to their synchronization requirements. Brief descriptions of three of these classes follow.

(a) *Synchronization of multiple periodic streams.*

The client requests the creation of a real-time bounded-jitter channel for each stream, and specifies for each channel the same delay bound and the same jitter bound. In the simplest case (e.g., that of a video stream and of its separate but co-stored sound track), all streams have the same source and the same destination. For (approximately) simultaneous arrivals of stream-frames from different streams at the destination, it is then sufficient that the stream-frames be input into the source node at the same time. The maximum difference between delivery times will be equal to the common jitter bound. Any phase difference between stream-frames at the input will be reproduced (with a maximum error equal to the jitter bound) at the output.

(b) *Synchronization of multiple aperiodic streams.*

This is a simple extension of class (a). If the rates of the streams are not constant but follow a more complicated time law, this law will be (approximately) reproduced at the output. This is, for example, what a scientist may need when comparing, for the purposes of validating and tuning a numerical model of a natural phenomenon, a video sequence recording the phenomenon and an image sequence produced by a simulator of the model [John91]. Initially, the speeds of the two streams will have to be matched, and this may require changing the rate of one or both of the streams. Then, the rates of both may have to be modified together according to a given law. Since our stream synchronization approach deals with each stream independently, all of these operations are straightforward: they only entail changing the input rates of those streams whose speeds have to be modified.

(c) *Simultaneous arrivals at different destinations.*

The method works (by definition of jitter) even when the streams to be synchronized have different destinations. This may be required in those applications in which (approximate) simultaneity is important; for example, in distributed music rehearsing, which has very strict delay and jitter requirements.

5. Conclusions

This paper has presented a distributed method for jitter control of wider applicability than the one proposed in [VeZF91]. The method requires synchronized clocks, but can tolerate small divergences between clock readings [Ferr91a]. Both methods can be used together, and may coexist with the absence of jitter control and with destination-based methods, in the same internetwork, the same node, and even the same channel [Ferr91a].

Distributed jitter control reduces the total buffer space requirement of a real-time bounded-delay channel, and spreads those total requirements much more evenly over the nodes of the channel's path. The presence of such a scheme will substantially reduce the maximum jitter perceived by the clients, and will facilitate further destination-based efforts to reduce end-to-end jitter. Careful design of the upper protocol layers and of operating system support mechanisms for real-time applications may make a jitter control service based on the distributed scheme sufficiently accurate for most of those applications.

Only experimentation with protocols based on these schemes in internetworking testbeds will allow us to confirm, or force us to modify, these conjectures. This is the obvious main goal of our future work in this area.

Acknowledgments

The author wishes to express his gratitude to all the members of the Tenet Group at the University of California at Berkeley and the International Computer Science Institute for the extensive discussions of the ideas and results presented in this paper. He is particularly indebted to Anindo Banerjea, Srinivasan Keshav, Massimo Maresca, Mark Moran, Dinesh Verma, and Hui Zhang for their suggestions, and to Professor Bernd Wolfinger of the University of Hamburg for his extremely careful reading of an earlier version of the manuscript and the many improvements that came from it.

References

[DeKS89] Demers, A., S. Keshav, and S. Shenker, "Analysis and Simulation of a Fair Queueing Algorithm," *Proc. ACM SIGCOMM Conf.*, Sept. 1989, 1-12.

[Ferr91a] Ferrari, D., "Distributed Delay Jitter Control in Packet-Switching Internetworks," Tech. Rept., International Computer Science Institute, Berkeley, October 1991.

[Ferr91b] Ferrari, D., "Real-Time Communication in an Internetwork," in preparation.

[FeVe90] Ferrari, D., and D. C. Verma, "A Scheme for Real-Time Channel Establishment in Wide-Area Networks," *IEEE J. Selected Areas in Communications*, SAC-8, April 1990, 368-379.

[John91] Johnston, W. E., personal communication, May 1991.

[KaKK90] Kalmanek, C. R., H. Kanakia, and S. Keshav, "Rate Controlled Servers for Very High-Speed Networks," *Proc. GlobeCom '90 Conf.*, San Diego, CA, Dec. 1990, 300.3.1 - 300.3.9.

[LiLa73] Liu, C. L., and J. W. Layland, "Scheduling Algorithms for Multiprogramming in a Hard Real Time Environment," *J. ACM*, 20, January 1973, 46-61.

[Mill89] Mills, D., "Measured Performance of the Network Time Protocol in the Internet System," Network Working Group, Request for Comments 1128, October 1989.

[Part91] Partridge, C., "Isochronous Applications Do Not Require Jitter-Controlled Networks," Network Working Group, Request for Comments 1257, September 1991.

[Ravi91] Ravindran, K., "Real-time Synchronization of Multimedia Data Streams in High Speed Packet Switching Networks," Tech. Rept., Dept. of Computing and Information Sciences, Kansas State Univ., February 1991.

[ShSa90] Shepherd, D., and M. Salmony, "Extending OSI to Support Synchronization Required by Multimedia Applications," *Computer Communications*, 13, 7, September 1990, 399-406.

[VeZF91] Verma, D. C., H. Zhang, and D. Ferrari, "Delay Jitter Control for Real-Time Communication in a Packet Switching Network," *Proc. TriComm '91 Conf.*, April 1991.

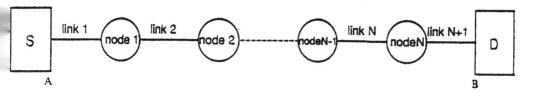

Fig.1. Model of a channel's path. Packet delays are measured between A and B. Since node 1 (node N) shares a host system with the source S (destination D), times at A (B) may be measured at the input (output) of node 1 (node N).

(1) Deterministic delay bounds on all level 1 links exist and are known or can be obtained by recursive application of an extension of the scheme in [FeVe90a].

(2) Level 1 nodes are store-and-forward.

(3) Each output link of a level 1 node and each host system is scheduled by a suitable discipline.

Fig.2. Sufficient conditions for guaranteeing end-to-end deterministic delay bounds in a hierarchical inter-network.

End-to-End Synchronization in Packet Switched Networks

N. Almeida*, J. Cabral*, A. Alves* **

** FEUP - Faculdade de Engenharia da Universidade do Porto

* INESC - Instituto de Engenharia de Sistemas e Computadores

Largo de Mompilher, 22, 4007 PORTO Codex, PORTUGAL

Abstract - The provision of real time services through a Packet Switched Network, requires the adoption of end-to-end synchronization methods. Several of these methods, are compared and a specific hardware solution is discussed. Results obtained through a specially designed Simulation Programme are included, for several network time delay jitter distributions and synchronization methods. These results are used to evaluate systems and design parameters.

1. INTRODUCTION

The communication services requiring a precise time relation between source and destination (e.g. circuit emulation, CBR and VBR video,...), when supported by a packet switched network (e.g. ATM, DQDB), require end-to-end synchronization methods with variable degrees of quality, according to the service and network performance in terms of delay jitter.

Recent activity in this area has been greatly oriented towards ATM networks, but it will also impact in other packet switched networks.

For circuit emulation, CBR, and VBR video services, the draft B_ISDN Recommendations (I.321, I.362, I.363) establish the end-to-end synchronization function as an ATM Adaptation Layer (AAL) function. The AAL is mapped between the ATM layer and the User layer (next higher layer). It enhances the services provided by the ATM layer in order to support the functions required by the next higher layer.

In the above recommendations and in the general literature, several end-to-end synchronization methods are being considered but very few and well characterized results are currently available for their evaluation.

A general study of several synchronization methods and their evaluation, is currently in progress, taking in consideration performance and implementation aspects. A simulation programme was developed, HW designed and, in order to make laboratorial tests of main concepts under different conditions, a flexible test system for CBR services is being developed.

The end-to-end synchronization methods most frequently referred to in the literature are:

- Adaptive clock (FIFO level);
- Time Stamping;
- Synchronization Patterns.

A slipping technique (plesiochronous clocks) may also be used, in some cases, where slips can be accommodated, as described in section 3.

This work was started in ESPRIT project 2054 (UCOL-UltraWideband Coherent Optical LAN), in relation to the development of high speed interfaces for CBR services [1]. UCOL is a coherent multichannel optical network of very high performance using a transfer mode based on ATM concepts [2]. These explains why much attention has been given to ATM standards in the current study.

However it must be stated that delay jitter characteristics and cell error rates of

UCOL are expected to be substantially different from B-ISDN specification.

In a first phase, the work was oriented towards a better specification of the selected end-to-end synchronization methods. A version of the adaptive clock method, based on buffer (FIFO) level, was the first selected for evaluating, as it showed advantages for CBR services. A preliminary design of all major blocks involved was made. The adaptive method (FIFO level) HW structure is described in section 4.

The simulation programme developed, which allows the analysis and study of the full system behaviour, has been used to optimize the main system parameters.

Simulation results are already available for the adaptive clock recovery (FIFO level) strategy. These results are presented in section 5.

2. GENERAL STRUCTURE OF A END-TO-END SYNCHRONIZATION SYSTEM

2.1. Structure

UCOL high speed interfaces contain two main subsystems: User interfaces (UI) and the ATM Adaptation Layer (AAL).

As mentioned before, the source clock frequency recovery function, is included in the AAL, which will be described in some detail.

The AAL is also the layer responsible for the adaptation of the service specific data format to the cell transport data format. The former is characterized by a continuous and isochronous delivery of information, and the later is of a bursty nature, as the information is being packed in cells for transmission.

2.2. Data formats

Recommendation I.362 (B-ISDN AAL functional description) provides a classification of services (based on the attributes of timing relation between source and destination, bit rate, and connection mode) in order to assist in the development and selection of suitable

methods to support the existing (and future) wide range of services. This classification (specific to AAL) has distinguished four classes, from A to D. Typical examples may be: class A-circuit emulation; class B-variable bit rate video and audio; class C-Connection Oriented data transfer and class D-Connectionless data transfer.

Recommendation I.363 (B-ISDN AAL specification) introduces four types of operation (1 to 4) needed to perform all the necessary functions for these service classes.

AAL type 1 and type 2 operation modes are, respectively, indicated for CBR and VBR services requiring end-to-end timing / synchronization.

To obtain a greater flexibility, AAL is also divided into two sublayers: the Segmentation and Reassembly (SAR) sublayer and the Convergence Sublayer (CS).

The SAR, and CS, Protocol Data Unit (PDU) formats for the AAL type 1, are shown respectively in figure 1 and 2.

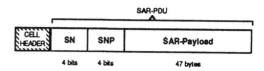

Figure 1. SAR-PDU format for AAL type 1.

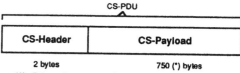

(*) Other sizes were also proposed to this field.

Figure 2. CS-PDU format for AAL type 1.

The CS-PDU format is not so well established, but some of the major guideline contributions, indicate the format shown above as a possible solution [3],[4]. The CS header will probably be two byte long, almost exclusively dedicated to synchronization purposes.

The system designed uses PDU formats as shown in figure 3.

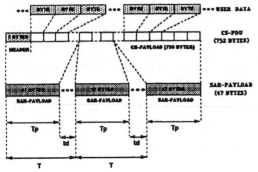

Figure 3. Sync. Test System PDU formats.

The value for **Tp** depends only on the network (line) bit rate. The average value for **T** equals the period service bit rate times 376 (47*8). **td** is the time between two successive cells.

2.3. Test System

In the stage of development of UCOL, there were still a number of uncertainties which recommended a flexible approach to be adopted. It was decided to design the system in such a way that several mechanisms could be easily implemented. On the other side it was decided to implement a test system to allow the evaluation of these subsystems in advance to the network being available.

The general structure of the End-to-End Synchronization Test System is shown in figure 4.

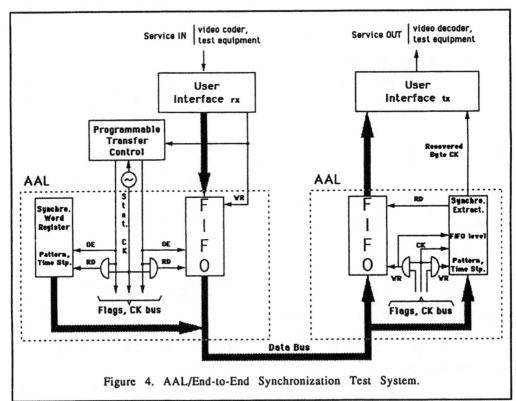

Figure 4. AAL/End-to-End Synchronization Test System.

It consists of user interfaces and AAL subsystems as designed for the real interfaces with the addition of a Programmable Transfer Control (PTC) unit.

The system operation is as follows: data coming out from the "user interface", in parallel, is introduced in a FIFO; a transfer control unit generates clock pulses moving data out of the FIFO in cells

(47 bytes long / no header), separated by programmable intervals; a remaining block on the transmitter side is used for the insertion of special words or patterns required by some of the synchronization methods; on the receiver side the reverse operations occur, with the communication between the two parts made via a parallel bus (19.4 Mbyte/s-155 Mbit/s).

The PTC unit acts as a traffic intercell interval generator. The delay jitter, resulting from the several nodes of the network is generated in the PTC, allowing the test and evaluation of the different methods of clock extraction.

The statistical characteristics of the delay jitter can be programmed. Figure 5 shows the functional structure of this unit.

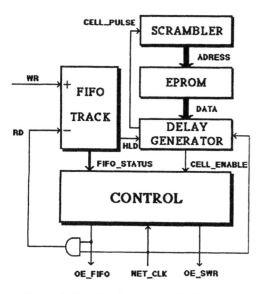

Figure 5. Functional structure of the PTC.

The FIFO_TRACK block saves the difference between the number of Write and Read pulses, in order to determinate the FIFO current level. In some cases it is possible that, as a consequence of the statistical distribution of the delay jitter, FIFO becomes empty. In this case, a Hold flag is activated, in order to delay the Read pulses.

The statistical distribution is generated in the Scrambler and EPROM blocks. The

Scrambler produce a pseudo-random sequence with Uniform distribution and with a period of 2^n (n-number of bits of the shift-register). Each value generated by the Scrambler block addresses the EPROM block which contains the inverse of Distribution Function of the chosen statistic, producing a pseudo-random sequence with a distribution according to the EPROM content. Thus it is possible to change easily the distribution function of the delay jitter just by changing the EPROM content.

The Delay Generator block receives the delay intercell interval information of the Data bus to produce the corresponding delay. This information is supplied to the Control block through the Cell_Enable signal which indicates the start of cell.

The Control block, by means of these signals, will generate the clocks and produce the pulses necessary to build the cell packets in the required format, affected with delay jitter following the selected statistical distribution function.

3. SYNCHRONIZATION METHODS

3.1. Slipping (plesiochronous clocks)

This method does not in fact perform any synchronization between the two-end terminals. The two clocks run independently with a predefined precision and the resulting slip are hidden, using an appropriate mechanism at a different level, so that to the user it looks as if the two systems are synchronized.

Indeed, services as video or audio, may accept periodic bit slips if these were forced at times not relevant to the service (e.g. non-active video lines, audio silence intervals). The faults will be subjectively null. These actions have to be related to the specific service and therefore not completely transparent.

For this reason it does meet UCOL requirements.

3.2. Adaptive clock (FIFO level)

This method adjusts the local clock with a digital phase-locked loop (DPLL) by observing the occupancy of an input buffer at the receiver end. This is the simpler method since no knowledge about the transmitted signal (e.g. detection of silence intervals or timing information) is required. Figure 6 shows the functional diagram of this DPLL.

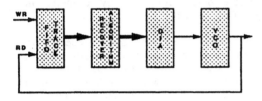

Figure 6. Functional diagram of the DPLL.

Since the write clock to the FIFO is bursty (no pulses when there are no packets arriving) and the read clock is continuous, the phase-detector is different from a conventional one, which compares the phase of a VCO output signal to the phase of a reference (input) signal.

Here the phase-detector is substituted by a FIFO-Level-Tracking (FLT) and a Recover-Algorithm (RA) block. The FLT block saves the current level of the FIFO occupancy, providing that information to the RA block. The RA based on the FIFO occupancy and its history will produce through the D/A, the appropriate control-voltage to VCO. The RA provides that the FIFO occupancy must be about half of its full capacity. This algorithm must avoid buffer over and underflow during the frequency acquisition process. Thus with a sufficiently sized FIFO and with a "good" algorithm the range of input delay jitters that the system may accept can be increased.

Section 4 describes, in more detail, studies made on this matter. The objective is to maximize the capacity of tracking the incoming frequency and to minimize buffer size.

3.3. Time Stamping

In the time stamping method, the transmitter writes an explicit time indication in the CS header. In the receiver, this time indication is compared with a local generated one. The result of the comparison is used to synchronize the locally clock.

Figure 7 shows one variant of the time stamping method. This variant does not take into account the possible existence of a common timing reference.

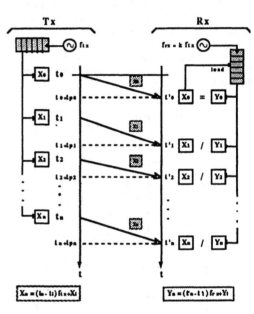

Figure 7. A time stamping operation mode.

The goal is to determine the value of **k**, in order to make the corrections which will equalise the receiver with the transmitter frequency. This may be difficult, taking into account the possible existence of significant network cell delay jitter and the fact that the reference clock used for stamping is not common, and great frequency precision is required.

Indeed, a better exploitation of the time stamp method requires a common timing reference (section 3.4). This represents a considerable drawback when considering networks without a reference clock, as UCOL network was designed.

3.4. Methods based on synchronized transmission links

In some cases a common clock is available distributed by the network. In this case it is possible to take advantage of the common timing reference present at the source and receiver nodes.

The operation of synchronization is achieved as follows: at the emitter the difference between the source and the network clock is coded and then transmitter to the receiver; the receiver, with this information, plus the available node network clock, can recover the source clock [5].

This method cannot obviously be used in most situations and was not considered as it is not applicable to UCOL.

These considerations led us to selected the adaptive clock method for our first implementation. This seems to be not very complex to implement and capable to the synchronization requirements.

4. ADAPTIVE METHOD (FIFO LEVEL)

A number of possibilities to implement the recovery algorithm are discussed below.

4.1. Direct approach

This method is based on the current position of the FIFO. The FIFO_LEVEL_TRACKING block keeps track of the FIFO occupancy, in order to provide the convertion block with the current level of the FIFO. The CONVERTION_TABLE converts this information in a correction word that, through the D/A, will give the error voltage to the VCO in order to track the incoming frequency. Figure 8 shows the functional structure of this method.

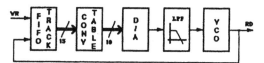

Figure 8. Functional structure of the Direct method.

4.1.1. Parameters and Procedures

The CONVERTION_TABLE is simply a table that converts the current level of the FIFO in a word that, through the D/A, produces the appropriate control voltage. The word generated by the CONVERTION_TABLE must be limited in order to keeps the recover frequency bound to the nominal range. The number of bits of the D/A must be sufficiently large in order to obtain a good resolution to the output frequency. The Low-Pass-Filter (LPF) main function is to smooth the D/A output voltage steps, between consecutive words.

4.2. Charge Pump approach

This method use a DPLL to recover the frequency of the transmitted clock. Here the Digital Phase Detector keeps track of the FIFO occupancy status and generates UP and Down signals to speed up or slow down the output frequency of the VCO. An error signal is generated by the Charge Pump circuit based on the UP/DOWN signals and then smoothed out by a loop low-pass-filter. Figure 9 shows the functional structure of this method.

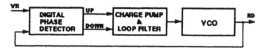

Figure 9. Functional structure of the Charge-Pump method.

4.2.1. Parameters and Procedures

The characteristics of the input delay jitter distribution and the allowed range of frequency variation , that the receiver terminal can accept, determine the size of the FIFO and the cutoff frequency of the LPF. This method establishes two thresholds, each one being close to the ends of the FIFO.

The FIFO occupancy can be determined by subtracting the WR and the RD pulses. If the subtracted value (FIFO occupancy) is between the two thresholds, both UP and DOWN signals are not asserted and the VCO output frequency is retained. However, if the value is below the lower threshold, Down signal is asserted and the VCO output frequency is decreased. Similarly, if the value is above the higher threshold, the read clock frequency is increased. Consequently, the read clock speed is adjusted dynamically to maintain FIFO occupancy between the two thresholds.

The cutoff frequency of the LPF, should be limited so that high frequency fluctuation of the FIFO occupancy will be filtered out and thus the frequency variation of the receiver clock is bound to the nominal range.

4.3. Hybrid approach

The adaptive method based on the FIFO occupancy, described on the two last preceding sections, shows two ways of extracting the source frequency of the incoming data. A hybrid method has been studied which comprises the techniques of both and incorporates some new procedures in order to improve the performance of the adaptive method. The functional structure of this method is shown on figure 10.

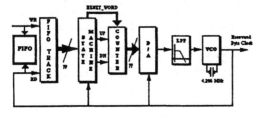

Figure 10. Functional structure of the hybrid method.

Here a FIFO_TRACKING block is also necessary and has the same characteristics of these in the previous sections. The state machine block perform the procedures based on the information which is provided by the FIFO_TRACKING block. The outputs of this block are the UP and DOWN signals to correct the frequency and a reset word to initialize the system to the nominal frequency in case of faults. The counter block is incremented or decremented in order to increase or decrease the read clock. The output of the counter feeds a D/A which converts the output in the corresponding control voltage to the VCO. A Low Pass Filter (LPF) is necessary to smooth the transitions between two successive words.

4.3.1 Parameters and procedures

The algorithm implemented in the state machine performs the corrections in the VCO based in two factors: the variation on the buffer filling level, between two or more time intervals, and the absolute level of the FIFO. The state machine, based on this two factors realize the correction of the read clock through the UP and DOWN signals.

The correction based on the variation of the buffer level is the only action that is performed while the FIFO_level status is between the two thresholds. When this limits are raised, only corrections based on this last factor are made.

The correction factor based on the variation of the buffer filling level is obtained by calculating the average through a set of values, corresponding to the difference between two consecutive FIFO level readings. The number of values in each average is one of the parameters evaluated under simulation.

5. SYSTEM SIMULATION PROGRAMME

In the previous sections some possible approaches to the adaptive method were discussed. The main idea is to consider a PLL to adjust the receiver clock, based on FIFO level. Although the main concepts behind this technique are simple, determination of the parameters involved (e.g. FIFO size, adjust speed, thresholds limits), is not a trivial task. In order to know how each one of these parameters affects the system performance it was decided to develop a specific simulation programme, described below.

5.1. Generalities

The simulation programme developed may be used to simulate any of the synchronization methods described in section 4.

In all simulations a programmable function is responsible for the generation of the intercell intervals. This will be the main input for the programme. Based on this it generates the write clock pulses.

The processes for generating the various network cell delay jitter patterns include, for the moment, Gaussian and Uniform distributions. Very little information is currently available on the statistics resulting from transmission in real networks. This implies that a considerable amount of work is still required in this area.

Basically, the programme outputs are: the minimum size of the FIFO, the mean frequencies of the read and write clocks, a histogram of FIFO occupancy and a histogram of the output frequency.

These histograms represent, respectively, the percentage of occupation of determinated FIFO zones and the frequency distribution of the recovered clock.

During the total simulation a table is generated, containing the values of some important parameters along the time axis, with programmable resolution.

5.2. Main system parameters

The main system parameters used in current simulations are:

- service bit rate (fo): 34 MHz
- network (line) bit rate (NetCk): 155 MHz
- byte period: Tbyte = (1/ NetCk)*8 = 51 ns
- output frequency range: fo ±20 ppm
- packet (cell) size: 47 byte

With this set of parameters the values in figure 3 are as follows:

Tp = 1 cell = 47 Tbyte

T (avg) = 211.5 Tbyte

td (avg) = 164.5 Tbyte

The standard deviation (STD) is also expressed in Tbyte.

5.3. FIFO size considerations

For each synchronization method the delay jitter distribution determines the minimum size of the receiver buffers necessary to its absorption. Large buffers, however, will have impact on implementation cost and on the total end-to-end delay.

The standardization bodies indicate that the total end-to-end delay (excluding transmission delay) should not exceed 20 ms for real-time interactive services. The provisional time for the transit nodes was established as 1 ms.

The simulations considered here take as standard a FIFO of 32Kbytes. Taking the average filling level as half of FIFO capacity, the average delay on the receiver buffer node is 3.81 ms for a 34 Mbit/s (4.296 Mbytes/s) transmission.

Nevertheless the results of the simulation, indicate that for large delay jitters, greater buffers will be necessary. A relation between the input intercell intervals standard deviation, for the Gaussian and Uniform distribution, and the FIFO size, was established, for the charge pump method. This relation is shown in figure 11.

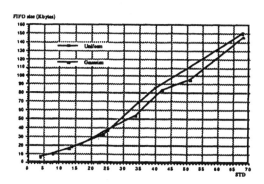

Figure 11. FIFO size versus intercell intervals standard deviation (STD).

5.4. Simulation results

The FIFO level histograms are divided in 34 zones, each one representing 1024 positions of the FIFO, excluding zones 1 and 34, which represent, respectively, the FIFO underflow and FIFO overflow.

The output frequency histograms cover the range of the nominal output frequency (fo) ±20 parts per million (ppm) [6].

Figure 12 shows the histograms for the Charge Pump method. For this method a digital approach was considered since this technique is mainly composed by analog parts.

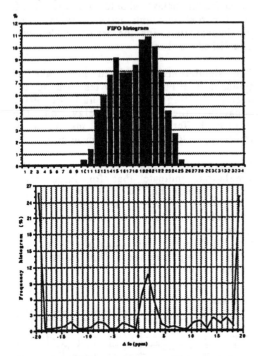

Figure 12. FIFO and frequency histograms for Charge Pump method (STD=15Tbyte).

Figure 12 shows a major disadvantage of this method: the tendency of the RD clock to oscillate between the VCO limits. These produce large oscillations on the RD clock which represent considerable frequency noise in the service provided.

Figure 13 shows a temporal diagram obtained using the Direct method.

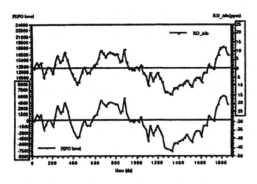

Figure 13. FIFO level and Read frequency temporal diagram, for Direct method (STD=15Tbyte).

This method has the drawback of presenting unnecessary fluctuation on the recovered clock, although FIFO size do not differ very much from the previous method.

Figure 14 shows the histograms for the Hybrid method.

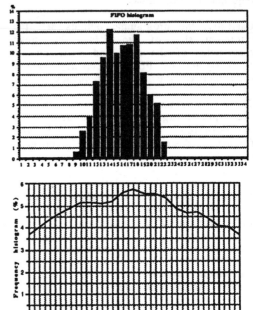

Figure 14. FIFO and frequency histograms for Hybrid method (STD=15Tbyte).

It shows an equilibrated frequency histogram and a reasonable FIFO size. The temporal diagram of figure 15, shows also smooth variations in the recovered clock.

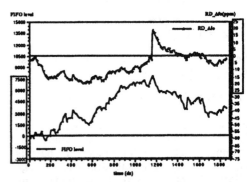

Figure 15. Temporal diagram for FIFO level and RD frequency (STD=15Tbyte).

This seams to be the most appropriated solution for the adaptive clock method. Nevertheless, further study is necessary using the appropriate delay jitter distributions to tune the main parameters and improve the system performance.

6. CONCLUSIONS

The adaptive clock method with an hybrid frequency control algorithm seems to be the best solution for the implementation of synchronization mechanisms for CBR services.

Results are still preliminary, as a better understanding of the system performance, may result from new simulations. This understanding could be achieved using different parameters, and above all, using more realistic delay jitter distributions.

Work will proceed with more simulation studies, with the refinement of the delay jitter distribution routines and the flexible hardware implementation will continue in order to allow the laboratorial testing of the results obtained.

7. REFERENCES

[1]   ESPRIT Project 2054, "Technical Annex", Section 2, January 1989.

[2]   N. Almeida, J. Cabral, M. Ricardo, E. Carrapatoso, A. Alves, "Rede Integrada de Multi-Serviços usando Técnicas ATM e Tecnologia Óptica Coerente", ENDIEL'91, Lisbon, May 1991.

[3]   ETSI/NA5, TD 90/118, "AAL for circuit emulation", Madrid, September 1990.

[4]   ETSI/NA5, TD 90/74, "AAL for VBR Videoservices", Rome, March 1990.

[5]   CCITT, Study Group XVIII, Temporary Document 38, "Report of SWP XVIII/8-3", Geneve, June 1991.

[6]   Rec. G.703, Red Book, Vol.III-Fasc. III.3, CCITT, Geneve, 1985.

Synchronization of Multimedia

Data Streams

in Open Distributed Environments

Peter Leydekkers, Bertjan Teunissen

PTT Research Tele-Informatics Netherlands

Abstract

This paper presents a study of synchronization mechanisms for real-time video, audio and text data streams. Synchronization is an important and complex issue when multimedia information, stored at geographically distributed locations, has to be transported to an end-system for presentation via various communication channels.

In this paper we briefly review the characteristics of multimedia data streams and the requirements that a network should support for these multimedia systems. An overview of synchronization mechanisms is presented and their applicability in an open distributed environment is discussed. Also the positioning of the synchronization mechanism in the OSI Reference Model is indicated.

0. Introduction

Conferencing systems are becoming more and more advanced. Formerly, only the exchange of voice between the participants was possible, at the moment video conferencing services are offered by PTTs and experimental multimedia desktop conferencing systems [MAS91] and other multimedia applications exist [HOP91]. Future conferencing systems will have the nature of a virtual conferencing room, allowing for the exchange of audio, full motion video and text independently of the geographical locations of the participants. The advance of multimedia conferencing systems is stimulated by the growing transport capacity of PTT networks [DOM91] and by new developments in multimedia workstation technologies .

Characteristic of multimedia information in conferencing systems is the variety of information which is stored, communicated and presented in an integrated way. In a multimedia conferencing system data may be stored locally, remotely, in singular or in multiple heterogeneous databases. A mechanism for combining various data elements from possibly distributed sources (at specified times) into one composed data object apt for storage, presentation and transmission to other multimedia systems is required [LIT90a], and is called the synchronization mechanism.

Important for the large scale realization of these advanced communication systems and services, is the availability and acceptance of international standards and recommendations. As already frameworks exist for communication in open distributed systems, it is important to seek solutions, e.g. for the synchronization problem, which are in line with these standards. Therefore, in this contribution, the problem of synchronization of multimedia streams is studied in the context of open distributed environments.

This contribution is organized as follows. Section 1 points out the basic user-requirements for a multimedia conferencing system followed by a summary of network requirements in section 2. Section 3 states the synchronization problem and Section 4 outlines possible synchronization methods for multimedia data streams. In Section 5 the positioning of the synchronization mechanism in the OSI Reference Model is discussed. At the end of this contribution, in Section 6, a summary is given.

1. User requirements for Multimedia Conferencing

This paper is concerned with the synchronization between audio, video and textual objects in an open distributed environment. The typical characteristics (e.g. accuracy) of the synchronization mechanisms used are determined by the requirements of the end user. In this paper we analyze these requirements for the case of a multimedia conferencing system.

A multimedia conferencing system in an open distributed environment might be composed of the following components:

- Workstation for presenting video, text and audio
- Camera
- Microphone
- Database (possibly distributed)
- Wide Area Network

An example of such a system is shown in figure 1. The system would enable a group of persons being apart from each other, to work together on a (multimedia) document, and also enables them to communicate in an almost natural way.

Multimedia conferencing systems require therefore the exchange of information like text, graphics and pictures (anisochronuous communication), as well as the synchronized, real time exchange of audio and video (isochronous communication). The transmission delay, synchronization accuracy and other requirements are determined by the participants, or, more precisely, by the limits of human perception. The requirements and properties of synchronized data streams are listed in Table 1 [HEH90].

Table 1 clearly shows large differences in the characteristics of multimedia data streams. These differences may be exploited in the design of communication protocols. Column one, in principle defining the concept "real-time perception", is especially important for synchronization since synchronized perception can be achieved by introducing extra delays in the data streams. Furthermore, delay jitter (column two) can be reduced ("isochronization") by storing data in buffers which are read out at a constant rate, thus also at the cost of extra delay.

	Max delay (s)	Max delay jitter (ms)	Average throughput (Mbit/s)	Acceptable bit error rate	Acceptable packet error rate
Voice	0.25	10	0.064	<0.1	<0.1
Video	0.25	10	100	0.001	0.001
Real-time data	0.001-1	--	<10	0	0

Delay between audio and video in the -20 ms and +40 ms range

Table 1 Data Streams

Another aspect of importance is the reliability of a multimedia conferencing system. The reliability requirements (column four and five) for real-time data transfer are much higher than those for voice and video transfer, since voice and video data streams contain highly redundant information. Real-time data contains typically information for cursor movements, region markings, document changes, etc. This information must be transmitted accurately and fast, in order to guarantee a acceptable response time.

2. Network Requirements to support MM Conferencing

The requirements mentioned in the previous paragraphs should be supported by the hard- and software of both end-systems and network. The quality of a multimedia conferencing system is determined by the qualities offered by the underlying network. Table 1 shows that, among others, the following Quality of Service (QoS) parameters need to be supported by the communication protocols:

- Throughput (bitrate)
- Maximum delay jitter
- Maximum delay
- Synchronization accuracy
- Bit error rate
- Packet error rate

The network should support multimedia communication and multipoint connections by providing the functionalities listed below [MAS91]. In this list we shortly check whether these functions are supported by B-ISDN based on ATM.

Flexible bitrate
In B-ISDN, having a capacity up to 150 Mbit/s, one can allocate virtual channels with the capacity needed.

Flexible channel speed division
B-ISDN provides virtual channels with a flexible bitrate.

Mesh connection capability
In order to facilitate a multipoint conference, all participant terminals should be logically connected to each other. In B-ISDN, the terminal will be connected by a physical transmission-line to the network which will include logical multiplexed subchannels for the other terminals involved.

Full duplex communication
For conferencing full duplex communication is absolutely necessary because of the interactive aspect of conferencing. Half duplex communication will degrade the human interface, due to real-time discrepancies between participants. B-ISDN supports full duplex communication.

Selective discard of data packets
Congestion of the network is possible, because a very large amount of data has to be transferred between participants. Due to certain aspects of real-time communication, retransmission is undesirable for audio and video. The network should have a selective packet discard function to clear and avoid congestion.

Multi-delivery of data
In the case of conferencing, data must be transmitted to several participants simultaneously. This function can be either carried out by the terminal or be supported by the network. If the network has a multi-delivery function, all terminals will commonly utilize this function. This feature is not supported by B-ISDN.

3. Synchronization problem

Suppose we wish to set up a conferencing session with two other participants in London and New York. Real-time information like full motion video, audio and text has to be exchanged, whereas the text has to be retrieved from a remote database located in Amsterdam. In this configuration, shown in figure 2, the following characteristics are used:

- Data streams are routed via different channels and possibly different nodes
- Multi-point connections
- Several multimedia objects are involved in the communication
- Several information sources located in different places are used

Due to the various delays in these channels, the data streams do not arrive at the same time. This demands a synchronization mechanism between the different data streams, in order to guarantee synchronized perception for e.g. audio- and video-objects. Depending on the application, several actions may be undertaken once synchronization is lost [STE90]. For instance, during a conference, the audio stream is unlikely to be hold in case of problems with the video stream, allowing the participants to continue their communication. On the other hand, such a failure occurring during the presentation of a multimedia document would probably cause the replay to be suspended.

4. Synchronization Methods

Several methods have been introduced to synchronize multimedia data-streams. From the communication point of view three locations can be distinguished at which the synchronization mechanism resides: at the source, at the network or at the receiver.

Synchronization at the source
Multimedia objects can be synchronized with other objects directly at the source. The relationship between different multimedia streams inside a multimedia object is and remains fixed during transportation and presentation of the object to the user. A buffer for holding incoming data is required so that small variations in network throughput and jitter can be compensated before presentation to the user. A CD-I signal is an example in which the audio and video information are woven together.

This synchronization method has several disadvantages. Different information types may have different characteristics, and thus different QoS requirements. For example, if text and voice are to be exchanged simultaneously, one would like to have an error free channel for the text and a channel with a constant delay and with no error checking for voice data. With this synchronization method only one QoS may be specified, and is determined by the QoS of the most demanding medium [NIC90]. Thus information type specific characteristics are not used to advantage, putting unnecessary heavy demands on the network. Another disadvantage is that the method does not apply when information forming a single document is stored at separate sites.

Synchronization at the network
Another approach is the use of a object called a 'synchronization manager' which provides clock pulses [RAY90]. Actions on other objects, like transmission of a video frame, take place at specified times. The synchronization manager is a functionality that is provided by the network and is in principle a handshake mechanism with a precisely defined timing. In advanced distributed applications, such as a multimedia conference, in which many data streams are involved, this centrally regulated handshake mechanism is inhibitory complex. Moreover, it can not be used in situations where several types of networks are combined, e.g. in a WAN.

Synchronization at the receiver
A third synchronization method is performed at the receiver, using information supplied by the sources. This information can be simple, as it is in the synchronization marker concept, or rather complex requiring a separate communication channel.

Synchronization Marker: In the synchronization marker concept synchronization markers are inserted by the sources in the data-streams marking sections in the data streams which should be simultaneously presented [SHE90]. Since the objects may use channels with different delays, the synchronization markers may arrive non-simultaneously (see fig. 3). By using buffers, the data from the different streams can be collected until all synchronization markers have arrived. Then, the re-synchronized data is passed to the recipient.

The disadvantage of this synchronization mechanism is that very large buffers must be used in order to compensate delays. It also modifies the data stream which implies that no direct attachment of in- and output devices is possible. This mechanism provides only basic synchronization services and always assumes a simultaneous presentation of information.

Synchronization Channel: A more complex synchronization method uses information, transmitted via a separate channel, to specify the presentation of information exchanged via other channels. The synchronization channel information contains references to data transported in the other channels, presentation and synchronization information.

With this mechanism it is possible to control whether the data is to be presented independently (uncorrelated), sequentially, simultaneously or in any combination of these [POS88]. The synchronization channel transmits synchronization information to the receiver when necessary.

In B-ISDN this could be performed in a virtual channel alongside the data stream. The question arises as to which capacity would be required for the synchronization channel. B-ISDN provides virtual channels with no fixed capacity. This implies that the granularity of synchronization may be determined by the user in such a way that the human perception requirements are fulfilled. For example, 9 bytes used on the 16 kbit/s synchronization channel implies that a 140 Mbit/s data-stream can be synchronized in units of 80 kbytes, i.e., every 4.5 ms [BLA91]. This fulfils the human perception requirement of variation of video and audio within the -20 ms and +40 ms range.

This synchronization channel method provides flexible means for synchronization and presentation of multiple streams in multimedia conferencing and other multimedia applications. Therefore it suited for the synchronization of multimedia data streams in an open distributed environment.

5. OSI and Synchronization

In order to enable the presentation of multimedia information, basic information types have to be integrated and synchronized. In the previous section we have argued that from the communication point of view, there are in principle three possible locations to place synchronization mechanisms. Two of them can be feasible in an open distributed environment. For these two methods we will try to point out the relationship with the OSI Reference Model here.

Synchronization at the source provides a pragmatic synchronization method for simple composite objects. It can be useful in an open distributed environment for dedicated applications using simple equipment for presentation, with few or no user-interaction. With respect to OSI, this synchronization method can be regarded as residing in the application or in the transport layer. In case the streams are multiplexed before being decoded the synchronization functionality resides in layer 7 and can be regarded as a Single Association Control Function (SACF) in the application layer of OSI. In the opposite case one can argue the synchronization functionality resides at the transport layer of OSI.

Synchronization at the receiver provides a flexible and powerful way for the provision of multimedia services in open distributed environments. The integration of multimedia information can be performed at three levels [LIT90b]. I.e. the user-interface-, service- and the transport-level. Integration at the transport level is concerned with multiplexing the data and stream synchronization. Integration at the service-level provides means for the composition of basic information types into documents and integration at the user-interface does consist of functions, like e.g. choice of undertitling and lip-synchronization. Only the integration at the service- and the transport-level can be related to OSI. Integration at the user-interface level is outside the scope of OSI.

Integration at the transport-level can be performed by the use of basic synchronization methods required to synchronize data in the lower layers of OSI. When for instance different streams are transported by the network, the transport service could provide mechanisms for inter- and intra-stream synchronization. The synchronization marker concept might be feasible here [SHE90]. Synchronization can e.g. be related to isochronous channels with very tight bounds on jitter and delay characteristics. It must be able to deal with ways of maintaining synchronization of data streams from real-time inputs, storage devices and communication channels. Buffering is needed to ensure the fulfilment of the requirements of isochronous channels. When several related isochronous media are present in the application, different isochronous channels need to be synchronized with respect to one another. Assuming that independent channels are used, synchronization mechanisms between related media must be performed at coarse intervals. To implement a synchronization mechanism, two parameters are necessary: the rate of change of jitter per sample and the instantaneous jitter for the current or most recent sample.In this situation, corrective re-synchronization actions would take place at coarse synchronization points.

Integration at the service-level can be established by applying the synchronization channel concept. The channel might transport meta-information revealing the structure of the object to the receiver. This information can be used by the application to compose a compound object from basic information types. In this case the control of synchronization is a function which is performed at the application layer.

According to the Application Layer Structure (ALS), the service provided by the application layer consists of several Application Service Elements (ASEs). Each ASE can provide a specific communication service to the application. For the coordination of different ASEs using one association, a layer seven connection, the Single Association Control Function (SACF) is defined. For coordination of activities among multiple associations of different ASEs, the Multiple Association Control Function (MACF) is applied. Therefore, in the exchange of multimedia information different ASEs may be involved, using different associations. One of the associations implements the synchronization channel. The coordination of the associations is provided by a specific Multiple Association Control Function (MACF). This functionality has to be standardized. Figure 4 shows the relationship between the MACF and the synchronization functionality in the OSI application layer. A similar approach is chosen by the authors in [SCH90].

The MACF seems, from an OSI Reference Model point of view, the most natural and logical place to perform this synchronization function, since it is already designed to perform the control of multiple associations. Therefore the MACF-functionality should be extended in order to include the synchronization function.

6. Summary

In this paper an overview has been given of a multimedia conferencing system and user and network requirements are discussed. The synchronization problem has been described and possible solutions are presented. A promising mechanism could be the use of a virtual channel to perform synchronization as it offers flexibility to support the requirements of multimedia conferencing systems. With respect to the OSI Reference Model the synchronization mechanism could be best performed in the application layer and in particular in the MACF. Further study is required to expand this concept.

Acknowledgements

We would thank dr. R.J. Meijer for his encouragement and assistance in preparing this contribution.

References

[BLA91]   G. Blair, D. Hutchinson, D. Shepherd (Lancaster University)
          Tutorial 4: 'Multimedia Systems' 3rd IFIP Conference on High
          Speed Networking, Berlin March 91.

[DOM91]   G. Domann, 'Two Years of Experience with Broadband ISDN
          Field Trial', IEEE Communications Magazine, January 1991.

[HEH90]   D. Hehmann, M. Salmony, H. Stuettgen, 'Transport services for
          multimedia applications on broadband networks', Butterworth &
          Co, vol.13 no.4, May 1990.

[HOP91]   A. Hopper, 'Design of High-speed networks in Multimedia
          Applications, 3rd IFIP WG6.4 Conference on High Speed
          Networking, Berlin March 91.

[LIT90a] T.D.C. Little, A.Ghafoor, 'Synchronization and Storage Models
          for Multimedia Objects', IEEE Journal on selected areas in
          communications, vol.8 no.3, April 1990.

[LIT90b] T.D.C. Little, A.Ghafoor, 'Network Considerations for
          Distributed Multimedia Object Composition and
          Communication', IEEE Network Magazine, November 1990.

[MAS91]   S. Masaki, 'Personal Multimedia-multipoint Teleconference
          system for Broadband ISDN', 3rd IFIP Conference on High
          Speed Networking, Berlin March 91.

[NIC90]   C. Nicolaou, 'An Architecture for Real-time Multimedia
          Communication Systems', IEEE Journal on selected areas in
          communications, vol.8 no.3, April 1990.

[POS88]   J.Postel et.al., 'An experimental multimedia mail system',
          ACM Transactions on Office Information Systems, vol.6 no.1,
          January 1988.

[RAY90]   M. Raynal, J. Helary, 'Synchronization and Control of
          Distributed Systems and Programs', Wiley series in parallel
          computing, 1990.

[SCH90]   G. Schuermann, U. Holzmann-Kaiser, 'Distributed Multimedia
          Information Handling and Processing', IEEE Network Magazine,
          November 1990.

[SHE90]   D. Shepherd, M.Salmony, 'Extending OSI to Support
          Synchronization Required by Multimedia Applications',
          Computer Communications, vol.13 no.7, Pages 399-406,
          September 1990.

[STE90]   R. Steinmetz, 'Synchronization Properties in Multimedia
          Systems', IEEE Journal on selected areas in communications,
          vol.8 no.3, April 1990.

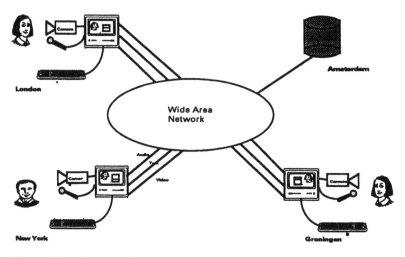

Figure 1 Multimedia Conferencing System

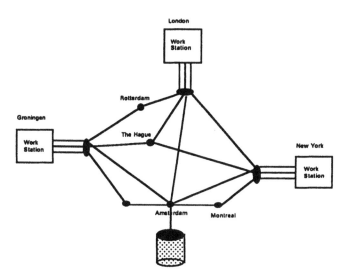

Figure 2 Synchronization problem

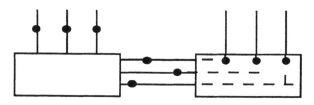

Figure 3 Synchronization markers

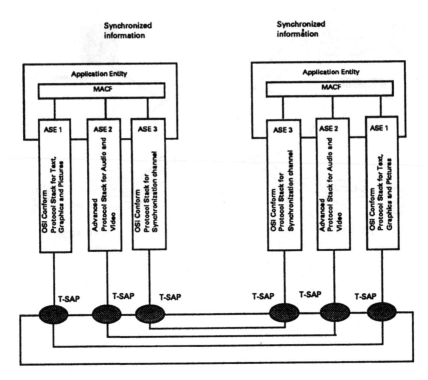

Figure 4 Multiple Association Control Function

Session IV: Synchronization II

Chair: Stephen Pink, Swedish Institute for Computer Science

The second session on synchronization focused more on operating system and runtime issues than on the network issues featured in the first session. A distributed multimedia system must deal with the asynchrony introduced by the host computer as well as the asynchrony of the network. Even if jitter, for example, could be eliminated from the network entirely, the scheduling of operating system tasks could easily reintroduce variations in the latency to the user. Thus, the designer of distributed multimedia systems must face the problem of real-time scheduling in the operating system from device driver handling up to presentation layer processing.

In the first talk, Dick Bulterman from CWI, summarized a paper co-authored by Robert van Liere called "Multimedia Synchronization and UNIX." The talk centered on a critical evaluation of UNIX as an operating system capable of providing synchronization services for multimedia applications. The authors also offered a solution to some of the problems inherent in using UNIX for multimedia by describing the design of a multimedia coprocessor.

Data location and data synchronization models are first described as part of the multimedia scenario that UNIX will have to serve. There are multiple locations that a multimedia source and sink can be in a distributed system. There can be single sinks and sources, multiple sinks and sources, as well as permutations and combinations of these. A similar complexity exists for data synchronization models. There are various synchronization classes (such as serial and parallel synchronization), different synchronization scopes (such as point and block synchronization), decisions on which of two streams is the synchronization master and which is the slave, and finally various levels of synchronization precision that are often set subjectively by the user.

The paper points out that some of the above combinations can put stress on an operating system such as UNIX that is built on a rather strict execution hierarchy. There are basically three ways in which code is run in a UNIX operating system environment. Processing can be done in user mode, where the thread and process abstractions are instantiated in UNIX, or in system mode. System processing is further divided up into kernel process execution and interrupt handling. Scheduling hierarchies in the UNIX kernel are well defined for these modes of execution, and it is generally recognized that UNIX was not designed to fit all the complex models of data synchronization, although for some models, such as serial synchronization, the UNIX system is adequate. For example, device interrupt handling is designed to be the most responsive part of system processing, yet is the most difficult for the user to control. Thus, the authors have suggested a new processing model.

The multimedia coprocessor (MmCP) will give UNIX and UNIX-like distributed operating systems such as Amoeba the flexibility to handle multimedia synchronization. The MmCP will be part of a multimedia workstation's architecture, sharing memory with the main UNIX processor. The MmCP will act as an intelligent device controller that will allow multimedia applications to have control of a low-level fast device to service the synchronization abstractions mentioned above.

Thomas Little from Boston University, in the second talk on joint work with Arif Ghafoor entitled "Scheduling of Bandwidth-Constrained Multimedia Traffic," presented a scheduling mechanism for smoothing the bandwidth requirements of multimedia applications that attempt to push the bandwidth limits of a network. A distributed multimedia system should be seen as a real-time system whose deadlines, or playout times, can be scheduled with a certain probability of failure (as opposed to having hard deadlines that must be met at all costs.) The scheduling algorithm presented in the paper assumes that the source for the multimedia data objects is data storage, as opposed to live media. The algorithm is most useful when attempting to schedule the transfer of multimedia data objects of varying sizes across a network whose limited capacity is being approached.

Thomas pointed out that since we have a priori knowledge of the characteristics of stored multimedia data, we can schedule our resources before playout occurs. Also, multimedia data can tolerate losses and data can arrive late without catastrophe, so playout schedules are probabilistic. By mathematically specifying two constraints, the minimum end-to-end delay of a multimedia data object through a network, and the finite capacity of a channel, a function is produced that describes a retrieval schedule (from storage) given the playout schedule of the data.

Two examples of retrieval schedules versus playout schedules were presented. The first was the transfer of compressed full-motion video images. The second was a mixed scenario of coordinated audio and video images. The results of applying the algorithm clearly showed that where maximum channel capacity was being temporarily exceeded, buffering was necessary to meet the scheduling constraints of the mixed audio and video traffic. The authors suggested that the algorithm presented in the paper would be useful in situations where digital audio and video are delivered to the home on constrained-bandwidth media, such as copper cable.

"Presentation scheduling" was the topic of the third talk in the session. Petra Hoepner of GMD-FOKUS in Berlin, in her talk on her paper "Presentation Scheduling of Multimedia Objects and Its Impact on Network and Operating System Support," defined presentation scheduling as the temporal ordering of actions and interactions in a multimedia application. The abstract units of these actions are called presentation frames. A presentation action may be "reconciled" to the presentation frame by conforming to the rules that define a presentation frame. These rules define the "synchronization strategy" of the actions and hence determine the schedule by which the actions in a multimedia application take place.

If there is more than one action in a presentation frame, then one must define the temporal relations of the actions. There are a number of presentation frame types, according to Petra, that occur frequently. Among these are the Sequentializer, Parallelizer, Splitter, Combiner and Brancher. Each of these presentation frame types contain identifiers, start and end event tags, as well as actions which form the rules of behavior for actions that are reconciled to presentation frames of these types.

The author gave two examples of presentation frame types: Parallelizers and Splitters. In the Parallelizer example, two events are executed at roughly the same time, but one could end before the other. The Parallelizer supplies a rule to determine what action should take place when the first event finishes. A concrete application of the Parallelizer is the mechanism that produces audio/video lip synchronization. The Splitter is similar to the Parallelizer except that more than two events must be managed. So, in addition to start and end event tags, there are also synchronization tags that the actions use to order the group of events temporally.

Network timeliness and availability, quality of service, workahead and buffering strategies are part of the operating system and network support needed for presentation scheduling. These issues were planned as topics for future research.

During the discussion, the author said that there are some similarities between the model of presentation scheduling described above and the work being done in the ODA and ODP areas. For example, there is a mechanism in ODP that has the same function as the Splitter. Petra also said that the goal of this research is to provide an environment for the a user to edit a presentation scheduling script that is independent of the host platform. The script could then be compiled or interpreted for a specific execution environment, which, after being linked with the runtime and network system, could be executed across a heterogeneous distributed system. Thus, one could design a multimedia medical document, for example, on one kind of workstation and make it available to users on different platforms.

Multimedia Synchronization and UNIX

Dick C.A. Bulterman
Robert van Liere
CWI: Centrum voor Wiskunde en Informatica
Amsterdam, The Netherlands

Abstract

One of the most important emerging developments for improving the user/computer interface has been the addition of multimedia facilities to high-performance workstations. Although the mention of multimedia I/O often conjures up visions of moving images, talking text and electronic music, multimedia I/O is not synonymous with interface bells and whistles. Instead, multimedia should be synonymous with the *synchronization* of bells and whistles so that application programs can integrate data from a broad spectrum of independent sources (including those with strict timing requirements). This paper considers the role of the operating system (in general) and UNIX (in particular) in supporting multimedia synchronization. The first section reviews the requirements and characteristics that are inherent to the problem of synchronizing a number of otherwise autonomous data sets. We then consider the ability of UNIX to support decentralized data and complex data synchronization requirements. While our conclusions on the viability of UNIX for supporting generalized multimedia are not optimistic, we offer an approach to solving some of the synchronization problems of multimedia I/O without losing the benefits of a standard UNIX environment. The basis of our approach is to integrate a distributed operating system kernel as a *multimedia co-processor*. This co-processor is a programmable device that can implement synchronization relationships in a manner that decouples I/O management from (user) process support. The principal benefit of this approach is that it integrates the potential of distributed I/O support with the standardization provided by a "real" UNIX kernel.

1. Introduction

One way of measuring progress in computer architectures is to study the evolution of the user/computer I/O interface. Ten years ago, the departmental minicomputer provided a dumb terminal and the occasional intelligent peripheral input or output device as the standard for user/computer interaction. Five years ago, the microprocessor-based personal workstation replaced the dumb terminal with a keyboard, a mouse and a high resolution bit-mapped display, as well as a network connection for accessing files and remote I/O devices. Currently, RISC-based workstations embody the standard of modern I/O support, adding 8-bit color displays, local disks, large main memories and a local I/O bus to the user's desktop. During this ten-year period, many of the implementation details of the user/computer interface have changed, providing faster generation and presentation of information through vastly increased processor speed, improved realism through special-purpose output technology, and (to a lessor extent) improved information precision through the use of enhanced data input facilities. Throughout this period, however, the fundamental user/computer I/O model has remained unchanged: information is first collected from one or more disjoint input sources, then transformed and filtered by a controlling application, and then passed on to one or more disjoint output destinations. The selection of I/O devices has also remained relatively static during the past decade, with input coming primarily from text- and pointer-based devices and output going to text- and picture-based devices.

While the expressive nature of the user/computer interface has seen little development during the past decade, it appears that its next evolutionary step *will* provide a fundamental change in the types of information that can be processed. The essence of this change is the introduction of temporal relationships among data sets. Consider that new workstations already provide the user with device-level access to DAT-quality stereo sound, 24-bit color images and live video input and output facilities. Taken alone, the independent manipulation of each separate information medium does not provide anything particularly new. Taken together, however, the integrated manipulation of this "multimedia" information provides the potential for dynamically creating new, user-directed composite data sets. (A review of the characteristics that can be expected from future multimedia workstations is provided in [1].) Unfortunately, the technology aspects of the user/computer interface have evolved more rapidly than the operating systems and programming languages interfaces for supporting dynamic temporal data relationships. As a result, it is clear that the essential problem of multimedia is not simply providing I/O support for sound and video. Rather, the essential problem of multimedia is that support must be provided for *synchronizing* otherwise autonomous data transfers within and across workstations.

To appreciate the scope of the multimedia support problem, consider that the development of programming languages, application-programmer interfaces, operating systems and device controllers have all been based on a model of user/computer interaction that treats I/O activity as a series of unrelated data movement operations on a set of independent (device) files. The UNIX[1] operating system is a good example of how I/O is currently managed. Here, each I/O request is passed to a device driver layer within the UNIX kernel; this layer is responsible for scheduling (possibly buffered) device transfers in such a way that the application in presented with a uniform sequential I/O model. The driver usually does not interpret any of the the data it is moving, leaving the coordination of multiple I/O operations within a process to the application layer. Since each driver is designed to manage activity within one device queue, low-level coordination of multiple resources is left to a first-come, first-serve contention strategy or to individual characteristics of the underlying hardware architecture. As a result of the UNIX I/O model, there is no way for an application to request coordinated input from multiple devices as an operating system primitive action, there is no way for the operating to conveniently coordinate separately specified I/O actions, and there is no way for a device controller to know how to effectively schedule itself in cooperation with other I/O devices. The situation becomes even more complicated when the sources and/or destinations of information are located on different workstations in a network. Here, no amount of *ad hoc* operating systems hacking will provide a comprehensive model for the synchronization of distributed time-based data.

In this paper, we consider the impact of supporting multimedia synchronization in a UNIX environment. (We use UNIX as a model for study because of it is representative of the types of operating environments that a multimedia user can expect to encounter.) We begin with a study of the scope of the problem by discussing two classes of multimedia data models: multimedia data location models and multimedia data synchronization models. We then consider these models in terms of the processing hierarchy provided by UNIX to see if UNIX is "multimedia ready". In so doing, we concentrate on current workstation-based user/computer I/O in local and distributed environments. We conclude by offering an alternative approach that we are studying as part of the CWI/Multimedia project [2,3] to better support distributed multimedia applications; this approach seeks to combine the benefits of providing a standard UNIX user interface model for common applications use with a co-equal distributed I/O subsystem that can be scheduled to implement multimedia data synchronization across a network of workstations.

---

[1] UNIX is a registered trademark of AT&T/USL in the United States and other countries.

2. Two Models of Data Interaction in Multimedia Systems

The broadest notion of multimedia encompasses the integration of arbitrary spatial and temporal data sets to encode and/or present information. In defining general support for this integration, it is important to model two aspects of information interaction. The first model defines the allowable data location relationships that exist among information sources and destinations; the second model defines the allowable synchronization relationships that exist among data sets. Both of these models are considered in the following sections.

2.1. Multimedia Data Location Models

The location of multimedia information determines the amount of operating systems support that is required to gather and route scattered data. Figure 1 reviews four general models that can describe the sources and destinations for multimedia information.

a) *Local single source*. This model has all data originating at a single source and routed to several destinations. An example is a CD-ROM that contains sound, text and picture information. This model provides the least synchronization complexity because all intra-sample synchronization is the responsibility of the source material designer. Inter-sample synchronization is implemented by fetching input blocks at a predefined rate and then routing them (by either the device controller, the operating system kernel or the application) to one or more output devices.

b) *Local multiple sources*. This model is similar to (a), except that source data is scattered across several devices. An example of this type of interaction is combining voice annotation with images in an electronic slide-show. The principal difference between single and multiple source data is the need for external (to the data) synchronization among the data streams.

c) *Distributed single source*. In this model, we assume that a single source of information exists that is located on a remote location relative to the workstation managing the user/computer interface. The single-source nature of data means that no multi-stream synchronization is necessary but that the remote location of the data will require compensation for transfer delays.

d) *Distributed multiple sources*. This is the most general model of data location. Information may be gathered from many sources on many workstations, and destinations may also be spread over several places. Synchronization problems include I/O scheduling across a set of workstation, transfer delays across several connections and processing delays at one or more sites.

The central issue in supporting the various location models is determining where synchronization is implemented in a system's processing hierarchy. (The hierarchy is: application code, operating systems code, device controller.) In the local single source model, the choice depends on the input sampling rate of the data. In the local multiple source model, the required synchronization can be placed in either user application code or within the operating system kernel; synchronization at the user level will yield more flexibility while synchronization at the kernel level will yield better performance. In the distributed single source model, the transfer-based synchronization can be accomplished by buffering data at the source or the destination; note that whatever option is used, data control is distributed over at least two kernels—the sending and receiving—as well as several protocol layers and at least one user layer. The distributed multiple source mode is the most interesting because it combines aspects of synchronization problems with transfer delays and raw information scheduling. The impact of data location on the processing hierarchy for UNIX systems will be discussed in detail in section 3, after we first consider exactly what we mean by synchronization.

2.2. Multimedia Data Synchronization Models

Regardless of the location of data, the data itself can contain synchronization information (that is, it can be self-synchronizing) or it may require synchronization through an external mechanism. The illustration in figure 2 can be used to get an intuitive feel for the general synchronization relationships that can exist among multimedia data streams. In this picture, we define five information streams that interact to provide a composite multimedia story. Each stream is made up of a number of blocks of information, with the timing of the presentation of each block dependent in some way on the presentation of information in other blocks. The details of the inter- and intra-stream relationships are beyond the scope of this paper. (Interested readers are referred to [3].) What is important is notion that synchronization concerns cover a broad spectrum. In this section, we consider four aspects that affect the partitioning of tasks among the application software, the OS and the device controller(s). These are: the basic type of relationship among data streams, the scope of synchronization information, the determination of the controlling party in a synchronization relationship, and issues regarding the precision of synchronization required.

a) *Synchronization classes.* There are two basic classes of synchronization within a multimedia framework: *serial* synchronization and *parallel* synchronization. Serial synchronization requirements determine the rate at which events must occur within a single data stream; this includes the rate at which sound information is processed, or video information is fetched, etc. Parallel synchronization requirements determine the relative scheduling of separate synchronization streams. In most non-trivial multimedia applications, each stream will have a serial synchronization requirement and a parallel relationship with other streams. Note that a special case of serial synchronization can be defined as *composite* or *embedded* synchronization; in this case, each serial block of data contains information for parallel output streams. In this case, the parallel synchronization among blocks is embedded in a serial stream.

b) *Synchronization scope.* We can distinguish between point and continuous synchronization. Point synchronization requires only that a single point of one block coincides with a single point of another. Continuous synchronization requires close synchronization of two events of longer duration. In general, point synchronization can be managed by the applications layer while continuous synchronization will need to be managed by a device controller or a high-performance, low-overhead portion of the operating system.

c) *Synchronization masters.* The third distinction regards the controlling entity in a (set of) stream(s). Sometimes we have two channels that are equally important, but sometimes one channel is the "master" and the other the "slave". It is also possible that an external clock plays the role of the master, either for all of the streams or for a subset of time-critical ones.

d) *Synchronization precision.* Finally, there are levels of precision. Stereo sound channels must be synchronized very closely (within 1 to 0.1 millisecond), because perception of the stereo effect is based on minimal phase differences. A lip-synchronous sound track to go with a video movie requires a precision of 10 to 100 milliseconds. Subtitles only require a 0.1 to 1 second of imprecision. Sometimes even longer deviations are acceptable (background music, slides). Note that in all cases the *cumulative* difference between the channels is what matters, not the speed difference.

In general, the diversity in individual device characteristics makes the level of support for a combination of media a challenging design issue. Most vendors of current commercial equipment use embedded synchronization that is mapped onto a serial stream of data. As a result, they need to consider only point-type synchronization scope with a single master device. The precision is determined by the characteristics of the input source and the system load; most of the synchronization precision is supported by managing interrupt contention between the input and output devices. While this approach can lead to dramatic results, it is not sufficient if the user is to be given more control over the data being processed or if information needs to be combined from several sources (either locally or from distributed points in a network).

3. The Impact of the UNIX I/O Subsystem on Multimedia Interaction

The previous section has characterized aspects of multimedia information. In this section, we consider the impact of multimedia information on UNIX (and vice-versa). We begin with a discussion of data location models and then consider synchronization topics.

3.1. Location Models

As illustrated in figure 3, processing within a typical UNIX system can take place at the following levels: within a physical I/O controller, within the operating system kernel, and within an application thread and/or process. Processing that occurs within an I/O controller is done in parallel with other activity in the system. Processing within the kernel is done in *system mode*, either in *process context* (that is, as part of the low-level support activity for a particular process) or in *interrupt context* (which is time that is not directly associated with the support of a particular process's threads). Processing within at the thread/process level is always done in *user mode* in the context of a particular process.

If controller-based I/O processing is supported, then information can be fetched, synchronized and output transparently within a particular device controller. (This assumes that the device controller can manage all the required data sources and destinations.) Operating system kernel I/O processing includes the conventional activity of device drivers, both in the interrupt and the process contexts. Thread/process I/O processing includes all user-level I/O synchronization. In general, controller I/O is immune from interrupt overhead. Kernel I/O will be immune from preemption (that is, a process switch resulting from the scheduler's determination that a higher priority process became ready to run), but it will typically not be immune from interrupts. Thread/process I/O is subject to both preemption and interrupts.

We can measure the effectiveness of the UNIX I/O system by considering the impact of the multimedia location models for each of the types of control considered above:

- *Controller-managed I/O*. In spite of its potentially attractive performance characteristics, controller-managed I/O can only play a minor role in general multimedia processing. The single benefit of this type of I/O control is that information can be quickly fetched, synchronized and routed without interference from other system activity. Unfortunately, since the entire point of providing multimedia services is to give the user the ability to integrate separate information streams, some measure of "interference" beyond start/stop/rewind/search will nearly always be required. In terms of our models of location, controller-managed I/O may be useful for local single source data that is self-synchronizing, but as soon as any management of separate streams is required, the limited scope of the controller will restrict its usefulness.

- *Kernel-managed I/O*. For reasons of performance, kernel-managed I/O can potentially play a dominant role in providing multimedia support. One example of this is the use of interrupt context processing to provide high-speed control and routing of incoming data. Another example is the use of multiplexing device drivers to coordinate the activity of a number of I/O streams. This model is especially useful for local multiple source data and (to a lessor degree) with distributed data. Unfortunately, kernel-managed I/O has a number of severe limitations, the most restrictive of which is that few application program builders have the option of writing new device drivers to cope with in-kernel I/O processing. This is especially true for applications that need to share I/O devices with other applications; in this case, driver modules simply cannot be unlinked and relinked efficiently enough to provide the flexibility required by several applications. Note that for distributed data, even in-kernel manipulation of data will be limited by the fact that there is no cross-kernel buffer sharing possbile among separate UNIX systems.

- *Thread-managed and process-managed I/O.* Application-based interaction in a multi-threaded model provides the most general form of support for all types of multimedia data processing. The application programmer can dispatch as many threads as is necessary to handle each type of data. Unfortunately, there is a cost for this flexibility: performance. The non-deterministic scheduling characteristics of UNIX systems make them unreliable at the thread level for collecting and processing information. To understand the limitations of processing at the thread level, consider that it takes about 40 microseconds for a 20-mHz processor to switch from user-mode to kernel-mode in executing a system call. (This is raw system call overhead; processing time is extra.) This means that even if we provide a set of device drivers with a great deal of memory to buffer incoming and/or outgoing data, an application still loses nearly 100 microseconds just in changing the modes necessary to initiate a data transfer between an input and an output device. Since each multimedia transfer will typically cause at least three systems calls for trivial I/O (one for fetching composite data and two for writing it out to two devices), this overhead can be substantial. Add to this the perilous scheduling situation that all threads must endure and the fact that at the thread level only limited resource management facilities exist in UNIX (such as memory locking or explicit control over kernel buffer management), then the situation at the thread level is not particularly encouraging.

Our consideration of the impact of data location on performance with the UNIX has concentrated on local data location models. The situation is even worse for distributed data, since here multiple thread layers and multiple kernel layers and multiple controllers must be transited by data as it moves from one machine to the other. We return to this point in section 4.

3.2. Synchronization Models

While the discussion above focused on the abstract gathering, processing and scattering of multimedia data, in this section we focus on the particular problems that arise in a UNIX environment for handling synchronization processing. The primary obstacle here is the UNIX scheduler. The priority-based scheduling mechanism offered by most kernel implementations is inflexible in responding to short-term constraints that can occur while synchronizing multiple data streams. This is a consequence of basic UNIX design; even so-called real-time scheduling classes within recent implementations of UNIX do not provide a user with a great deal of dynamic scheduling control to respond to transient critical conditions. Although the scheduler could conceivably be changed, most users will not have this as an option.

In terms of our detailed list of synchronization types, we can make the following observations about the ability of UNIX to support multimedia processing:

- *Synchronization classes:* UNIX can do reasonably well in supporting serial synchronization of data if the sampling rates are sufficiently low to not cause a burden on the system. The block-oriented fetching of data can significantly increase the number of samples processed by an application, although the limited scheduling control of each thread will not ensure the constancy required by high-bandwidth devices. For parallel synchronization, the prospects are less promising: the sequential nature of UNIX I/O will result in either a loss of data resolution or in a limit on the number of parallel tracks that can be processed. One reason for this is the form of the generic I/O system call; all I/O is done on a single file descriptor at a time, with separate file descriptors requiring separate system calls. It may be possible to build multiplexing drivers to combine I/O on a number of file descriptors, but this will not offer a general solution to most applications builders. Another possibility may be the development of multi-file I/O system calls (with particular synchronization semantics defined in the system call argument list), but even *if* these were to become accepted by the growing list of standards organizations, most languages would be unable to cope with the notions of parallel I/O accesses. For the time being, the best one can hope to do is to provide either an applications-based multi-threaded scheduling solution to parallel stream synchronization (with all of the performance limitations discussed above) or to rely on smarter controllers to by-pass the CPU altogether.

- *Synchronization scope:* Of the two types of synchronization scope defined above, point synchronization can be relatively well managed by the thread level, but continuous synchronization can only be managed if the input and output data rates are sufficiently low. Once again, the scope of the synchronization is not only restricted by the implementation concerns of the UNIX I/O system, but also by the ability of applications code to flexibly access data at a low-enough layer in the system.

- *Synchronization masters:* the easiest way to support synchronization within a UNIX environment is to have a master clock regulate the gathering of samples and the dispatching of samples to various output devices. In order for such a clock to function, it will need to be able to influence processing in a number of threads in the same way that real-time clock can influence the scheduling of various real-time processes. (The problems are, of course, not simply similar, they are identical.) Unfortunately, the level of real-time support in UNIX systems has never been particularly good. As for peer-level synchronization, the problems with guaranteed scheduling time under UNIX once again limit the amount of coordinated processing that can be realistically accomplished.

- *Synchronization precision:* depending on the level of precision, processing can be implemented at any of the five layers in the UNIX hierarchy. If stereo channels need to be synchronized, then it can only occur at the controller or interrupt level (unless the data need only be resynchronized at a much lower rate). If, on the other hand, subtitles need to be added to a running video sequence, then this can easily be done at the thread level.

The general dilemma of processing multimedia data remains that those applications requiring the most processing support are probably the least likely to get it in a general UNIX environment. This is not really surprising: manufacturers of high-performance output devices (such as graphics controllers or even disk subsystems) have long realized that the only way to really improve over-all system performance is to migrate this processing out of the UNIX subsystems. Unfortunately, doing so is difficult for multimedia applications, since the type of processing required over a number of input and output streams is usually beyond the scope of the implementation of any one special-purpose I/O processor.

3.3. Is UNIX Multimedia Ready?

The discussions in the preceding sections can lead us to two initial conclusions:

(1) The fastest and most responsive layers in the UNIX I/O hierarchy are the device controller and the interrupt layers; these layers enjoy high-priority scheduling and can be invoked with relatively little overhead. In terms of processing efficiency, it can be argued that once you reach either the normal kernel or thread/process layers, it is

probably too difficult to provide efficient and deterministic multimedia processing.

(2)   It can be assumed that for all but the most trivial types of fetch-and-deposit multimedia operations, it is both desirable and necessary to provide a layer of applications support to manage the synchronization interactions among the various incoming and outgoing data streams. (Recall that the entire reason for having computerized multimedia systems is the measure of control a user can have over the sequencing and presentation of pieces of data.) This type of processing is "easily" done at the thread/process layers, it is possible (but often impractical) at the device driver layer, it is improbable at the interrupt layer and it is usually totally unavailable at the controller layer.

The net effect of these conclusions is that it is desirable to supply a new programmable layer in the UNIX hierarchy that combine the performance benefits of the existing lower layers with the flexibility of the existing upper layers. In providing this layer, it is probably not useful to simply steal cycles from the CPU—doing this is, in effect, only replacing the existing UNIX scheduler with a semi-real-time one. If we assume that all of the normal services available to a user must continue in addition to multimedia processing, then some form of co-processing will be required to satisfy both the UNIX user and the multimedia application.

In the next section, we provide a brief description of an approach being studied at CWI for providing multimedia applications support. This approach, which is based on a distributed I/O and processing architecture, is a generalization of existing approaches for offering high-performance graphics and computation processing on a workstation: the special-purpose co-processor.

4. An Alternative Approach to Providing Multimedia Synchronization Support

Supporting multimedia synchronization on a local workstation requires a balance between I/O data rates, system scheduling and user interaction control. In a local environment (such as in the local single source and local multiple source models), it is possible that a combination of clever implementation techniques, creative kernel hacking and low user expectations can produce interesting multimedia results. (The wealth of personal computer multimedia applications prove this to be true.) Unfortunately, for many users, clever kernel hacking is not an option; they expect to run standard applications on standard systems that exhibit standard (if not always exciting) functionality. Even if it were possible to modify local UNIX implementations to improve I/O performance, the result can be the isolation of data and user/computer functionality just at a time when data sharing across applications in a networked environment is at the heart of modern computer evolution.

Decentralized data sharing requires a level of communication and coordination that surpasses that which can be cleverly added to a single kernel. This coordination may consist of bandwidth reservation algorithms for efficient network use or intelligent algorithms for information transfer. An example of the latter type of algorithm may be a transport-style communications layer that knows to bias its service towards one type of media—such as audio—at the expense of others—such as video—if bandwidth become limited during a transmission. If one were to try this in a typical UNIX kernel, then the process and mode switching time may well be longer than the adaptive period of transmission delay!

In order to address the twin issues of decentralized data sharing and local UNIX standardization, we have been investigating the development of alternative workstation support for multimedia synchronization. Figure 4 illustrates the placement of a *multimedia co-processor* (MmCP) as a component of a workstation architecture. The MmCP is assumed to be a programmable device that can be cross-loaded from the master UNIX processor. It is assumed that the MmCP can execute arbitrarily complex processing sequences, and that it will have access to all or a part of the workstation's memory. As with arithmetic co-processors, a simple interface should exist to control information flow from the UNIX processor. Unlike normal co-processors, however, we assume that the MmCP will be driven by a separate distributed operating system that will provide communications support between its hosting workstation and other workstations in a network environment. This distributed support (Fig. 4b) will provide for coordination among the various sources and destinations in the local and distributed location models discussed in section 2.

The development of the MmCP is driven by the following three observations:

(1)   It should be clear from the sections above that the general motivation for a programmable, high-performance processing layer exists. It may be argued that this need will be satisfied at the thread/process layer by faster processors, although we feel that such processors will only stimulate the requirements for even higher processing rather than satisfying it.

(2)   The considerable standardization effort currently being undertaken for UNIX (and UNIX-like) systems demonstrates the fundamental importance of providing a compatible user interface for running a variety of application programs. While a totally new operating system could be developed with a better scheduler, a lighter-weight I/O system and a more flexible user/system interface, such a system would inevitably need to be loaded down with a UNIX support layer to gain general acceptance. Other approaches, such as assuming that UNIX will go on a functionality diet and be transformed into a lean-and-mean multimedia operating system, have no basis is recent history; in fact, all recent versions of UNIX have gotten functionally fatter as a part of the standardization process.

(3)   A parallel development that encourages our work is the rapid development of multiprocessor workstation architectures. Although many of these systems are little more than trade-press rumors, several systems already provide moderate-cost multiprocessor workstations coupled with a wide array of input and output subsystems. There is no inherent reason why these systems can not simultaneously support multiple operating systems (one or more for the MmCP and the rest for UNIX processing.)

The MmCP provides a reasonable balance between keeping the benefit of nearly fifteen years of UNIX development and providing new, high performance services for tasks such as synchronization.

If we examine the location models in section two, we can see that the MmCP will probably achieve the greatest benefit for distributed data. Its impact on controlling local I/O can also be substantial, however. In fact, one way of viewing the MmCP is as a very intelligent (and programmable) device controller rather than a second operating system. As far as synchronization support is concerned, mechanisms will be necessary to provide the required primitives for application program support, but these primitives can be developed in a much more flexible environment than that of the UNIX kernel.

Our work is currently centered around evaluating the use of the Amoeba operating system as the basis for an MmCP [4,5]. Amoeba has two main advantages in our research: first, it has excellent communications characteristics that appear to make it suitable for light-weight protocol development; second, it is mature but relatively unused—meaning that it is still an open, experimental system that is unencumbered by hundreds of users or thousands of standardization committee members. It should be pointed out that we are investigating *basing* our work on Amoeba, not adding multimedia functionality to Amoeba. It is expected that various tasks that can be performed by the UNIX processors need not necessarily be duplicated in the MmCP. Also, unlike other operating systems research projects [6,7], we are not intending to develop a "micro-kernel" as such (that is, a kernel with core services for use in controlling activity on a workstation), but rather something which could be called a "nano-kernel": a kernel that handles a particular subset of services that can be allocated to one or more users of on a general workstation. (Figure 5.) In this sense, our work is aimed at replacing the partitioned intelligence in device controllers with a layer of shared intelligence at a super-controller level. This has the advantages of providing a full (and standard) UNIX environment plus a programmable interface layer for high-performance support.

5. Summary

We have attempted to argue that the conventional UNIX environment for workstation computing—as useful as it is for many applications—may not be ideally suited for high-performance multimedia computing. Although some of the factors that constrain UNIX are technology dependent, much of this problem lies with fundamental design issues that were a part of the original uniprocessor, sequential serial I/O model developed for UNIX in the 1970's. The approach of the multimedia co-processor that we have presented here is an attempt to overcome many of these problems without sacrificing the positive aspects of a uniform UNIX interface.

6. Acknowledgments

Members of the CWI/Multimedia group (notably Guido van Rossum, Dik Winter and Jack Jansen) have played important role in the development of the ideas in this paper. In particular, Guido van Rossum contributed the synchronization class definitions in section 2.2. An early version of this paper was presented at the 1991 EurOpen Autumn technical conference in Budapest, Hungary.

7. References

[1] *SIGGRAPH '89 Panel Proceedings,* "The Multi-Media Workstation," Computer Graphics, Vol. 23, No. 5 (Dec 1989), pp 93-109.

[2] *Bulterman, D.C.A.,* "The CWI van Gogh Multimedia Research Project: Goals and Objectives," CWI Report CST-90.1004, 1990.

[3] *Bulterman, van Rossum and van Liere,* "A Structure for Transportable, Dynamic Multimedia Documents," Proceedings of the Summer 1991 Usenix Conference (Jun 1991), pp 137-156.

[4] *Mullender, van Rossum, Tanenbaum, van Renesse and van Staveren,* "Amoeba: A Distributed Operating System for the 1990s," IEEE Computer Magazine, Vol. 23, No. 5 (May 1990), pp 44-53.

[5] *van Renesse, van Staveren and Tanenbaum,* "Performance of the World's Fastest Distributed Operating System," Operating Systems Review, Vol. 22, No. 4 (Oct 1988), pp 25-34.

[6] *Accetta, Baron, Bolosky, Golub, Rashid, Young and Tevanian,* "Mach: A New Kernel Foundation for UNIX Development," Proceedings of the Summer 1986 Usenix Conference (Jul 1986).

[7] *Dale and Goldstein,* "Realizing the Full Potential of Mach," OSF Internal Paper, Open Software Foundation, Cambridge MA (Oct 1990).

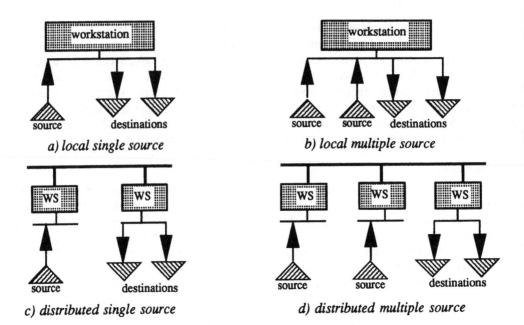

Figure 1: Four location models for multimedia data.

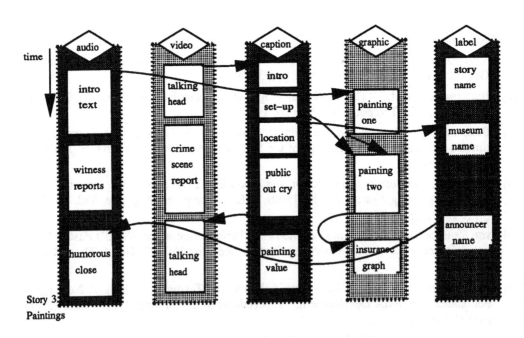

Figure 2: Synchronization in a composite multimedia document [3].

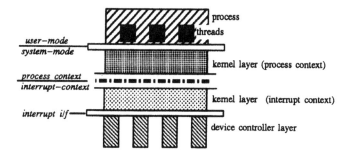

Figure 3: Elements of the UNIX processing hierarchy.

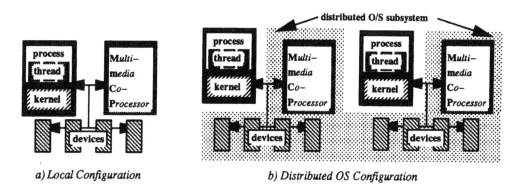

a) Local Configuration *b) Distributed OS Configuration*

Figure 4: The Multimedia Co-Processor (MmCP).

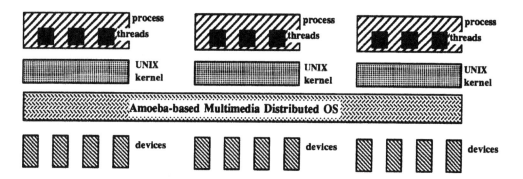

Figure 5: Multimedia support using an embedded distributed operating system.

Scheduling of Bandwidth-Constrained Multimedia Traffic

T.D.C. Little
Department of Electrical, Computer and Systems Engineering
Boston University, Boston, MA 02215 USA
tdcl@buenga.bu.edu

A. Ghafoor
School of Electrical Engineering
Purdue University, West Lafayette, IN 47907 USA

Abstract - Multimedia applications describe unique requirements that must be met by computer network and operating system components. In particular, the time-dependencies of multimedia data require mechanisms to ensure timely and predictable delivery of data from their sources to destinations. For single medium applications which have relatively constant bandwidth utilization, connections from source to destination can be tailored to moderate ranges of data rates. On the other hand, due to the large variation in multimedia object sizes and concurrency in object presentation, multimedia applications can require a correspondingly large variation in required bandwidth over the life of a connection. Data of these types may not arrive in time to meet the intended playout schedule when the capacity of the channel is exceeded. In this paper we present an approach to remedying this situation by effectively smoothing the bandwidth requirement over time via a scheduling mechanism.

1 Introduction

One of the unique requirements of a multimedia information system (MMIS) is the capability to present time-dependent data to the user. Time dependencies of multimedia data can be implied during creation, e.g., sequences of live video images. Data can also have time dependencies formulated explicitly, e.g., a pre-orchestrated slide presentation [1]. In either case, these time dependencies must be characterized and supported by the system. A MMIS must also be able to overcome any system delays caused by storage, communication, and computational latencies in the procurement of data for the user. These latencies are usually random in nature, being caused by shared access to common system resources. In a distributed multimedia information system (DMIS), multiple data sources including databases, video cameras, and telephones can be connected to user workstations via high-speed packet-switched networks. System latencies in a DMIS are problematic since several streams originating from independent sources can require synchronization to each other in spite of the asynchronous nature of the network.

To support the presentation of time-dependent data, scheduling disciplines associated with real-time operating systems are necessary. However, the real-time requirements can be relaxed since data delivery can tolerate some occasional lateness, i.e., catastrophic results do not occur when data are not delivered on time. For multimedia data, real-time deadlines consist of the playout times for individual data elements that can be scheduled based on a specified probability being missed. The design of a system to support time-dependent media must account for latencies in each system component used in the delivery of data, from its source to its destination. Therefore, specific scheduling is required for storage devices, the CPU, and communications resources.

Most recent work supporting time-dependent data is in the areas of live data communications [2-10], data storage systems [1, 11-17], and general distributed systems [1, 8, 18-22]. The work in data communications has usually been applied to live, periodic data sources such as packet audio and video [2-10] and for applications that do not exceed the capacity of the communication channel. These applications typically use a point-to-point configuration since there is seldom any need to deliver synchronous data streams from multiple independent live sources. Packet delay variations at the receiver are effectively eliminated by buffering and the introduction of a constant time offset resulting in a reshaping of the distribution of arriving packets to reduce delay variance.

For stored-data applications, time-dependencies must also be managed [1, 11-17]. Unlike live data, which are typically periodic during acquisition and presentation, stored data can require aperiodic playout and can be retrieved from storage at arbitrary times prior to presentation to the user. Audio and video data types require very large amounts of storage space and will exist primarily in secondary storage. When data originate from storage, the system has more flexibility in scheduling the times at which data are communicated to the application. Since data are not generated in real-time, they can be retrieved in bulk, well ahead of their playout deadlines. This is contrasted with live sources, e.g., a video camera, that generate data at the same rate as required for consumption. Unlike live sources, stored data applications in a DMIS can require synchronization for data originating from independent sources and simple data sequencing cannot provide intermedia synchronization.

Our work is intended for stored-data applications which, due to the heterogeneity and concurrency of multimedia data, the channel capacity can be exceeded for some intervals. We propose the use of *a priori* knowledge of data traffic characteristics to facilitate scheduling, when available, rather than providing on-line scheduling of dynamic input data based on its statistical nature. In essence, the dynamic bandwidth requirements of a multimedia object are fit into a finite channel capacity rather than the available bandwidth dictating the feasibility of the application, i.e., the inherent burstiness of the object is smoothed via our approach. The result is that delays are traded-off for the satisfaction of a playout schedule, even when the capacity of the channel is exceeded by the application.

In the remainder of this paper we describe an approach to the scheduling of time-dependent multimedia data retrieval under limited bandwidth conditions. In Section 2, we discuss the properties of time-dependent multimedia data. In Section 3, we describe our approach to scheduling time-dependent traffic under a bandwidth constraint, and provide several examples. Section 4 describes practical considerations for applying the derived schedules. Section 5 concludes the paper.

2 Properties of Time-Dependent Data

For management of time-dependent data in a computer system we are confronted with sequences of data objects which require periodic and aperiodic computation and communication that is oriented towards presentation of information to the user. In this section we describe several approaches to characterizing these time-dependencies through temporal modeling. These models can then facilitate real-time scheduling of data presentation when system latencies are present (Section 3).

2.1 Temporal Models

A common representation for the temporal component of time-dependent data is based on temporal intervals [12, 23, 24] in what is called temporal-interval-based (TIB) modeling. This technique associates a time interval to the playout duration of each time-dependent data object. For example, the playout of a sequence of video frames can be represented by using temporal intervals as shown in Fig. 1. A binary relationship called a *temporal relationship* (TR) can be identified between pairs of intervals. A temporal relationship can indicate whether two intervals overlap, abut, etc. [24]. With such a TIB modeling scheme, complex timeline representations of multimedia object presentation can be represented by application of the temporal relations to pairs of temporal intervals. Each interval can also be represented by its *start time* and *end time* via instant-based modeling. Therefore, for a sequence of data objects, we can describe their time dependencies as either a sequence of related intervals or as a sequence of start and end times. In the latter, the start times can be interpreted as *deadlines* as required for real-time scheduling. Fig. 2 shows the symbols that we associate with the intervals and deadlines in a time-dependent data specification. For each interval i we indicate its start time π_i and duration τ_i. The relative positioning of multiple intervals represents their time dependencies τ_δ^i. Their overall duration is defined by τ_{TR}.

A disadvantage with the instant-based approach is that modifications, including deletions or insertions, to a sequence of time-dependent data require changing subsequent time instances. For a TIB representation, the deadlines are implied by the precedence relationships described by the temporal relations. However, time instances in the form of deadlines are more suitable for the real-time scheduling during data playout [25]. To satisfy both needs we need the ability to derive the deadlines from the TIB representation.

2.2 *n*-ary Temporal Relationships

Binary temporal relations are sufficient for describing any simple or complex time-dependent data. However, we can generalize the binary temporal relations and ultimately simplify the data structures necessary for maintaining temporal semantics. Furthermore, by restricting the temporal relationships to have certain characteristics, we can simplify the process of scheduling data retrieval of time-dependent data. We therefore propose a new kind of homogeneous temporal relation for describing time-dependent data. The new temporal relation on n objects, or intervals, is defined as follows:

Definition 1: Let P be an ordered set n of temporal intervals such that $P = \{P_1, P_2, ... P_n\}$. A temporal relation, TR, is called an n-ary temporal relation, denoted TR^n, if and only if

$$P_i \ TR \ P_{i+1} \ \forall i \ (1 \leq i < n).$$

Like binary temporal relations, there are thirteen possible n-ary temporal relations which reduce to seven cases after eliminating their inverses [12]. When $n = 2$, the n-ary temporal relations simply become the binary ones, i.e., when $n = 2$, $P_1 \ TR \ P_2$. If we restrict our discussion to n-ary relations with the property that

$$\pi_i \leq \pi_j, \ \forall i, j, \ i < j, (1 \leq i, j \leq n),$$

then the deadlines of each object in a given sequence will be monotonically increasing

[26]. This property is satisfied by the temporal relations of *before, meets, overlaps, during*[1], *starts, finishes*[1], and *equals*. By using this subset of the thirteen temporal relations we do not restrict our ability to model time-dependencies [26].

In the process of presenting time-dependent multimedia data, we can use precedence relationship between related intervals [12], or we can use a deadline-driven approach. Since the n-ary temporal representation uses precedence relationships among intervals, to apply a deadline-driven approach we must be able to generate the exact playout deadline of each element based on the intervals and their relationship. This task is necessary for real-time scheduling of object retrieval in the presence of non-negligible delays [25]. The following theorem describes the relative playout time (deadline) for any object or start point of a temporal interval.

Theorem 1: The relative deadline π_k for interval k for any n-ary temporal relation, is determined by

$$\pi_k = 0, \text{ for } k = 1$$

$$\pi_k = \Sigma_{i=1}^{k-1} \tau_\delta^i, \text{ for } 1 < k \leq n$$

Proof: Since timing is relative to the start of the set of intervals, we let $\pi_1 = 0$. $\pi_2 = \pi_1 + \tau_\delta^1$ since $\tau_\delta^1 = \pi_2 - \pi_1$, by definition of delay for the binary case. Suppose that $\pi_m = \Sigma_{i=1}^{m-1} \tau_\delta^i$, for some m. For intervals P_m and P_{m+1}, $P_m \; TR \; P_{m+1}$ by Definition 1. Therefore, we can conclude the hypothesis by applying induction on π_m. ‖

Theorem 1 describes the efficient conversion of an interval-based representation to an instant-based one. In the following section we investigate the implications of real system latencies on the retrieval of time-dependent data and the satisfaction of the temporal specification.

2.3 Playout Timing with Delays

In the presence of real system latencies, several additional delay parameters are introduced as shown in Fig. 3. In order to properly synchronize some data element x with a playout time π, sufficient time must be allowed to overcome the latency λ caused by data generation, packet assembly, transmission, etc., otherwise, the deadline π will be missed. If we choose a time called the *control time* T such that $T \geq \lambda$, then scheduling the retrieval at T time units prior to π guarantees successful playout timing. The *retrieval time*, or packet production time ϕ is defined as $\phi = \pi - T$. The case of synchronizing one event is equivalent to the problem of meeting a single real-time deadline, a subset of the real-time scheduling problem. For streams of data, the multiple playout times and latencies are represented by the sets $\Pi = \{\pi_i\}$ and $\Phi = \{\phi_i\}$.

Fig. 4 illustrates the effect of introducing a control time to reduce the delay variation on the elements of a sequence of data called a *stream*. Here, elements of stream are generated at a source and experience a random delay with a distribution described by $p(t)$ before reaching their destination. They then require synchronization based on a playout schedule with deadlines π_i and playout durations τ_i. A suitable control time can be found for a stream of data and target percentage of missed deadlines based on such a delay characteristic [2-5]. Variation in both λ_i, the latencies of individual stream elements, and τ_i, their playout durations, can result in short term average or instantaneous need for buffering which is accommodated by T. For live sources, if ϕ_i are the packet production times, then $\phi_i = \pi_i - T$, $\forall i$, where π_i are the playout times. The playout sequence is merely shifted in time by T from the generation times. Furthermore, playout durations are typically uniform ($\tau_i = \tau_j, \forall i \, j$), and delay variation governs the buffering requirement. In the general case, we are presented with arbitrary streams characterized by π_i, λ_i, and τ_i and would like to determine the amount of buffering necessary to maintain playout timing with a given probability of failure.

One of the assumptions in the earlier analyses is that the generation of packets is at a rate equal to the consumption rate as is typical for live sources. When we consider a database as the source of a packetized stream instead of a live source (e.g., a camera), it is permissible for the playout schedule to specify retrieval of data exceeding the available channel capacity, and the earlier scheduling policies are deficient. In essence, data storage sources give us more freedom in controlling the packet production times. This is particularly important when we require concurrency in the playout of multiple media.

3 Scheduling with a Bandwidth Constraint

Since it is assumed that the channel capacity can be exceeded, we seek a method for smoothing the overloaded capacity during the life of a connection interval yet still maintaining playout timing. In this section we present our proposed scheduling approach and illustrate it with several examples. The approach generates a schedule that utilizes the full capacity of the channel by providing buffering at the destination based on *a priori* knowledge of the playout schedule Π.

3.1 Data Retrieval Timing

To develop an approach to scheduling data retrieval we must first characterize the properties of the multimedia objects and the communications channel. The total end-to-end delay for a packet can be decomposed into three components: a constant delay D_p corresponding to propagation delay and other constant overheads, a constant delay D_t proportional to the packet size, and a variable delay D_v which is a function of the end-to-end network traffic. D_t is determined from the channel capacity C as $D_t = S_m/C$, where S_m is the packet size for medium m. Therefore, the end-to-end delay for a single packet can be described as

$$D_e = D_p + D_t + D_v.$$

Since multimedia data objects can be very large, they can consist of many packets. If $|x|$ describes the size of some object x in bits, then the number of packets r constituting x is determined by $r = \text{ceiling}(|x|/S_m)$. The end-to-end delay for an object x is then

$$D_e = D_p + rD_t + \Sigma_{j=1}^r D_{v_j}.$$

Like the non-bandwidth constrained approaches, a control time can be identified for either single packets or for complete objects given a selected probability of receiving a late packet (object) called $P(fail)$. Control time T_i is defined as the skew between putting a packet (object) onto the channel and playing it out ($T_i = \pi_i - \phi_i$). Given $P(fail)$, and the cumulative distribution function F of interarrival delays on the channel, we can readily find T_i for either packets or objects as shown below.

$$T_i = D_P + S_m/C + F^{-1}(1 - P(fail))$$
$$T_i = D_P + r_i S_m/C + F_r^{-1}(1 - P(fail))$$

In these equations, $F^{-1}(p)$ determines the delay d for $p = F(d) = P(D_v \leq d)$ and $F_r^{-1}(p)$ determines the delay d for $p = F_r(d) = P(\Sigma_{j=1}^r D_{v_j} \leq d)$. Since $F_r(d)$ requires a convolution on r random variables, it is computationally impractical for an arbitrary distribution on D_v when r is large. However, a Gaussian approximation for the distribution of a sum of r independent random variables is reasonable for large r, and can be realized given the mean μ, and variance σ^2, of the interarrival distribution.

3.2 Delay and Capacity Constraints

In Section 3.1, only single packets and objects are considered. When multiple objects are retrieved, it is possible to exceed the capacity of the channel, and therefore we must consider the timing interaction of the objects. Transmission of objects is either back-to-back or introduces slack time, depending on their size and time of playout. We define an optimal schedule for a set of objects to be one which minimizes their control times. Two constraints determine the minimum control time for a set of objects. These are the minimum end-to-end delay (MD constraint) per object, and the finite capacity (FC constraint) of the channel. The MD constraint simply states that an object cannot be played-out before arrival. This constraint must always be met. The FC constraint describes the relation between a retrieval time and its successor when the channel is busy and accounts for the interarrival time due to transit time and variable delays. It also represents the minimum retrieval time between successive objects. These constraints are summarized mathematically below.

$$\pi_i \geq \phi_i + T_i \qquad \text{MD constraint}$$
$$\phi_{i-1} \leq \phi_i - T_{i-1} + D_P \qquad \text{FC constraint}$$

The following result combines the MD and FC constraints and describes the characteristics of an optimal schedule (see Fig. 5).

Theorem 2: An optimal schedule for any object i has the following characteristics:

$$\forall i, \phi_i \geq \pi_{i-1} - D_P \Rightarrow \phi_{i-1} = \pi_{i-1} - T_{i-1}, \text{ and } \phi_i < \pi_{i-1} - D_P \Rightarrow \phi_{i-1} = \phi_i - T_{i-1} + D_P$$

Proof: Any object in the set can find the channel idle or with a backlog of items. If it is slack, the equality of the MD constraint is optimal, by definition. Similarly, when the channel is busy, the equality of the FC constraint is optimal. Clearly the channel is slack when ϕ_i occurs after π_{i-1}, but it can also be slack when $\phi_i \geq \pi_{i-1} - D_P$ due to the "pipeline" delays. Therefore, the state of the channel, idle or slack, can be identified. ‖

We now show how to construct such a schedule. Given the characteristics of the channel and of a composite multimedia object $(D_v, D_P, D_t, C, S_m, \pi_i, |x_i|)$, a schedule can be constructed. We begin by establishing an optimal retrieval time for the final, mth object, i.e., $\phi_m = \pi_m - T_m$. The remainder of the schedule can be determined by iterative application of the MD and FC constraints for adjacent objects. The resultant schedule indicates the times at which to put objects onto the channel between source and destination, or it can be used to establish the worst-case buffering requirement. This approach is embodied in the following scheduling algorithm shown below.

$$\phi_m = \pi_m - T_m$$
for $i = 0 : m - 2$
   if $\phi_{m-i} < \pi_{m-i-1} - D_P$
      $\phi_{m-i-1} = \phi_{m-i} - T_{i-1} + D_P$
   else
      $\phi_{m-i-1} = \pi_{m-i-1} - T_{m-i-1}$
   end
end

It is also possible to determine the number of buffers required when using the playout schedule Φ, again noting that it is necessary to obtain an element prior to its use. For each fetch time ϕ_i, the number of buffers K_i required is equal to the size of the objects with elapsed playout deadlines between ϕ_i and π_{i-1} [25].

From this methodology we can determine the buffer use profile, ϕ_i versus K, and the maximum delay incurred by buffering, which can be used for allocation of storage resources. We now show several examples of the application of the scheduling approach.

3.3 Examples

Fig. 6 illustrates the time-dependencies of a sequence of full-motion video images represented by the OCPN [12]. The video is assumed to be compressed and has a nominal playout rate of 30 frames/sec. ($|x_i| = 1 \times 10^6$ bits and $\tau_i = 1/30$ sec.). The following conditions are assumed for the communications channel: $C = 1.5$ Mbit/sec., $S_m = 8192$ bits, $D_v = 50$ μsec., and $D_p = 100$ μsec. The results of the scheduling algorithm are:

$$\Pi = \{0.0, 0.033, 0.067, 0.10, 0.13, 0.17\} \text{ sec.}$$
$$\Phi = \{-0.028, 0.0057, 0.039, 0.072, 0.11, 0.14\} \text{ sec.}$$
$$K = \{0, 0, 0, 0, 0, 0\} \text{ bits}$$

These results indicate that there is ample time to transmit the specified video traffic without buffering (K buffers) beyond the object in transit, i.e., the size and playout timing is not affected by the channel capacity constraint.

Fig. 7 shows a more complex example which exceeds channel capacity. In this case, audio (64 Kbits/sec.), video and images (1024 x 1024 pixels x 8 bits/pixel) are coordinated in a multimedia presentation. Video and audio data elements are retrieved in 5 and 10 sec. segments, respectively. The results of the scheduling algorithm applied to the playout sequence are:

$$\Pi = \{0, 0, 0, 5, 10, 10, 15, 20, 20, 20, 25, 30, 30, 35\} \text{ sec.}$$
$$\Phi = \{-9.5, -6.1, -5.7, 0.0, 0.0, 3.4, 3.8, 7.2, 10.5, 11.0, 16.7, 21.6, 26.2, 26.6\} \text{ sec.}$$
$$K = \{0, 50, 100, 184, 50, 100, 184, 184, 100, 184, 184, 50, 50, 100\} \times 10^5 \text{ bits}$$

In this case, the scheduling algorithm produces a retrieval schedule based on scheduling adjacent objects in the channel. The results indicate that an initial delay of 9.5 sec. is required prior to beginning playout at the receiver.

We now consider practical aspects of the use of the scheduling approach including determination of network delays and application of the derived schedule.

4 Implementation Issues

In this section we show how the derived schedule Φ can be used in two ways. First, the sender can use it to determine the times at which to put objects onto the channel. The receiver then buffers the incoming objects until their deadlines occur. In this case the source must know Φ, the destination must know Π, and the overall control time is T_1 (the computed delay of the first element in the sequence). This scheme can be facilitated by appending π_i onto objects as they are transmitted in the form of time stamps. The second way to use Φ is to determine the worst-case skew between any two objects as $T_w = \max(\{\pi_i - \phi_i\})$, and then to provide T_w amount of delay for every object. Transmission scheduling can then rely on a fixed skew T_w and the Π schedule, that is, it can transmit all items with π_i in the interval $(t \le t + T_w)$.

The first method minimizes both buffer utilization and delay since the schedule is derived based on these criteria. Furthermore, required buffer allocation need not be constant but can follow the buffer use profile. The second method provides unnecessary buffering for objects, well ahead of their deadlines, and requires constant buffer allocation. However, it provides a simpler mechanism for implementation. Consider a long sequence of objects of identical size and with regular (periodic) playout intervals. For this sequence, $T_1 = T_w$, and $\phi_i = \pi_i - T_w$, assuming the channel capacity C is not exceeded. In this case, the minimum buffering is also provided by the worst-case delay. Such a se quence describes constant bit rate (CBR) video or audio streams which are periodic, producing fixed size frames at regular intervals. Rather than manage many computed deadlines/sec. (e.g., 30/sec. for video), a transmission mechanism more simply sequentially transmit objects based on sequence number and T_w.

When data originate from live sources, the destination has no control over packet generation times, and sufficient channel capacity must be present to preserve the real-time characteristic of the streams. In this case, the control time can be determined from the size, delay, and channel capacity requirements of a representative data object, e.g., a CBR video frame, and only provides jitter reduction at the receiver. For variable bit rate (VBR) objects, the source data stream is statistical in size and frequency of generated objects, and we apply a pessimistic worst-case characterization of the largest and most frequent VBR object to determine the control time. $\{\phi_i\}$ defines the packet production times, and playout times are $\pi_i = \phi_i + T$, as done in previous work. This service can be supported by appending timestamps, in the form of individual deadlines, π_i, to the transmitted objects.

In the general case, a playout schedule is aperiodic, consisting of some periodic and aperiodic deadlines. Typically, during class decomposition, these are isolated, and one of the two scheduling schemes above can be applied. If both exist in the same playout schedule Π (e.g., if classes are not decomposed), then the derived schedule Φ can be used as follows: choose a new control time T_E such that $(T_1 \le T_E \le T_w)$, and drop all deadlines ϕ_i from Φ such that $T_E \ge \pi_i - \phi_i$. The result is a culled schedule Φ reflecting the deadlines that will not be satisfied by simply buffering based on T_E. By choosing T_E to encompass a periodic data type (e.g., video), the burden of managing periodic deadlines is eliminated, yet aperiodic objects, requiring extensive channel utilization, can be dealt with using $\phi_i - T_E$, where $\phi_i - T_E > 0$.

When multiple delay channels are cascaded, e.g., channel delays plus storage device delays, we see that the end-to-end path capacity is reduced to the value of the slowest component. To facilitate scheduling of each resource, we can apply the derived schedule

Φ of the first channel as the playout schedule of the succeeding channel, while moving towards the source. In this manner, each resource merely meets the schedule stipulated by the preceding resource in the chain, with the initial schedule as Π. For this scheme, however, to meet the same *P(fail)*, the failure probability used for computing Φ on each link must be divided between each component [27].

5 Conclusion

Multimedia applications require the ability to store, communicate, and playout time-dependent data. In this paper we present an approach to satisfying timing requirements in the presence of real system delays when a limited bandwidth component such as a communications channel or storage device is present. The approach utilizes a timing specification in the form of a set of monotonically increasing playout times which are shown to be derivable from a temporal-interval-based timing specification. From the timing specification a retrieval schedule is computed based on the characteristics of the limited-capacity resource. The examples presented illustrate the utility in the proposed mechanism for applications which would normally not be viable due to timing specification violation during multimedia object retrieval.

6 References

[1] Little, T.D.C., Ghafoor, A., "Network Considerations for Distributed Multimedia Object Composition and Communication," *IEEE Network*, Vol. 4, No. 6, November 1990, pp. 32-49.

[2] Barberis, G., Pazzaglia, D., "Analysis and Optimal Design of a Packet-Voice Receiver," *IEEE Trans. on Comm.*, Vol. COM-28, No. 2, February 1980, pp. 217-227.

[3] Barberis, G., "Buffer Sizing of a Packet-Voice Receiver," *IEEE Trans. on Comm.*, Vol. COM-29, No. 2, February 1981, pp. 152-156.

[4] Montgomery, W.A., "Techniques for Packet Voice Synchronization," *IEEE J. on Selected Areas in Comm.*, Vol. SAC-1, No. 6, December 1983, pp. 1022-1028.

[5] De Prycker, M., "Functional Description and Analysis of a Video Transceiver for a Broad Site Local Wideband Communications System," *Esprit '85: Status Report of Continuing Work*, The Commission of the European Communities, Ed., North-Holland, New York, 1986, pp. 1087-1108.

[6] De Prycker, M., Ryckebusch, M., Barri, P., "Terminal Synchronization in Asynchronous Networks," *Proc. ICC '87* (IEEE Intl. Conf. on Comm. '87), Seattle, WA, June 1987, pp. 800-807.

[7] Naylor, W.E., Kleinrock, L., "Stream Traffic Communication in Packet Switched Networks: Destination Buffering Considerations," *IEEE Trans. on Comm.*, Vol. COM-30, No. 12, December 1982, pp. 2527-2524.

[8] Adams, C., Ades, S., "Voice Experiments in the UNIVERSE Project," *IEEE ICC '85 Conf. Record*, Chicago, IL, 1985, pp. 29.4.1-29.4.9.

[9] Ades, S., Want, R., Calnan, R., "Protocols for Real Time Voice Communication on a Packet Local Network," *Proc. IEEE INFOCOM '87*, San Francisco, CA, March, 1987, pp. 525-530.

[10] Ma, J., Gopal, I., "A Blind Voice Packet Synchronization Strategy," *IBM Research Rept.* RC 13893 (#62194) Watson Research Center, July 1988.

[11] Salmony, M.G., Sheperd, D., "Extending OSI to Support Synchronization Required by Multimedia Applications," *IBM ENC Tech. Rept.* No. 43.8904, April 1989.

[12] Little, T.D.C., Ghafoor, A., "Synchronization and Storage Models for Multimedia Objects," *IEEE J. on Selected Areas in Comm.*, Vol. 8, No. 3, April 1990, pp. 413-427.

[13] Postel, J., Finn, G., Katz, A., and Reynolds, J., "An Experimental Multimedia Mail System," *ACM Trans. on Office Information Systems*, Vol. 6, No. 1, January 1988, pp. 63-81.

[14] Ghafoor, A., Berra, P., and Chen, R., "A Distributed Multimedia Database System," *Proc. Workshop on the Future Trends of Distributed Computing Systems in the 1990s*, Hong Kong, September 1988, pp. 461-469.

[15] Gemmell, J., Christodoulakis, S., "Principles of Delay Sensitive Multi-Media Data Retrieval," *Proc. 1st Intl. Conf. on Multimedia Information Systems '91*, Singapore, January 1991, McGraw-Hill, Singapore, pp. 147-158.

[16] Yu, C., Sun, W., Bitton, D., Yang, Q., Bruno, R., Tullis, J., "Efficient Placement of Audio Data on Optical Disks for Real-Time Applications," *Comm. of the ACM*, Vol. 32, No. 7, July 1989, pp. 862-871.

[17] Wells, J., Yang, Q., Yu, C., "Placement of Audio Data on Optical Disks," *Proc. 1st Intl. Conf. on Multimedia Information Systems '91*, Singapore, January 1991, McGraw-Hill, Singapore, pp. 123-134.

[18] Nicolaou, C. "An Architecture for Real-Time Multimedia Communication Systems,"*IEEE J. on Selected Areas in Comm.*, Vol. 8, No. 3, April 1990, pp. 391-400.

[19] *Proc. 1st Intl. Workshop on Network and Operating Support for Digital Audio and Video*, Berkeley, CA, November 1990, (ICSI Tech. Rept. TR-90-062).

[20] Anderson, D.P., Tzou, S.Y., Wahbe, R., Govindan, R., Andrews, M., "Support for Continuous Media in the Dash System," *Proc. 10th Intl. Conf. on Distributed Computing Systems*, Paris, France, May 1990, pp. 54-61.

[21] Anderson, D.P., Herrtwich, R.G., Schaefer, C., "SRP: A Resource Reservation Protocol for Guaranteed-Performance Communication in the Internet," *ICSI Tech. Rept.* TR-90-006, February 1990.

[22] Anderson, D.P., Govindan, R., Homsy, G., Abstractions for Continuous Media in a Network Window System," *Proc. 1st Intl. Conf. on Multimedia Information Systems '91*, Singapore, January 1991, McGraw-Hill, Singapore, pp. 273-298.

[23] Herrtwich, R.G., "Time Capsules: An Abstraction for Access to Continuous-Media Data," *Proc. 11th Real-Time Systems Symp.*, Lake Buena Vista, FL, December 1990, pp. 11-20.

[24] Allen, J.F., "Maintaining Knowledge about Temporal Intervals," *Comm. of the ACM*, November 1983, Vol. 26, No. 11, pp. 832-843.

[25] Little, T.D.C., Ghafoor, A., "Multimedia Synchronization Protocols for Broadband Integrated Services," *IEEE J. on Selected Areas in Comm.*, December 1991.

[26] Little, T.D.C., Ghafoor, A., Chen, C.Y.R., "Conceptual Data Models for Time-Dependent Multimedia Data," submitted to the *1992 Workshop on Multimedia Information Systems (MMIS '92)*, Phoenix, AZ.

[27] Ferrari, D., "Client Requirements for Real-Time Communication Services," *IEEE Comm. Magazine*, Vol. 28, No. 11, November 1990, pp. 65-72.

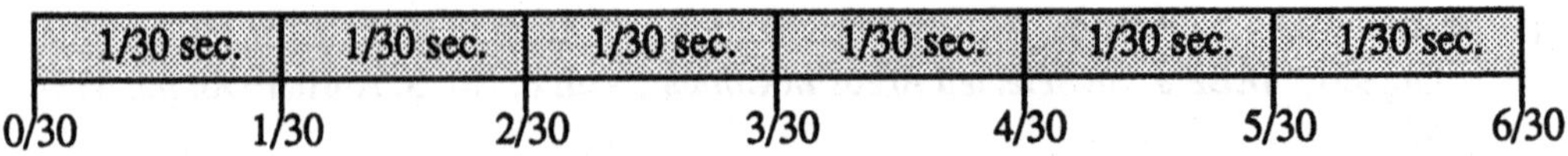

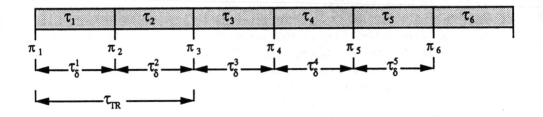

1/30 sec.
1/30 sec.
1/30 sec.
1/30 sec.
1/30 sec.
1/30 sec.
0/30
1/30
2/30
3/30
4/30
5/30
6/30

Fig . 1. Temporal-Interval-Based Specification for Video Frames

τ_1
τ_2
τ_3
τ_4
τ_5
τ_6
π_1
π_2
π_3
π_4
π_5
π_6
τ_δ^1
τ_δ^2
τ_δ^3
τ_δ^4
τ_δ^5
τ_{TR}

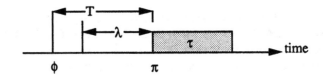

Fig. 2. Timing Parameters

T
λ
τ
time
ϕ
π

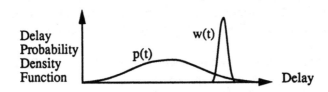

Fig. 3. Timing Parameters Including Latency

Delay
Probability
Density
Function
w(t)
p(t)
Delay

Fig. 4. Reduction of Delay Variance

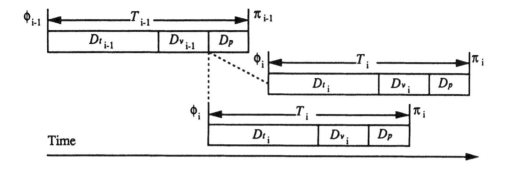

Fig. 5. Scheduling Constraints

Fig. 6. Video Frame Timing Using the OCPN

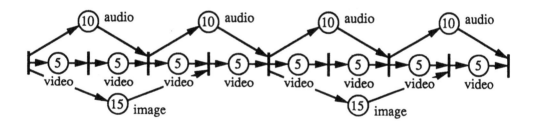

Fig. 7. Audio, Video, and Image OCPN Timing Specification

Presentation Scheduling of Multimedia Objects and Its Impact on Network and Operating System Support

PETRA HOEPNER

GESELLSCHAFT FÜR MATHEMATIK UND DATENVERARBEITUNG (GMD)
FORSCHUNGSZENTRUM FÜR OFFENE KOMMUNIKATIONSSYSTEME (FOKUS)
HARDENBERGPLATZ 2
D - 1000 BERLIN 12

The presentation of a multimedia object is concerned with the conveyance of information for a human user. The information conveyed depends not only on the contents presented, but also on their layout in time and space, i.e. on a specific presentation behaviour. The specification of the presentation behaviour in a system- and application-independent way provides for a compatible presentation of the multimedia object at different locations of a distributed, heterogeneous system. The paper concentrates on the temporal aspects of an abstract presentation behaviour description, i.e. the synchronization and control of temporally related presentation actions. The operating system and network support required for the execution of a presentation schedule is outlined.

1 Introduction

The *presentation of multimedia objects* is concerned with the conveyance of information for a human user. The information conveyed is not based solely on common representation types for the presentation [1] (e.g. character text, geometric graphics, moving pictures/images, audio/speech, raster graphics/images), but also on the resources/devices involved, on the transformation of the application-specific representation types (e.g. product definition data, trade data, animation) into a human perceptible form and on the control of spatial, temporal and hypermedia (hyperlinks) relations. In addition, an active participation of the user in the presentation process (interaction) must be possible.

Regarding different multimedia information/data models (e.g. [2], [3], [4], [5], [6], [7], [8], [9]) the gradation of presentation behaviour specification varies from the prescription of only the basic information/data objects associated with some content to composite information/data objects imposing also spatial, temporal and/or hypermedia constraints on the presentation of the information/data. The description of the intended presentation behaviour is required by diverse multimedia applications such as multimedia mail systems [10], educational software [11], etc.

If a specific presentation layout in a distributed heterogeneous environment is intended, an exchange format for multimedia information/data has to provide for content exchange as well as for conveying structuring information. Spatial, temporal and/or hypermedia structuring capabilities can thereby be integrated in the information/data model used, such as proposed for the integration of temporal relations into the 'Open Document Architecture (ODA)' [12], [13], can be defined separately in, for example, a script or can be internal to the data as, for example, in the data format of DVI.

This paper is mainly concerned with the specification of the temporal aspects of the presentation behaviour and the requirements imposed on system behaviour.

Temporal relations can be generated at the creation time of data (termed 'live synchronization' [12]) or set up artificially for stored data (termed 'synthetic synchronization' [14]). The temporal relations for both, live and synthetic synchronization, can be specified by a *synchronization rule*. The synchronization rule determines the goal of synchronization [15]; in the following context this goal is the temporal ordering and control of presentation actions and interactions termed *presentation scheduling*. Detaching presentation scheduling and its specification from any specific application, information/data format or concrete synchronization or scheduling mechanism, advocates the so-called 'separation of concerns', [16] proposed for the specification of hard real-time systems to achieve advantages such as the flexibility to change and the detection and correction of errors.

Presentation schedules of multimedia objects to be performed in heterogenous, distributed environments do strongly rely upon network and operating system support. The general requirements for such a support will be outlined.

2 Presentation Schedule

The presentation of a multimedia object comprises a set of temporally related actions and interactions, as defined by the originator, user or some other instance. Different approaches for the modelling and the specification of temporal relations and their synchronization exist, such as [3], [7], [12], [13], [17], [18].

An extraction of the basic concepts provided within these approaches leads to the following modelling requirements regarding the synchronization and control of the presentation:

- *Actions.* The presentation of a multimedia objects can be decomposed into a set of actions, such as present picture X, play audio comment Y, show menu Z, get menu selection.

- *Temporal Relations.* Actions are temporally related to each other thereby defining a specific presentation order, such as sequential, parallel.

- *Reconciliation.* The performance of an action may be modified or adapted to accomplish the presentation order specified, such as clip, fill, repeat.

- *Control.* The performance of one or more actions may be influenced by external control functions, such as stop, pause, resume.

The specification of a *presentation schedule* takes into consideration the requirements listed above. Interpreting or compiling a presentation schedule leads to a specific arrangement of processes in a distributed system which behave according to the specification.

In [13] a synchronization model for the presentation was described, based on actions and their temporal relations. An *action* is generally the "representation of something which happens" as defined in the 'Reference Model of Open Distributed Processing (ODP-RM)' [19]. A presentation action consists of events which construct a time interval. Some events are significant for the synchronization of actions, the *synchronization events.* Start- and endpoints of actions are synchronization events. Depending on the (non)existence of synchronization events between the start- and endpoint of an action, 'atomic actions' and 'composed actions' can be differentiated. Composed actions are actions that are composed from atomic actions or composed actions. The participating actions of a composed action have to be synchronized to fulfill a defined order of execution. An extension of the model with reconciliation and control concepts for the specification of presentation schedules is introduced in the following.

The temporal extent of a presentation action can be defined by its inherent duration (e.g. 'play video X') or can be delimited externally (e.g. 'show picture P one minute', 'show picture P until the end of the music M'). This temporal extent is termed *presentation frame* in the sequel. The extent of a presentation frame can be defined internally or externally, for example, by end of the contained actions, by a specified duration with respect to the start of the presentation frame (this can be also an absolute duration) or by external events such as the start of another presentation frame or a termination event issued by the user at the control interface.

An action can be reconciled to the extent of the presentation frame by so-called extent reconciliation strategies. Strategies to accomplish the extent reconciliation are (based on HyTime [4]):

Alignment   A specific alignment strategy for the action can be specified with respect to the presentation frame extent (e.g. fading, stretching, shrinking, repeat number etc.) (prior to vamping, filling, clipping or center).

Vamping   The action is repeated until the presentation frame extent is reached.

Filling   The rest of the extent is filled with a 'malleable action', i.e. one that can be scaled arbitrarily without unacceptable distortion, such as pause, or with an alternative action (concept of 'restricted blocking' introduced in [17]).

Replacement   Potential replacement actions are defined for example for the best fit.

Clipping   If an action is larger than the presentation frame extent, it will be clipped.

Center   If the presentation frame extent is known, the action will be centered within the extent (e.g. accompanied by a 'filling' strategy).

Offset   The action is shifted 'offset' time units from the startpoint of the presentation frame (e.g. accompanied by a 'filling' strategy).

Duration   The action is performed for a specified 'duration'. This is especially useful for actions with no inherent timing.

A presentation frame may not only delimit the extent of one action, but also the extent of two or more actions. Therefore a temporal relation must be associated with the presentation frame, which defines the synchronization strategy for the contained actions. In Section 3 different types of presentation frames are introduced, in each case prescribing a temporal relation (based on the concepts described in [19] and [13]).

The termination of a presentation frame ends all contained actions, i.e. the actions are no longer performed. Note, that 'not performing' an action logically can imply physically a negating action, such as presenting a picture and removing the picture.

With respect to a specific presentation schedule a presentation frame is identified uniquely by a 'presentation-frame-id'. Actions contained in a presentation frame can be referenced by the following notation 'presentation-frame-id.action-id'.

3 Presentation Frame Types

3.1 Sequentializer

A *Sequentializer* contains one or more actions which are performed in sequence (ODP-term: chain of actions). The start of the sequentializer starts the first action in sequence. The end of the sequentializer may be defined by the end of the last action in the sequence or by a specific presentation frame extent. Additional constraints for the sequential performance can be specified, that is continuation or non-continuation. Continuation requires an exact sequential performance, i.e. the endpoint of one action corresponds to startpoint of the next action in sequence. Non-continuation does not impose such a rigorous constraint, i.e. sequential actions start as soon as possible one after each other, for example, between pictures there may be a delay for the time the picture is prepared for presentation.

EXAMPLE:

```
FRAME        presentation_frame_name
TYPE         sequentializer
START_EVENT  event_from_other_frame
END_EVENT    end_presentation_frame_name = end_action_C
CONSTRAINT   continuation
ACTION       A  with  RECONCILIATION duration(5)
ACTION       B
ACTION       C
```

The presentation frame 'presentation frame name' defines the sequential performance of the actions 'A', then 'B', then 'C'. The start of the presentation frame is signalled by the event 'event from other frame'. The end of the presentation frame is signalled to the environment by event 'end presentation frame name', which is internally determined by the end of action 'C'. The endpoint of one action is exactly the startpoint of the next action. Action 'A' lasts 5 time units (e.g. a picture shown for 5 minutes). For actions 'B' and 'C' no duration is specified, which means either that the actions perform for a certain time, such as audio and video actions, or that they are terminated externally (see Section 4).

3.2 Parallelizer

A *Parallelizer* contains two or more actions. All actions start relative to the beginning of the parallelizer (similar to the ODP-term: forking action). The end of the presentation frame can be defined by (1) a specific action, i.e. a master action, by (2) the end of a non-deterministic action, i.e. the first, second, ..., last action in time, or by (3) a specific presentation frame extent. Extent reconciliation strategies are possible, for example, all actions are clipped if the master action terminates.

Tight synchronization requirements, such as lip synchronization between an audio and a video action, may be modelled by a fine granularity of parallelizer presentation frames containing atomic actions on audio/video frame basis (as modelled in [7] with Petri-Nets).

EXAMPLE:

```
FRAME        presentation_frame_name
TYPE         parallelizer
START_EVENT  event_from_other_frame
END_EVENT    end_presentation_frame_name = end_master
CONSTRAINT   master is A
ACTION       A
```

```
ACTION      B  with  RECONCILIATION alignment(shrink OR stretch)
ACTION      C  with  RECONCILIATION offset(2) AND clip
```

The presentation frame 'presentation frame name' defines the parallel performance of the actions 'A', 'B' and 'C'. The start of the presentation frame as signalled by the event 'event from other frame' starts the actions 'A' and 'B'; the action 'C' is shifted two time units from this startpoint. The end of the presentation frame is signalled to the environment by the event 'end presentation frame name', which is internally determined by the end of the master action 'A'. Action 'B' is exactly aligned to the presentation frame extent by shrinking or stretching its extent, the action 'C' is clipped, when the presentation frame ends.

3.3 Splitter

A *Splitter* contains two or more actions. All actions start relative to the beginning of the splitter. The difference from the parallelizer is, that more than one synchronization event can be issued (ODP-term: dividing action). A splitter therefore splits the presentation into independent branches, i.e. independent parts of the presentation, because different presentation frames may be scheduled to the various synchronization events. The presentation frame ends when all synchronization events have occurred. For example, two actions A and B are started at the same time. However, the end of the first action in time starts presentation frame X and the end of the last action in time starts presentation frame Y. Extent reconciliation strategies have to be chosen carefully because, for example, an aligning of endpoints may be contradictory to the semantics of this presentation frame type.

EXAMPLE:

```
FRAME       presentation_frame_name
TYPE        splitter
START_EVENT event_from_other_frame
SYNC_EVENT  sync_event_1 = end_first
            sync_event_2 = end_action_A
END_EVENT   end_presentation_frame = sync_event_1 AND sync_event_2
ACTION      A
ACTION      B  with  RECONCILIATION clip
```

The presentation frame 'presentation frame name' defines the parallel performance of the actions 'A' and 'B'. The start of the presentation frame as signalled by the event 'event from other frame' starts the actions 'A' and 'B'. The synchronization event 'sync event 1' is signalled to the environment as soon as the first action in time terminates (event 'end first'). Independently, the end of action 'A' is signalled to the environment by the event 'sync event 2'. Therefore two different cases of presentation ordering can be differentiated: (1) action 'B' is the shorter action in time, action 'A' is then continued until it terminates ('sync event 2' and 'end presentation frame' are issued simultaneously) and (2) action 'A' is the shorter action in time, all three synchronization events happen at the same time and action 'B' is clipped.

3.4 Combiner

A *Combiner* joins two or more independent branches of the presentation schedule (ODP-term: joining action). The start of the presentation frame therefore depends on more than one synchronization event, i.e. the synchronization events of two or more preceding presentation frames. A combiner may contain no action or one action. Ending the presentation frame however emits only one synchronization event.

EXAMPLE:

```
FRAME       presentation_frame_name
TYPE        combiner
START_EVENT end_frame_X AND end_frame_Y
END_EVENT   end_presentation_frame
ACTION      none
```

The presentation frame 'presentation frame name' ends as soon as the the presentation frames 'X' and 'Y' have ended.

3.5 Brancher

A *Brancher* is a special presentation frame that represents the possibility of several alternative futures (e.g. a 'choice' construct in programming languages). A possible presentation schedule may be regarded as a tree; only one branch of the tree will be selected and performed. The brancher always contains an action which has the ability to determine a selection value, necessary for the decision of the branch. As soon as the value is determined the brancher ends, i.e. the contained action is finished with the end of the presentation frame. Following presentation frames depend not only on the end of the brancher but also on the selection value this presentation frame determined. Start constraints for the following presentation frames have to be specified according to that rule. It is also possible to hide the value from the environment if the brancher is able to determine which presentation frame waits for which value and emits a value-specific synchronization event.

EXAMPLE:

```
FRAME       presentation_frame_name
TYPE        brancher
START_EVENT event_from_other_frame
END_EVENT   end_presentation_frame = end_action_A
VALUE       input_value   = choice of 1,2,3
```

```
        default_value = 1
ACTION      A  with  RECONCILIATION duration(20)
```

The start of the presentation frame as signalled by the event 'event from other frame' starts the action 'A'. The action 'A' determines a value (1, 2 or 3). The default value '1' is adopted if no input value was determined. Action 'A' either terminates as soon as the selection is done or after 20 time units.

4 Presentation Frame Interfaces

Presentation frames can be regarded as objects enforcing the above-specified behavioural constraints. The construction of a specific presentation schedule as well as the active participation of the user in the presentation requires the provision of interfaces by presentation frames allowing for different types of interactions, namely

- interactions between presentation frames
- interactions between a user and a presentation frame

The interactions between presentation frames are determined by the presentation schedule based on the specification of the synchronization events offered and accepted by the various presentation frames. These interactions have to be supported by the operating system.

The interactions a user can participate in depend on the interface that a presentation frame exports, i.e. the methods provided. These methods may either influence the presentation frame as a whole, i.e. the performance of all contained actions, or influence only the performance of a specific action. Nevertheless both types of interactions may influence the temporal behaviour of the presentation frame. For example, the operation 'stop' on a presentation frame will terminate the presentation frame, thereby terminating all contained actions. Performing the operation 'stop' on a specific action will terminate this action in the presentation frame; this, however, may influence other contained actions and also the presentation frame, if, for example, the action is a master action.

5 Impact on Network and Operating System Support

The above-described presentation schedule specification requires the distribution and instantaneous reaction (ideal case) to synchronization events. This concept is the basis of a "reactive system" (as introduced by Harel and Pnueli). A reactive system is based on a strong synchronism hypothesis: any reaction (processing) in response to an event occurrence is instantaneous, and any reaction is synchronous with the event that produced it. [20] considered distributed multimedia applications in general as a collection of interacting objects, organized around particular reactive objects, called "reactive kernels". Analogous to this, the real-time aspects (control and synchronization) of a multimedia presentation can be handled by *reactive presentation kernels*.

For the programming of reactive systems some formalisms have been proposed, such as ESTEREL [21] and Statecharts [22]. These formalisms are based on the concept of an instantaneous broadcast communication mechanism as, for example, modelled by finite-state machines. Synchronization events are therefore broadcast immediately, i.e. instantaneously. Based on the synchrony hypothesis, the distribution of synchronization events of presentation frames or actions takes "no time".

The concept of a reactive system leads to a problem regarding the distribution aspect of the presentation [20]. A distributed presentation system has to cope with asynchronous behaviour as introduced by communication objects. To enable synchronous reactions to synchronization events during the presentation, the timeliness of information/data availability has to be supported by real-time communication services. Performance requirements therefore have to be specified [23]. Based on these requirements, a meta-scheduler [24] may be used that reserves components guaranteeing a certain quality of service. As far as buffering capabilities are available at the destination (location of presentation) and the requested data is persistent, a 'work ahead' in transmitting presentation data is possible; if no buffering capabilities are available, startpoint-point synchronization and real-time transmission of presentation data is required.

The execution of a presentation schedule also has to take into account the allocation of resources and the temporal behaviour of presentation devices regarding their interface for the aquisition and restitution of continuous media data. Real-time scheduling strategies [25] have to be applied in dependance on the available resources and the processes to be scheduled.

As outlined, the synchronization and control of the presentation is not a local feature, but has to be supported at different architectural levels of an open, distributed system, examined for example in [14] and [26].

The enhancement of presentation schedules with features supporting the specification of performance requirements (for example Quality of Service (QOS) parameters) for the communication as well as with resource characteristics is a subject for further study.

6 Conclusion

The presentation of a multimedia object consists of temporally related presentation actions. The synchronization and control of these presentation actions, i.e. the temporal behaviour, is specified in a presentation schedule. Presentation actions are arranged in presentation frames; presentation frames are composed into a specific presentation schedule. Presentation frames are responsible for the reconciliation of actions to the extent defined and for the provision of interaction interfaces to other objects, for example, other presentation frames or the user.

The execution of a presentation schedule by a reactive presentation kernel supporting the instantaneous broadcasting of synchronization events was proposed. The adherence to the synchronization rules defined by the presentation schedule also imposes temporal requirements on the transmission of data streams in the communication system.

The presentation of multimedia objects in open, distributed systems requires the support of several services, such as communication services, retrieval services, transformation services. All these services have to offer certain qualities to handle real-time information/data transfer and processing. Further research has to be done in the evaluation and realization of presentation schedules regarding the quality characteristics of supporting services in a distributed application platform [27].

References

[1] "BERKOM Reference Model – Application-Oriented Layers", ed. GMD-FOKUS, DETECON Technisches Zentrum Berlin, Germany (May 1991), Version 3.2.

[2] ISO 8613, "Information Processing – Text and Office Systems – Office Document Architecture (ODA) and Interchange Format" (1989).

[3] ISO/IEC JTC1/SC2/WG12, "Coded Representation of Multimedia and Hypermedia Information, Working Document for the Future MHEG Standard", Version 3 (November 1990).

[4] ISO N2948, "Information Technology - Hypermedia/Time-based Structuring Language (HyTime) Committee Draft 10744" (April 1991).

[5] R. Steinmetz, R. Heite, J. Rückert, and B. Schöner, "Compound Multimedia Objects - Integration into Network and Operating System", in: *First International Workshop on Network and Operating System Support for Digital Audio and Video*, International Computer Science Institut (ICSI): TR-90-062 (November 1990).

[6] F. Oguet, C. Schwartz, F. Kretz, and M. Quere, "RAVI, A Proposed Standard for the Interchange of Audio/Visual Interactive Applications", *IEEE Journal on Selected Areas in Communications*, vol. 8, no. 3, pp. 428–436 (April 1990).

[7] T.D.C. Little and A. Ghafoor, "Synchronization and Storage Models for Multimedia Objects", *IEEE Journal on Selected Areas in Communications*, vol. 8, no. 3, pp. 413–427 (April 1990).

[8] G. Blakowski, "Konzeption für eine Sprache zur Beschreibung von Transport- und Darstellungseigenschaften multimedialer Objekte", in: *Telekommunikation und multimediale Anwendungen der Informatik*, ed. J.Encarnacao (Hrsg.), pp. 465–474, Proc. of the GI'91, October 14-18 1991, Darmstadt, Informatik-Fachberichte 293, Springer Verlag (1991).

[9] S. Lehert and E. Moeller (eds.), *Data and Information Modelling*, Proc. of the BERKOM Workshop, Annelsbach - Hoechst/Odenwald, July 9-13, 1990, GMD Report no. 196, R. Oldenbourg Verlag, München/Wien (1991).

[10] J.B. Postel, G.G. Finn, A.R. Katz, and J.K. Reynolds, "An Experimental Multimedia Mail System", *ACM Transactions on Office Information Systems*, vol. 6, no. 1, pp. 63–81 (January 1988).

[11] M.E. Hodges, R.M. Sasnett, and M.S. Ackerman, "A Construction Set for Multimedia Applications", *IEEE Software*, no. 1, pp. 37–43 (January 1989).

[12] R.G. Herrtwich and L. Delgrossi, "ODA-Based Data Modeling in Multimedia Systems", International Computer Science Institut (ICSI): TR-90-043 (1990).

[13] P. Hoepner, "Synchronizing the Presentation of Multimedia Objects - ODA Extensions -", *ACM SIGOIS Bulletin*, vol. 12, no. 1, pp. 19–32 (July 1991).

[14] T.D.C. Little and A. Ghafoor, "Network Considerations for Distributed Multimedia Object Composition and Communication", *IEEE Network Magazine*, vol. 4, no. 6, pp. 32–49 (November 1990).

[15] P.Hoepner, G.Schürmann, and K.-H.Weiss, "Synchronization, Isochronous and Anisochronous Communication", *BERKOM-Projekt: "Multi-Media-Dokumente im ISDN-B"*, DETECON, Technisches Zentrum Berlin, Voltastr. 5, 1000 Berlin 65, Germany (März 1991).

[16] S.R. Faulk and D.L. Parnas, "On Synchronization in Hard-Real-Time Systems", *Communications of the ACM*, vol. 31, no. 3, pp. 274–287 (March 1988).

[17] R. Steinmetz, "Synchronization Properties in Multimedia Systems", *IEEE Journal on Selected Areas in Communications*, vol. 8, no. 3, pp. 401–412 (April 1990).

[18] ISO/IEC JTC1/SC18 N2028, "Information on French Proposal on Audiovisual Interactive Applications" (September 1989).

[19] ISO/IEC JTC1/SC21 N6079 CD Text, "Basic Reference Model of Open Distributed Processing Part 2: Descriptive Model" (July 1991).

[20] L. Hazard, F. Horn, and J.-B. Stefani, "ISA Project - Notes on Architectural Support for Distributed Multimedia Applications", Centre National d'Etudes des Telecommunications (1991).

[21] G. Berry, P. Couronne, and G. Gonthier, "Synchronous Programming of Reactive Systems: An Introduction to Esterel", INRIA, Institut National de Recherche en Informatique et en Automatique, Technical Report No. 647 (1987).

[22] D. Harel, "Statecharts: a Visual Formalism for Complex Systems", *Science of Computer Programming*, vol. 8, no. 3 (June 1987).

[23] D. Ferrari, "Client Requirements for Real-Time Communication Services", *IEEE Communications Magazine*, vol. 28, no. 11, pp. 65–72 (November 1990).

[24] D.P. Anderson, "Meta-Scheduling for Distributed Continuous Media", University of California, Computer Science Division, Technical Report No. UCB/CSD 90/599 (October 1990).

[25] R.G. Herrtwich, "An Introduction to Real-Time Scheduling", International Computer Science Institut (ICSI): TR-90-035 (July 1990).

[26] C. Nicolaou, "An Architecture for Real-Time Multimedia Communication Systems", *IEEE Journal on Selected Areas in Communications*, vol. 8, no. 3, pp. 391–400 (April 1990).

[27] R. Popescu-Zeletin, V. Tschammer, and M. Tschichholz, "'Y' distributed application platform", *Computer Communications*, vol. 14, no. 6, pp. 366–374 (July/August 1991).

Session V: Communication II

Chair: Domenico Ferrari, University of California at Berkeley and ICSI

This session continued the discussion of network support for continuous media. While the first talk described a scheme for communication with performance guarantees in a packet-switching environment, the other three presented real-time protocol designs or implementations.

The first talk was given by Brian Field of the University of Pittsburgh, co-author with Taieb Znati of the paper "Alpha-Channel, A Network Level Abstraction to Support Real-Time Communication." The scheme proposed in it is based on the concept of "alpha-channel," a connection with one parameter, alpha, which represents the fraction of packets that must be delivered on time, i.e., before a client-specified deadline. Alpha = 1 corresponds to deterministic delay service, alpha = 0 to best-effort service, and alpha between 0 and 1 to statistical delay service. Like other approaches previously described in the literature, this is based on worst-case arguments, per-node rate control, and deadline scheduling with priorities (no best-effort packet is transmitted by a node if there is even a single deterministic packet in it).

The main differences between the scheme and, for example, the one on which Berkeley's Tenet protocols are based are the traffic characterization (which in the alpha-channel method has no burstiness parameter), and the establishment procedure (the source chooses the path, and asks the nodes on it to verify that they have sufficient resources for the new connection), while the rate control mechanism coincides with the one proposed by the Tenet Group at the Berkeley Workshop to control jitter in simple networks.

Comparisons with other schemes, that is, with Virtual Clock and Stop-and-Go, were the subject of some discussion at the end of the presentation. The question was also raised of what applications might want a value of alpha between 0 and 1. Clearly, no client would prefer statistical to deterministic delay service if the two services had identical prices: Statistical bounds can only be tolerated in certain less demanding applications if they cost less than deterministic ones. Another topic that was debated at some length was the usefulness that the application mark packets as droppable or undroppable by the network. Considerable skepticism was expressed by some participants about the assumptions that most applications have packets that are more important than the others, that they know which packets these are, and that the network can really choose which packets to drop.

The second paper in the session, "An Implementation of the Revised Internet Stream Protocol (ST-2)," by Craig Partridge and Stephen Pink, was presented by Stephen Pink of the Swedish Institute of Computer Science. The speaker told the audience about the experiences he and his co-author gathered while implementing ST-2 in the 4.3BSD kernel. The philosophy and the good and bad features of ST-2 were evaluated; among the good features, the speaker listed the information (about alternative requests that could be met) the network gives a client whose request cannot be met, and the ability a source has to specify the destinations of multicast messages; those listed among the problems were the almost complete separation of data and control paths (which may cause major difficulties when the server is recovering from certain failures), and the very large and extendible number of performance parameters to be specified.

The authors' statement that the need for state information to be kept along the path of a real-time connection, which is a fundamental aspect of ST-2's philosophy, is still "a subject of hot debate in the networking community" was a bit surprising, as this need was not seriously challenged by any of the workshop's attendees, unlike what happened in Berkeley during the previous edition of the workshop.

The discussion provided further details about the implementation, and some of the difficulties that were encountered in doing it within 4.3BSD (in particular, the modifications that had to be made to the "connect" system call). The code consisted of about 6,500 lines of C, in spite of the absence of the resource management functions (which ST-2 does not specify); the implemented version is expected to run on top of a protocol that provides performance guarantees and to capitalize on the resource management mechanisms of that protocol.

The last two talks in the session were concerned with two of the four protocols included in the prototype version of the Tenet real-time protocol suite being developed at the University of California at Berkeley and ICSI. The first of the talks was given by Anindo Banerjea of the University of California at Berkeley, and was based on the paper "The Real-Time Channel Administration Protocol" by him and Bruce Mah. The speaker described the structure and some of the features of the implementation of RCAP, the protocol in the suite to be used to set up and tear down real-time channels. A channel in the prototype Tenet suite is characterized by a minimum and a minimum-of-averages throughput bounds, a deterministic or statistical delay bound, an optional deterministic delay-jitter bound, and a loss probability bound. These bounds are specified by the client, and do not have to obey any non-obvious constraints in order to be acceptable by the service.

RCAP receives the values of these performance parameters and traffic specifications (coinciding with the throughput bounds) from the client, and tries to build a connection from source to destination that will provide the requested guarantees. A mechanism that establishes such a connection, or determines that it cannot be created along that path, in a single round-trip even in a complex internetwork was described in the talk, which covered the other functions of RCAP (channel teardown, status reporting) as well.

The question of call-blocking and of the vastly insufficient information that a yes/no answer by the establishment procedure gives the requesting client were discussed by several participants. An approach similar to the one proposed by ST-2, which provides the client with an alternative set of values for the performance parameters, was felt to be more convenient by some of the attendees. The speaker pointed out that what he had described was the prototype implementation of RCAP, which has been purposely kept simple for ease of experimentation. There is nothing in the design of RCAP that prevents the later introduction of much more sophisticated dialogues and negotiations between client and server.

The second of the Tenet talks in the session, which was also the last, was given by Bernd Wolfinger of the University of Hamburg. He co-designed the CMTP protocol while on sabbatical leave at ICSI, and described it in a paper co-authored with Mark Moran ("A Continuous Media Data Transport Service and Protocol for Real-Time Communication in High Speed Networks"), on which his talk was based. CMTP is one of the two transport protocols in the Tenet real-time suite; it is oriented towards the transmission of continuous-media (voice, high-quality sound, video) streams, while its companion protocol, RMTP, is message-oriented.

The service that CMTP provides, called CMTS, consists of the unreliable, periodic, and sequenced transfer of continuous-stream data units with guaranteed throughput, delay, and loss probability bounds. The client interface is fundamentally different from that of RMTP; the performance parameters and traffic characterization have purposely been designed for continuous-media applications, including descriptions of the burstiness due to compression and the allowed workahead.

CMTP translates the client-specified traffic and performance parameters into those that the underlying real-time network protocol, RTIP, expects. RTIP is relied on to provide the performance guarantees; since the nature of real-time communication precludes the use of the mechanisms for error control and flow control that account for most of the overhead and the size of traditional transport protocols, CMTP is a fairly lightweight protocol.

An aspect of CMTP that attracted quite a bit of attention during the discussion was the mechanism to start, sustain, and stop the transfer of stream data units. Questions were concerned with buffer management, synchronization, the desirability that the receiver communicate with the sender before and/or during transfers, and the handling of unexpected situations, such as a full buffer on the sender's side or an empty buffer on the receiver's. For most of these questions, the answers were easy: CMTP relies on the real-time services offered by RTIP, and can ask for guarantees that most of these situations will not occur, or will occur only with a very small frequency; those situations that depend on the behavior of the application can usually be handled by making allowances in the design for the inevitable uncertainties about that behavior; finally, recovery from failures of the underlying service will heavily rely on the recovery facilities of that service.

Not discussed, because of the shortage of time, were a number of intriguing issues; for example, the best transport protocol arrangement for real-time traffic: should there be a single universal transport protocol (e.g., TP++), a few protocols to handle different types of traffic (e.g., RMTP for message-oriented traffic, and CMTP for continuous-media traffic), or media-specific protocols (e.g., one for video, such as PVP, one for voice, one for high-fidelity sound, and so on)?

Note: Stephen Pink's and Craig Partridge's paper is not contained in these proceedings. For information about their work contact steve@sics.se.

α-Channel, A Network Level Abstraction
To Support Real-Time Communication

Brian Field and Taieb Znati† *
Computer Science Department
(† Telecommunications Program)
University of Pittsburgh
Pittsburgh, PA 15260

Abstract

In this paper, we propose a network level abstraction called α-channel to support the requirements of real-time applications, based on resource reservation and fixed routing. An α-channel represents a simplex, end-to-end communication channel with guaranteed service characteristics, between two entities. Based on the value of α, we define three classes of α-channels, namely a *deterministic* α-channel (α = 1), a *statistical* (0<α<1) and a *best-effort* (α=0). The semantics of these classes derive from the percentage of on-time reliability that need to be supported by the α-channel.

The primary attribute supported by the α-channel is the *on-time* reliability. At the α-channel establishment phase, the sender specifies the *end-to-end delay* of each packet, the *maximum number* of packets to be generated during this delay, and the on-time reliability required. We described the basic scheme that our model uses to verify the feasibility of accepting a new α-channel with guaranteed performance. A formal description of the proposed model and a sketch of the proof of its correctness is provided. We also present the preliminary results of a simulation experiment implementing the basic functionalities of the proposed scheme.

1 Introduction

Recent advances in VLSI technology led to the development of innovative packet switching architectures such as Fast Packet Switches and Photonic Switches, and created opportunities for a new class of *real-time* applications [1].

The performance requirements of these applications span a wide range, in terms of throughput, reliability, jitter bounds and delay characteristics. The "best-effort paradigm" offered by the Internet architecture, which proved to be very successful in the realization of a universal network in a heterogeneous environment is not adequate to support real-time traffic. The network protocol offers no guarantees about timely, reliable, or ordered packet delivery [2]. Circuit switching, which is the standard method for providing real-time performance, does not optimize the utilization of the network resources, and may be very inadequate for short transmissions, where the connection setup overhead is prohibitive with respect to the duration of the transaction. Furthermore, the scheme offers no provision for specifying data rates or limits on delays, and supports no guaranteed performance.

More recently, a number of schemes which aimed at providing performance guarantees based on virtual-circuits, resource reservation and fixed routing have been proposed [3,4,5,6]. A rate based network control system which provides guarantees for average throughput was proposed by Zhang [5]. The mechanism used is based on the concept of the "virtual clock", and aims primarily at regularizing the data flows according to the average throughput. The scheme does not provide guarantees to support end-to-end deadlines, and does not support an adequate buffer management scheme to limit packet loss due to buffer overflow. Another project, Multiple Congram Oriented High Performance Internet Protocol (McHIP), aims at providing variable grade services with guarantees for communication across subnets with diverse capabilities [4]. Subnets are characterized by a set of performance parameters, and their interconnection is achieved based on high speed gateways with all data delivery functions implemented in hardware. Furthermore, each subnet may have a designated resource server which keeps record of the established channels and the available resources. The network makes explicit resource allocation decisions at the time of connection establishment based on the requirements specified by the user. The authors, however, do not provide a clear definition of the connection semantics and the scheme used for implementing these connections.

The DASH communication system provides an abstraction called Session Reservation Protocol (SRP) [6]. An SRP is a simplex stream characterized by a set of performance, reliability, and security parameters. The model, designed to be independent of the transport protocol and compatible with the IP protocol, is based upon the "linear bounded arrival process". The DASH resource model describes a set of primitives that allow a user to express workload characteristics and performance requirements. A negotiation algorithm embodied in the SRP allows the user to negotiate reservation of the distributed resources.

Ferrari and Verma have proposed a real-time channel that guarantees end-to-end performance. The real-time channel is viewed as a binding contract between the client and the network. The client describes the characteristics of the traffic and specifies a set of performance requirements to be supported by the network. Based on these specifications, the network attempts to create a connection with fixed routing and performance bounds for each node along the path from source to destination. This is accomplished by computing the best performance guarantees for each node along the path, and making tentative resource reservations. The computation of the offered guarantees is based upon processing power, buffer capacity and the bandwidth of the links, and takes into consideration the requirements of the currently guaranteed real-time channels. Based on the performance guarantees accumulated and the guarantees requested, the destination channel determines whether the new channel is accepted or rejected. In the former case, the destination node adjusts the guarantees of each node along the path to the required guarantees to support the new channel, and sends a confirmation message along the same path to reserve the required resources. By controlling admission of new channels, the network ensures that it can meet its guarantees.

To support the requirements of real-time applications, we propose a network level abstraction called α-channel. An α-channel is a communication channel between a source and a destination, characterized by a set of performance attributes which reflect the requirements of the application. Upper layers, which view the α-channel as an end-to-end abstraction, specify the requested performance characteristics in quantitative terms when requesting the establishment of the connection. Based on the performance specifications requested by the upper layers and the current status of the network, the underlying delivery system either accepts and guarantees the specified performance attributes or denies the request.

Our scheme differs from the scheme described above primarily in terms of the specifications characteristics of the real-time channel, the mechanism used to establish a new channel, and the mechanisms used to maintain the specified rate among the node along the routing path.

In the first section of the paper, we introduce the concept of the α-channel. The proposed resource reservation model used to guarantee the requested performances is described in section 2. Issues related to the the management of the queues of a switching node are discussed in the last part of this section. The last section is dedicated to the description of the preliminary result of a simulation experiment.

2 α-Channel, A Network Level Abstraction

Without proper characterization of the network components and without any resource allocation, it is extremely difficult, if not impossible, to provide predictable performance [11]. Two approaches are used to provide network performance guarantees. The first approach is to over-engineer the network to the extent that an application is certain to get the resources it needs. Thus, the applications can be unconstrained in its resource usage and still receive the guaranteed level of performance. In a high speed network, this approach would require prohibitively large amount of resources.

The second approach involves monitoring and controlling resource allocation and usage for each application. This approach requires that an application specifies its resources needs a priori, and unless its needs can be met, it is blocked or rejected. Once started, mechanisms are provided to ensure that the application does not use more resources than requested. Because every application uses only its share of resources, all performance needs can be met. This approach is adopted in our definition of the α-channel.

2.1 Specification of the Semantics of α-channels

An α-channel represents a simplex, end-to-end communication channel between two entities, a *sender* and a *receiver*. The sender requests an α-channel by specifying the real-time performance characteristics required by the application. The primary performance attribute supported by the α-channel is the *on-time reliability*. This attribute represents the minimum percentage, α, of this channel's packets that must be delivered on time, assuming that no failures occur among the nodes of the fixed routing path. The value $(1 - \alpha)$ represents, therefore, the percentage of traffic that need not be delivered on time. In our scheme, this percentage of traffic will be considered best effort and will only be serviced if the current conditions of the node permits.

Based on the value of α, we define three classes of α-channels. The semantics of these classes derive from the percentage of on-time reliability that need to be supported by the α-channel. The first class, *deterministic* α-channel ($\alpha = 1$), requires that all of the generated traffic is delivered on time. The second class, *statistical* ($0<\alpha<1$), requires that only a percentage α of the generated traffic is delivered on time, while the rest of the traffic will only be considered if the current conditions of the node permit. Finally, the third class, *best-effort* ($\alpha=0$), is characterized by the absence of guarantees.

In the following section, we describe the resource reservation model used to guarantee the requested performance.

2.2 Resource Reservation Model Specification

At the α-channel creation time, a sender specifies the *end-to-end delay* of each packet, the *maximum number* of packets to be sent during this delay, and the *on-time reliability* required.

For each request, the network must identify a path from the source to the destination that can support the requested α-channel. This identification process consists of two parts. In the first part, a possible set of paths from the source to the destination is identified. The second part of this process consists of verifying that each node on a possible path can support the real-time requirements of the new α-channel without violating those already guaranteed. In this paper, only the second issue is considered. The issues related to efficiently identifying possible paths to handle real-time traffic is currently being investigated. We assume that the network can generate a list of possible paths from the source to the destination. Each node along the path is required to support exactly α percent of the source generated traffic. The switching nodes rely on the source node to appropriately tag α percent of the traffic as requiring deterministic service.

Given a possible path, the proposed α-channel verification algorithm is then run on the nodes to determine the feasibility of accepting the new α-channel. This is achieved in our scheme by the *Channel Acceptance* process. All accepted α-channels are then managed by the *Run-Time Support* process.

2.3 Channel Acceptance Process

This process is responsible for determining whether to accept or reject a newly generated α-channel. Based on the specifications of the α-channel, the process determines whether the new α-channel can be serviced such that its real-time requirements are met, and all existing α-channels requirements continue to be guaranteed.

As stated above, the only information provided to the node specifies the maximum number of arrivals per node delay and does not provide any specification of the arrival pattern. The per node delay is the end-to-end delay divided by the length of the routing path. Consequently, the Channel Acceptance process must consider the worst case arrival pattern. This occurs when all packets arrive simultaneously at the node, and all α-channels are operating at their maximum rate. Under these conditions, the Run-Time Support process must be capable of processing the offered traffic without violating the deadline requirements.

2.4 Verification Model

The main idea underlying our verification scheme is based on the relationship among the maximum number of packets received by a given node over a given time interval, the per-packet processing time and the delay a packet may suffer at that node. This relationship may be stated as follows:

The knowledge of both the maximum number of packets received by a given node over a given time interval and the per-packet processing time determines the maximum delay a packet may suffer at that node.

Given that our scheduling policy is driven by the earliest deadline packet first, the process of determining whether an α-channel is acceptable consists of verifying that the above relationship holds in the case where:

- The given time interval is the per-node delay of the α-channel,
- The maximum number of packets is the total number of packets received from all α-channels over this time interval that require deterministic service, and
- The packet processing time for a given α-channel.

In the following, a more formal description of the scheme is provided.

2.4.1 Formal Specification of the Model

Let c be an α-channel going through node n. The α-channel c is characterized by :

- The maximum amount of time, Δ_c, any packet from α-channel c may be delayed at node n,
- The maximum number of packets, $N_c(\Delta_c)$, generated by α-channel c during any interval of size Δ_c. Therefore, $\alpha_c \times N_c(\Delta_c)$ represents the number of packets from α-channel, c, that need to be serviced by node n during any interval of size Δ_c.

The deadline of packet p from α-channel c may be define as $\delta_c(p) = a_p + \Delta_c - \rho$, where a_p is the arrival time of packet p at node n, and ρ is the processing time of a packet from α-channel c.

Based on the above characteristics, the model used to verify the feasibility of accepting an α-channel is based on the following theorem.

Theorem:

Let $1, 2, \cdots N$ be the α-channels currently supported by node n such that $\Delta_i \leq \Delta_j$ if $i < j$, and $N_k(\Delta_k) \times \rho \leq \Delta_k$, for all $k = 1, 2, \dots N$. Furthermore, assume that the packet scheduling algorithm is based on the earliest deadline. Then, if for all $k = 1, 2, \cdots N$,

$$\sum_{i=1}^{k-1} W(i,k) + \alpha_k \times N_k(\Delta_k) \times \rho + \rho \leq \Delta_k,$$

then

$$delay_{p^k} \leq \Delta_k, \text{ for all } k = 1, 2, \dots, N,$$

where $delay_{p^k}$ is the delay observed by a packet, p^k, from α-channel k, and $W(i,k)$ ($i < k$), represents the amount of time to service the maximum number of deadlines from α-channel i that occur during any interval of size Δ_k, in the worst case.

The worst case pattern of arrivals of two α-channels i and k occurs when the packets from α-channel i arrive in a way such that the deadline of the first packet coincides with the arrival of a packet from α-channel k and continue to arrive at their maximum predefined rate, as illustrated by Figure 1. Based on this observation, the quantity $W(i,k)$ may be computed by counting the maximum number of deadlines that can occur during an interval of size Δ_k. The number of deadlines in this interval is distributed among the three regions, R_1, R_2, and R_3. The first region, R_1, contains $\alpha_i \times N(\Delta_i)$ deadlines. The cumulation of servicing these deadlines specifies the size of the region. The second region, R_2, contains the number of intervals of size Δ_i completely embedded in an interval of size $\Delta_k - R_1$. The amount of time required to service the deadlines that may occur within R_2 may be computed as $\left\lfloor (\Delta_k - R_1) / \Delta_i \right\rfloor \times N_i(\Delta_i) \times \alpha_i \times \rho$. Finally, R_3 can be expressed as $R_3 = \Delta_k - R_1 - R_2$. Thus a deadline from α-channel i will only need to be serviced during R_3 if $\Delta_i - \alpha_i \times N_i(\Delta_i) \times \rho \leq R_3$, in which case the number of deadlines from α-channel i that will require service during R_3 is $\left\lfloor R_3 - (\Delta_i - \alpha_i \times N_i(\Delta_i)) \right\rfloor + 1$. Therefore, the value of $W(i, k)$, in the worst case, during an interval of size Δ_k, may be expressed as:

$$W(i, k) = R_1 + \left\lfloor (\Delta_k - R_1) / \Delta_i \right\rfloor \times N_i(\Delta_i) \times \alpha_i \times \rho + \left(\left\lfloor R_3 - (\Delta_i - \alpha_i \times N_i(\Delta_i)) \right\rfloor + 1 \right) \times \rho$$

Based on the above theorem, the Acceptance Algorithm may be stated as follows. Let C_n = the set of α-channels currently supported by n. Assume that $|C_n| = N - 1$. Furthermore, let k be a new α-channel requesting to be established and assume that $\Delta_1 \leq \Delta_2, \cdots \leq \Delta_{k-1} \leq \Delta_k < \Delta_{k+1} \cdots \leq \Delta_N$. The new α-channel k is accepted if and only if the two following conditions are satisfied.

$$\sum_{i=1}^{k-1} W(i, k) + \alpha_k \times N_k(\Delta_k) \times \rho_k + \rho \leq \Delta_k \tag{i}$$

and

$$\sum_{i=1}^{k+j-1} W(i, k+j) + N_{k+j}(\Delta_{k+j}) \rho_{k+j} + \rho \leq \Delta_{k+j}, \text{ for all } j = 1, 2, \cdots (N - K). \tag{ii}$$

Sketch of the proof:

According to the previous theorem, if conditions (i) and (ii) are verified, then any channel i $\in C_n$, $delay_{p^i} \leq \Delta_i$ for all $p^i \in i$. Consequently α-channel k must be accepted. Otherwise admission of the new α-channel into the network causes the guarantees of at least one α-channel to be violated. Consequently, the new α-channel must be rejected.

In the next paragraph, we describe the run-time support process to service deterministically packets in a switching node.

2.5 Run-Time Support Process

In order to guarantee real-time requirements, the Run-Time Support process must provide deterministic packet service. This can be accomplished if the process can determine the order in which all packets are to be serviced, given any arrival pattern. In the following sections, a description of the basic scheme used to manage different queues in the switching node and the policy used to allocate the CPU to these packets is provided.

2.5.1 Processing Node Configuration

Each packet switching node contains a *service queue* and a set of α-*channel rate regulator queues*. These queues are managed by two entities, namely a *service queue manager*, and an α-*channel rate regulator*. The configuration of the switching node is illustrated in Figure 2.

The service queue manager waits on the service queue. This queue contains all packets that are currently eligible to be serviced by the CPU. The α-channel rate regulator monitors the rate regulator queues. The switching node provides a rate regulator queue for each α-channel currently supported by that node. The rate regulator is responsible for moving packets from the regulator queues into the service queue. In the next sections, a more detailed description of the basic operations performed by these two entities and the policies used to move packets from the rate regulator queue into the service queue is provided.

2.5.2 Service Queue Manager

As stated previously, the service queue contains all packets that are eligible to be serviced. These packets are ordered based on their deadlines. Packets with shorter deadlines are serviced before those with larger deadlines. When a packet has finished receiving service from the CPU, the service queue manager provides the CPU with the packet currently holding the shortest deadline in the queue. This packet may be labeled as best-effort, in which case the service queue manager must first verify that no packets requiring deterministic service is currently awaiting service. If there are no deterministic packets in the service queue, then the best-effort packet is serviced. Otherwise, the best-effort packet is dropped, and the next packet in the queue is considered. This continues until the next most eligible deterministic packet is located. Consequently, best-effort packets are only serviced if no deterministic packets are awaiting service.

2.5.3 α-channel Rate Regulator

The primary responsibility of the flow regulator is to maintain the flow rate of each channel at each switching node. This is necessary to ensure that the number of packets arrival a node may see does not exceed the prescribed rate of the α-channel. The rate regulation can be done in one of several ways. One way to monitor the rate of the α-channel is to maintain a list of departure times for the last several packets from this channel. The first packet in the regulator queue is moved into the service queue only when servicing it, and therefore sending it to the next node, will not violate the arrival rate at the next node. This can be achieved if the current history is maintained and used to determine when the first packet can be moved into the service queue. Maintaining the current history list, however, may necessitate the recording of a large number of packets an α-channel over its prescribed interval, thereby making the technique very expensive. This problem can be alleviated by observing that the source must obey its prescribed arrival rate. Consequently, maintaining the original relative gaps of these packets throughout the nodes along the routing path ensures that the prescribed arrival rate is not violated at any intermediate node. This method was adopted in our scheme.

The network entry point of the α-channel keeps track of the elapsed time between each two consecutive packets. Each packet entering the network is then tagged by the corresponding elapsed time. This gapping information is used by the flow regulator to maintain the original time gap between two consecutive packets. This ensures that bursts of packets from the same α-channel can not develop within the network. Consequently, the packet at the head of the the regulator queue is moved to the service queue only after the appropriate amount of gap time has elapsed since the previous packet of the same α-channel was serviced. This mechanism ensures that the observed gaps among adjacent packets never decrease below their original values, thereby guaranteeing that no packet bursts that violates the prescribed arrival rate can develop.

3 Simulation Implementation

In order to test the validity of the proposed scheme, a simulation model was developed. The simulation was written using CSIM, a simulation package that combines the C programming language and a library of routines to provide basic simulation functionalities [12]. The simulated network testbed consisted of an arbitrarily topology were each node had the same processing capacity. In this study, the links connecting nodes were assumed to have 0 propagation delay.

For each simulated node, two CSIM processes were created: one to handle the service queue maintainer requirements, and one to handle the flow regulator requirements. Additionally, each accepted channel was implemented as a CSIM process. Each channel was configured to simulate packet arrivals in one of the following ways: bursty, were all packets were sent one after the other, periodic, were packets were distributed evenly over the delay interval, and exponentially, where packets arrivals where generated from an exponential distribution.

The simulation experiment was run with several network topologies. Different α-channels with various real-time requirements and various arrival pattern were tested. These requirements as well as the arrival pattern were changed to exercise different load on the network and verify the on-time reliability of the scheme. The results obtained for the network configuration specified by Figure 3 are reported in Table 1. The results show that the required guarantees of the supported α-channels were not violated. Furthermore, the observed guarantees of the statistical α-channels most of the time exceeded their required guarantees. We are currently in the process of extensively investigating the performance of the scheme and the per node utilization. Further work is aimed at modifying the basic scheme to support packets of variable size. In addition, we are currently investigating models to characterize the excess capacity of the switching nodes and ways of incorporating this information in determining the set of nodes that can support the α-channel.

4 Conclusion

In this paper, we introduced a network level abstraction called α-channel to support the requirements of real-time applications. An α-channel represents a simplex, end-to-end communication channel with guaranteed service characteristics, between two entities. We described the basic scheme that our model uses to verify the feasibility of accepting a new α-channel with guaranteed performance. A formal description of the proposed model and a proof sketch of its correctness was provided. In addition, the preliminary results from a simulation experiment implementing the basic functionalities of the proposed scheme and were presented. The results show that the guarantees of the supported α-channels were preserved.

5 Bibliography

[1] Chen, T.M., and D.G., Messerschmitt, "Integrated Voice/Data Switching", IEEE Commun Mag., vol. 26, pp. 16-26, June 1988.

[2] Postel, J., "Internet Protocol; DARPA Internet Program Protocol Specification", RFC 791, September 1981.

[3] Ferrari, D., and Verma, D., "A Scheme for Real-Time Channel Establishment in Wide-Area Networks", IEEE Journal on Selected Area of Communications, Vol. 8, No. 3, April 1990.

[4] Parulkar, G.M., and Turner, J.S., "Towards a Framework for High Speed Communications in a Heterogeneous Networking Environment", in Proc. INFOCOM, Ottawa Canada, April 1989, pp. 655-688.

[5] Zhang, L., "A New Architecture for Packet Switching Network Protocols", PhD. Thesis, Massachussetts Institute of Technology, July 17, 1989.

[6] Anderson, D.P., Herrwich, R., and Shaefer, C., "SRP: A Resource Reservation Protocol for Guaranteed-Performance Communication in the Internet", Tech. Rept. No. TR-90-006, International Computer Science Institute, Berkeley, February 1990.

[7] Field, B., and Znati, T., "Experimental Evaluation Of Transport Layer Protocols For Real-Time Applications", in Proceedings of the 16th Annual Conference on Local Computer Networks, Minneapolis, Minnesota, 1991.

[8] Schwetman, H.D., "CSIM: A C-Based, Process-Oriented Simulation Language", *Microelectronics and Computer Technology Corporation*, Technical Report, PP-080-85.

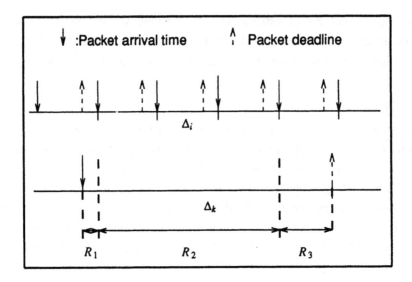

Figure 1- Worst case packet arrival pattern (channel i and k).

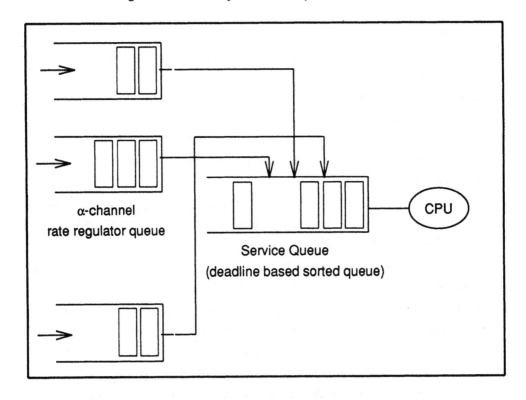

Figure 2- Configuration of a switching node.

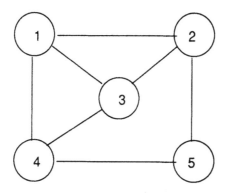

Figure 3- Network Configuration.

α-channel id	$N_{id}(\Delta_{id})$	Per node delay	Path	Required α	Observed α
0	25	65.0	1	0.90	0.94
1	13	54.0	2	0.80	0.86
2	18	47.0	3	1.00	1.00
3	16	99.0	4	1.00	1.00
4	22	47.0	5	0.84	0.96
5	15	51.0	2 -> 5	0.80	0.80
6	16	56.0	3 -> 4	1.00	1.00
7	17	49.0	2	1.00	1.00
8	11	171.0	1 -> 2 -> 5	1.00	1.00

Table 1- Performance Results for Various α-channels

The Real-Time Channel Administration Protocol

Anindo Banerjea
email: banerjea@tenet.berkeley.edu
Bruce A. Mah
email: bmah@tenet.berkeley.edu

The Tenet Group
Computer Science Division
University of California, Berkeley
and
International Computer Science Institute
Berkeley, California

ABSTRACT

The Real-time Channel Administration Protocol (RCAP) provides control and administration services for the Tenet real-time protocol suite, a connection-oriented suite of network and transport layer protocols for real-time communication. RCAP performs per-channel reservation of network resources based on worst-case analysis to provide hard guarantees on delay, jitter, and packet loss bounds. It uses a hierarchical approach to provide these guarantees across a heterogeneous internetwork environment.

In this paper, we outline our assumptions and approaches to real-time communication. We then describe the service provided by RCAP, the protocol itself, our plans for implementation, and current status of our research.

1. Introduction

The Real-time Channel Administration Protocol (RCAP) [BanMah91][Low91] provides control services for the Tenet real-time protocol suite, which consists of the Real-time Internet Protocol (RTIP) [VerZha91], the Real-time Message Transport Protocol (RMTP) and the Continuous Media Transport Protocol (CMTP) [WolMor91]. The protocol suite is intended to provide packet based data delivery services with guaranteed delay and jitter (delay variance) bounds, bandwidth guarantees, and bounded packet loss.

We use the term *real-time* to denote network services which provide such guarantees, especially guarantees on delay and jitter bounds. Applications which require such services include digital video and audio, interactive systems, and remote control systems. Current networks based on packet-switching provide no such guarantees. Networks based on

This research was supported by the National Science Foundation and the Defense Advanced Research Projects Agency (DARPA) under Cooperative Agreement NCR-8919038 with the Corporation for National Research Initiatives, by AT&T Bell Laboratories, Hitachi, Ltd., Hitachi America, Ltd., the University of California under a MICRO grant, and the International Computer Science Institute. Bruce A. Mah was also supported by a National Science Foundation Graduate Fellowship. The views and conclusions contained in this document are those of the authors, and should not be interpreted as representing official policies, either expressed or implied, of the U.S. Government or any of the sponsoring organizations.

circuit-switching do provide these guarantees, but at the expense of blocking resources regardless of whether they are being used or not. The Tenet real-time protocol suite is designed to provide guarantees by reserving resources for channels which require guarantees, but making the resources available to non-real-time traffic if they are not actually being used at the time.

RTIP is a connection-oriented network layer protocol which provides delivery of fixed size packets over a simplex channel, with client-specified bounds on packet delay, jitter, and loss rate, at a guaranteed bandwidth which is also chosen by the client. RMTP is a simple transport protocol which uses RTIP as a underlying network layer protocol to provide delivery of arbitrarily-sized messages with guarantees on delay, jitter, loss rate, and bandwidth. CMTP is a more complex transport layer protocol that provides a stream-like interface for continuous media clients that generate data at regular intervals, such as a digital video source. CMTP also uses RTIP as an underlying layer, and provides guarantees similar to RMTP.

RCAP provides the services of channel establishment, channel teardown, and status inquiry for these delivery protocols. It manages resources at nodes along the path of a channel, and provides admission control, to enable the network to provide end-to-end performance guarantees.

Designing and developing RCAP as a separate protocol has the advantages of making both RCAP and the data delivery protocols simpler and easier to debug. This division into control and data delivery portions also allows us to make changes in one without necessarily requiring changes in the other.

2. Assumptions

The Berkeley approach to real-time communication [FerVer89] makes some assumptions about the network, which must be satisfied in order for the guarantees to be met.

1. The network consists of nodes interconnected by logical links, which are either physical links (with bounded latencies) or subnetworks (see Figure 1). The delays across the subnetworks must also be boundable.

2. The nodes are store-and-forward nodes. The logical links may be non-store-and-forward networks (such as circuit-switched networks) provided their total delay can be bounded.

3. The loss rate on the physical links (due to noise) is low enough to be negligible. RCAP provides means of bounding the loss rate due to buffer overflow.

4. To provide jitter guarantees, either it must be possible to bound the jitter on each logical link, or the clocks on the nodes must be at least loosely synchronized.

The Berkeley approach can be used to provide real-time communication on any network which satisfies the above assumptions.

The first assumption may seem rather restrictive, because we seem to be assuming that the delays on the subnetworks are already bounded. However, RCAP provides the mechanism to bound the delays on the subnetworks as well. It is useful to think of the network as consisting of a hierarchy of different levels. The top level network is level 1; in the rest of the paper we refer to this level as the internetwork level, since it is potentially composed of a number of subnetworks linked together by gateways. These subnetworks can be considered level 2, and may in turn be composed of lower level networks, to an arbitrary depth of levels. RCAP provides a mechanism to recursively apply the approach on subnetworks to bound their delays, by structuring its control messages in a hierarchic way to model the hierarchic nature of the network. Once the delay on the subnetwork has been bounded, this bound can then be used to provide an end-to-end bound on the delay in the higher level network.

3. Approach

The Berkeley approach to real-time communication is connection-oriented and involves resource reservation during connection establishment. Connection establishment consists of a control message originating from the sending client and making one round trip along the path of a simplex channel.

During the forward pass the best possible resources are reserved for the channel. The lowest possible delay and jitter, sufficient bandwidth, and a corresponding amount of buffer space are reserved at each node at this time. This means that the network does its best to see if the channel can be established. However, it also means that the resources reserved during this stage are likely to be far better than the requirements of the channel.

The reverse pass allows the network to relax the reservations depending on the client's performance requirements and the actual performance attained by the reservations made on the forward pass. For example, if the end-to-end delay achieved by the forward pass is lower than the client's requirement, the delay reservations can be relaxed on the reverse pass. Jitter reservations are also relaxed, and buffer space may be released. This relaxation is needed to correct the over-allocation of resources during the forward pass.

The Berkeley approach proposes multi-class Earliest Due Date (EDD) as the scheduling mechanism of choice for the level 1 network nodes. This has many advantages. EDD is very amenable to fine grain resource allocation. Bandwidth and delay are decoupled as resources, so that low delay channels do not necessarily also require high bandwidth allocation. The scheduling mechanism is also amenable to worst-case analysis, so that hard guarantees can be made about delay, jitter, and packet loss bounds.

In addition, EDD allows the coexistence of channels having probabilistic guarantees (statistical channels) and best-effort traffic on the same network with channels having hard guarantees (deterministic channels). This allows higher utilization of the network resources, in addition to providing more flexible price versus performance choices for clients.

The subnetworks can use EDD scheduling or any other mechanism, as long as the delay across the subnetworks can be bounded. This means that the network may be heterogeneous, with subnetworks based, for example, on circuit-switching or Asynchronous Transfer Mode (ATM). All that matters is that there should be some mechanism to bound the delay on the subnetwork.

This bounding usually involves an establishment phase and resource reservation. We believe it is a good idea to use the same establishment process for subnetwork levels as for the internetwork level. By doing so, we can preserve the single round-trip establishment scheme and extend it to an arbitrary number of levels of subnetworks. These goals are achieved through the hierarchic structure of RCAP control messages.

4. RCAP Service Description

We now define the services that RCAP must provide to the client. We do this in the context of a simple client-server model, where the client asks the local RCAP entity to provide channels and to perform actions on those channels.

RCAP performs channel administration functions (setup, teardown, and status inquiry) for the data delivery protocols in the Tenet real-time protocol suite. The data delivery protocols are in the network layer (RTIP) and transport layer (RMTP and CMTP).

RCAP exports a number of primitive functions to client programs:

establish_request   The sending client invokes this primitive to request a channel. The client provides its performance requirements, traffic characteristics, and addressing information. RCAP returns a unique channel identifier if the request succeeds.

status   The sender can request information about the state of a channel by invoking the *status* primitive.

register   A receiving client uses the *register* primitive to indicate to RCAP that it is ready to receive connections on a given port.

receive_request   This primitive, when invoked by the receiving client, causes the receiving client to wait until an establishment request arrives from a sending client. It then returns the establishment information from that request, allowing the receiver to accept or deny the request.

accept   The receiving client invokes the *accept* primitive to indicate acceptance of an establishment request.

deny   The receiving client uses the *deny* primitive to reject an establishment request.

unregister   This primitive is used by the receiving client to indicate that it will no longer be accepting requests on a port.

close   Either the sending or receiving client can invoke the *close* primitive to close a channel and release the resources of that channel.

5. Protocol Description

To provide the services described above, RCAP entities on each node in the network need to communicate using control messages. These must be structured hierarchically to capture the hierarchic nature of the network. We now describe our protocol in the above context and elucidate with an example.

The RCAP entities in the network interact by exchanging RCAP control messages. At each node along the path of a channel, the local RCAP entity communicates with the network layer (RTIP) entity to reserve or release resources; at the channel endpoints, RCAP communicates in a similar fashion with the appropriate transport layer entity.

RCAP entities exchange a number of different control messages:

```
establish_request
```
    This control message contains channel parameters provided by the sending client. This message causes RCAP entities along the channel path to reserve resources for the channel, if possible.

`establish_accept`	The `establish_accept` control message indicates that a channel was successfully established. It propagates along the reverse path from the receiver to the sender and causes the RCAP entities in the nodes to confirm, and possibly relax, their resource reservations.
`establish_denied`	This control message indicates that a channel could not be established. It propagates backwards along the channel path towards the sending client and causes the RCAP entities to release resources allocated to the channel.
`status_request`	Sent forward along the channel's path, this message collects information from each node about its status and resource allocation.
`status_report`	This `status_report` message returns the status information from a `status_request` message to the sending client.
`close_request_forward`	This message is a request from the sending client to close the channel. The channel's resources are deallocated.
`close_request_reverse`	This message is a request from the receiving client to close the channel. The channel's resources are deallocated. The effects of the two `close_request` messages are very similar; they are distinguished in order to associate an implicit direction (forward or backward) with each type of control message.

RCAP is designed for use in an internetwork; it uses a hierarchical view in which the links and nodes in a subnetwork are abstracted into a single logical link in the internetwork. Such an approach allows RCAP to utilize the characteristics peculiar to an individual network in order to provide guarantees, yet hide the underlying details of the networks whenever possible. The RCAP control messages reflect this hierarchical view when applicable.

This hierarchical approach is most clearly seen in the `establish_request` RCAP control messages generated by the *establish_request* primitive (see Figure 2). Each message is composed of various records:

Header Record	The Header Record (HR) contains end-to-end parameters for the transport layer entities. There is exactly one HR in each `establish_request` message.
Network Subheader Record	A Network Subheader Record (NSR) marks the start of the records for a network. Each NSR contains end-to-end network-dependent parameters for its associated network and is followed in the message by zero or more Establishment Records.
Establishment Record	An Establishment Record (ER) carries network-dependent, per-node, local parameters.

The `establish_request` control message is created by the RCAP entity on the sending host; the message initially contains only an HR followed by an NSR for the inter-network level. As the control message travels along the channel path, records are added and deleted. Upon the entrance to a lower-level network, the RCAP entity on the entering router adds an NSR for the network; ERs appropriate to that network are then added by the nodes in that network. When exiting a network, the RCAP entity on the exiting router saves the NSR and ERs collected in that network and summarizes the information they contain into a new ER for the next higher-level network. In this way, end-to-end information about lower-level networks is abstracted for higher-level networks.

Figure 3 illustrates a sample `establish_request` message about to exit a network and enter another, as shown. The RCAP establishment message in (i) has traversed the first two networks; the establishment information for those networks has been summarized in $ER_{inet,I}$ and $ER_{inet,II}$. The progress of the message across the third network is recorded in NSR_{III}, $ER_{III,1}$, and $ER_{III,2}$. Upon arrival at the exiting router of the third network, $ER_{III,3}$ is added to describe the last hop as shown in (ii). (iii) shows the entire path through network III summarized in $ER_{inet,III}$. Before entering network 4, NSR_{IV} is added to the message, as shown in (iv).

The `establish_accept` message, sent by the *accept* primitive to confirm resource reservations and indicate acceptance of a channel, has a similar hierarchical structure. ERs are removed from the message by the individual nodes as they confirm the resource reservations. Strings of records that were removed and summarized on the forward pass are processed and reattached to the `establish_accept` message using the state stored in the routers.

Some control messages, such as `establish_denied` and the two `close_request` messages, do not need such a hierarchical structure, as they can convey all necessary information without needing to refer to network-specific parameters. Accordingly, they take on a much simpler structure, containing only the channel identifier and the action to be performed.

The `status_request` and `status_report` control messages have a structure similar to that of the `establish_request` control message in that they are all composed of records which are added or removed by the nodes along the channel path. The major difference is that while the channel establishment process abstracts details of lower-level networks, the status reporting process requires these details to be made available to the client process. Therefore, the status information from the nodes is not summarized or abstracted in any way.

RCAP does not depend on any particular delivery mechanism for its control messages. An ideal implementation would utilize some service providing in-order, reliable message delivery.

6. Implementation Plans

The Tenet Group's implementation plans involve two stages. At first, the protocol suite will be implemented on a local testbed with a small number of nodes and a simple topology. After some experience with this version, the protocols will be ported to the Experimental University Network (XUNET II).

6.1. Local Testbed

The first implementation is targeted for a local testbed with three workstations connected by FDDI links. The FDDI subnetworks, being one hop subnetworks, can be regarded as physical links for the purposes of the protocol. Thus, the links between the nodes are all simple physical links with bounded delays. Only the queueing delays in the nodes are variable and need to be bounded by RCAP. This provides the simplest environment in which to build and test the protocols.

6.2. XUNET II

The Tenet Group plans to implement the protocol suite on the Experimental University Network (XUNET II) in collaboration with AT&T Bell Laboratories. XUNET II will provide a heterogeneous topology [Fer91]. The backbone will be an Asynchronous Transfer Mode (ATM) based wide area network, consisting of XUNET II switches connected by T3 (45 Mb/s) links. At the periphery of the backbone network will be routers, with one or more FDDI rings attached to each router, and host computers attached to the rings.

Thus a typical path on XUNET II can be viewed as composed of the following logical links:

1. Host to router (subnetwork - FDDI)

2. Router to router (subnetwork - ATM backbone)

3. Router to host (subnetwork - FDDI)

As before, the FDDI subnetworks can be treated as physical links for the purposes of the protocol. However, the ATM backbone needs to be treated in a hierarchical manner.

7. Status

As of the date of this writing, the design of the protocol suite is complete and implementation for the local testbed is in progress. The RCAP software has been partially written. Software for RTIP and RMTP has been written and is being tested. The CMTP software is in the process of being written.

At this time, the long-haul fibers for XUNET II are in place and rudimentary routers have been deployed. Switches and more advanced routers are being tested, and are expected to be installed by the end of 1991.

8. Summary

RCAP provides the functions of connection establishment, teardown, and status reporting for the real-time data delivery protocols of the Tenet real-time protocol suite. The protocols in the suite provide data delivery with hard guarantees on the delay and jitter of the packets. Packet loss due to buffer overflow can be eliminated or bounded by correct calculation of buffer requirements.

Separating the control functions from the data delivery functions gives the advantage of simpler design for each of the protocols and a certain amount of insulation, preventing changes in the implementation of one set of functions from adversely affecting the other functions.

The hierarchical nature of RCAP makes it a very flexible protocol that can work in a large variety of environments with little modification. It preserves a single round-trip establishment scheme in an internetwork composed of gateways linking together a number of heterogeneous subnetworks. It preserves a one round trip establishment scheme across heterogeneous networks.

The approach is based on worst-case analysis. It allows the provision of hard guarantees with very general assumptions about the underlying network. For applications that do not require hard guarantees, multi-class EDD allows statistical channels and best-effort traffic on the same network as deterministic channels, with correspondingly higher utilization of network resources.

9. References

[BanMah91]   A. Banerjea and B. Mah, "The Design of a Real-Time Channel Administration Protocol," unpublished report, University of California at Berkeley and International Computer Science Institute, Berkeley, California, May 1991.

[FerVer89]   D. Ferrari and D. Verma, "A Scheme for Real-Time Channel Establishment in Wide-Area Networks," Report TR-89-036, International Computer Science Institute, Berkeley, California, May 1989.

[Fer91]   D. Ferrari, "Tailoring the Tenet Real-Time Protocols to XUNET II," unpublished report, University of California at Berkeley and International Computer Science Institute, Berkeley, California, May 1991.

[Low91]   C. Lowery, "Protocols for Providing Performance Guarantees in a Packet-Switching Internet," Report TR-91-002, International Computer Science Institute, Berkeley, California, January 1991.

[VerZha91]   D. Verma and H. Zhang, "Design Documents for RTIP/RMTP," unpublished report, University of California at Berkeley and International Computer Science Institute, Berkeley, California, May 1991.

[WolMor91]   B. Wolfinger and M. Moran, "A Continuous Media Data Transport Service and Protocol for Real-Time Communication in High Speed Networks," *Proc. 2nd Intl. Workshop on Network and Operating System Support for Digital Audio and Video*, Heidelberg (November 1991).

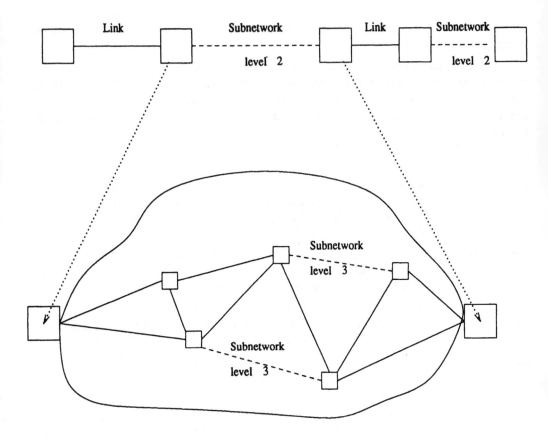

Figure 1. Hierarchic view of the internetwork. The network can be viewed as composed of nodes joined together by links which are either physical links or subnetworks.

The internetwork in this example consists of two physical links and two level 2 subnetworks. One of the level 2 subnetworks is shown in greater detail. It composed of physical links and two level 3 subnetworks.

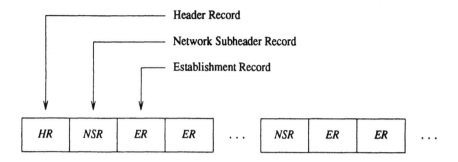

Figure 2. Structure of an `establish_request` message. The start of the message is to the left. The Establishment Records (ERs) following a Network Subheader Record (NSR) contain establishment information on nodes within the network associated with the NSR.

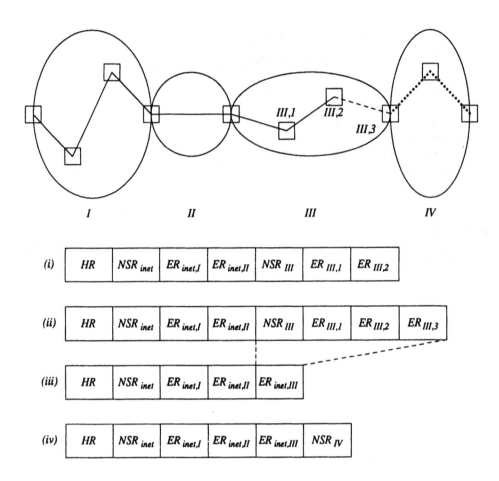

Figure 3. An establish_request control message at various points along a path through networks *I, II, III,* and *IV.* Ellipses represent networks and squares represent nodes along the path. A node in two adjacent networks is a router, which considered to be a part of both networks.

Within the representations of the establishment messages, *HR* denotes a header record. NSR_i represents a Network Subheader Record; with *i* indicating the applicable network (the subscript *inet* indicates that the NSR is for the internetwork level). $ER_{i,j}$ stands for an establishment record for node *j* in the network *i*. Note that in higher levels of the hierarchy, $ER_{i,j}$ can contain router-to-router establishment information for the subnetwork preceding node *j* in network *i*.

(*i*) shows the message just after leaving node *III*, 2. (*ii*) shows the message just after arrival at node *III*, 3; the real-time establishment information computed over the last hop through network *III* has been added in $ER_{III,3}$. In (*iii*), the establishment information for network *III* has been summarized in $ER_{inet,III}$. (*iv*) shows the message just prior to entering network *IV*; NSR_{IV} has been attached with the router-to-router parameters for network *IV*.

A Continuous Media Data Transport Service and Protocol for Real-Time Communication in High Speed Networks *

Bernd Wolfinger [1]
Mark Moran

The Tenet Group
Computer Science Division, Department of EECS,
University of California, Berkeley
and
International Computer Science Institute
Berkeley, CA 94720, USA.

ABSTRACT

An important class of applications with real-time data transport requirements is defined by applications requiring transmission of data units at regular intervals. These applications, which we call continuous media (CM) clients, include video conferencing, voice communication, and high-quality digital sound. The design of a data transport service for CM clients and its underlying protocol (within the XUNET II project) is presented in this paper. The service makes use, in particular, of an a priori characterization of future data transmission requests by CM clients.

First, we will give a few examples of CM clients and their specific data transmission needs. From these clients, we then extract a generalized list of data transport requirements for CM and describe the basic features of a service designed to meet these requirements. This service provides unreliable, in-sequence transfer (simplex, periodic) of so-called stream data units (STDUs) between a sending and a receiving client, with performance guarantees on loss, delay, and throughput. An important feature of the solution is the use of shared buffers to eliminate most direct client/service interactions and to smooth traffic patterns, which may be bursty due to fluctuations in the arrival process of data and variability of network delays. The paper concludes with some aspects of implementation.

1. Introduction

Applications with real-time data transport requirements fall into two categories: those which require transmission of data units at regular intervals, hereafter referred to as *continuous media (CM) clients*, and those which generate data for transmission at relatively arbitrary times, hereafter referred to as *(real-time) message-oriented clients*. Examples of the former are video conferencing, in which video frames (of fixed or variable length) are sent from source to destination once per frame time (e.g. 33 ms), voice communication, playback of high-quality digital sound, and transmission of sensor data that is measured and transferred with strict periodicity. Examples of real-time message-oriented clients are those which require urgent messages or transactions, and mail service with guaranteed delivery latency.

It is generally accepted that dedicated transport protocols are necessary for high speed networks. Adaptation of existing transport protocols, originally designed for lower speed networks (such as OSI Transport Protocol Class 4 or TCP), to high speed environments is not straight-forward, and may not provide satisfactory performance to transport

* This research was supported by the National Science Foundation and the Defense Advanced Research Projects Agency (DARPA) under Cooperative Agreement NCR-8919038 with the Corporation for National Research Initiatives, by AT&T Bell Laboratories, Hitachi, Ltd., the University of California under a MICRO grant, and the International Computer Science Institute. The views and conclusions contained in this document are those of the authors, and should not be interpreted as representing official policies, either expressed or implied, of the U.S. Government or any of the sponsoring organizations.

[1] On sabbatical leave from the University of Hamburg, Computer Science Dept., Bodenstedtstr.16, D-2000 Hamburg 50

service users of future networks. Therefore, considerable research has been conducted in the design of completely new transport protocols to support high speed end-to-end communication between users. Surveys of the general requirements these protocols must satisfy, and of existing protocol proposals can be found in e.g., [DDK90], [LPS91], [WrT90], [Zit91]. In these publications, it is suggested that new algorithms be developed to support basic transport protocol functionality (such as flow control, error detection and correction, connection management, etc.). In addition, the use of specific implementation techniques, e.g. parallel processing, is advocated. Current transport protocols designed for high speed networks include, e.g., Delta-t Transport Protocol, cf. [Wat89], Network Block Transfer Protocol (NETBLT), cf. [CLZ87], Versatile Message Transaction Protocol (VMTP), cf. [ChW89], Express Transport Protocol (XTP/PE), cf. [Che88], and the protocol designed by Netravali et al. and described in [NRS90].

The literature suggests general agreement among network designers that transport protocols should be tailored to meet the various transport service requirements of end users. Requirements of various users can be supported by using rather general transport services and providing options to flexibly adapt the service to the differing requirements of users (e.g. during establishment of a transport connection). Alternatively, it is also possible to split the transport service *a priori* into two (or more) different, cleanly separated, services. Each service would support a class of users with similar data transport requirements. This second solution is chosen in this paper, as we will describe a transport service designed for continuous media clients, which we expect to coexist with a transport service designed for message-oriented clients. For surveys on the requirements of continuous media applications for data transport, the reader is referred to [HSS90] and to [ITC91]. Requirements of video transfer, in particular, can be found in [LiH91]. Section 2 details the arguments in favor of a separate transport service for CM clients.

2. Necessity of a Dedicated Transport Service for Continuous Media Applications

Because CM clients are better able to characterize their future behavior than message-oriented clients, a dedicated CM data transport service can potentially provide them with a more cost-effective service by characterizing the future resource demands of such clients a priori.

In addition to the efficiency advantages mentioned above, a dedicated CM data transport service could provide better functionality for CM clients than a traditional message-oriented transport service, because of the many incompatible functional requirements of these two classes of applications. For example, a dedicated transport service could provide the abstraction of *logical (data) streams* between CM clients. A *stream* here denotes a continuous sequence of data units (continuous only with respect to some limited granularity in time) provided and to be transmitted during a given time interval, which corresponds to the duration of the stream. As we also assume in-sequence delivery of the data units of a stream, we can consider such streams to be generalization of Lixia Zhang's "flows" (defined as a "stream of packets that traverse the same route from the source to the destination and that require the same grade of transmission service" in [Zha91]). At any instant in time, each established connection between two CM clients is used by at most one stream. By making such streams visible to the data transport service, network and system resources could be conserved between streams. More importantly, some connection parameters may be redefined for the duration of the stream, allowing conservation of resources and providing better cooperation between sender, receiver, and the transport service (e.g. the current stream could be stopped and a new, slower stream started to support video playback with "freeze frame" and "slow motion"). Streams also provide a natural mechanism for synchronizing data from different connections (e.g. video and audio). Such streams would be difficult to implement on top of a message-oriented transport service, because of the requirement that streams (and their associated parameters) be visible to the transport service. Since this functionality would not be used by most message-oriented clients, implementing a new service is preferable to adding this functionality to a message-oriented service.

A dedicated CM transport service can also provide error-handling mechanisms that are more suitable for CM clients. Although most message-oriented clients cannot use data which is only partially correct, many CM clients can tolerate limited data loss, and some can even utilize corrupted data. [HTH89]

Finally, while message-oriented clients cannot predict the time of their next data transfer, and so must explicitly inform the system to initiate each transfer, the time of each data transfer for a CM client can be deduced; therefore, the requirement that data transfer be initiated via an explicit interaction with the system introduces unnecessary overhead. This situation is exacerbated when data is provided for network transfer by a DMA-like device, e.g. a hardware coder-decoder (codec) for compressed video, since a (user) process must intervene between data generation and transmission.

The above differences in data transport requirements between message-oriented and CM clients justify design of a dedicated CM data transfer service which could provide better service for CM applications in four ways:

(1) a better traffic model for characterizing CM traffic and specifying performance requirements;

(2) the abstraction of (logical) streams, which are visible to the transport service;

(3) CM specific error handling, including delivery of all correctly received data and (possibly) of corrupted data, and of optionally replacing corrupted/lost data with dummy data; and

(4) the elimination of the need for a rendezvous (e.g. via a system call) between the client and the service for each data transmission.

At this point, we would like to emphasize that there is no fundamental reason a message-oriented transport service could not offer a service that included (1) - (3); however, as we argued above, these capabilities are neither required nor desirable for most message-based applications, and hence it seems wiser to implement a new service to provide them to CM clients.

3. Data Transport Requirements of Continuous Media Clients

In this section, we will give a few examples of CM clients and their data transport requirements. From these clients we extract a generalized list of data transport requirements for CM. In section 4 we will describe the basic features of a service designed to meet these requirements. All the applications listed require a strict upper bound on delay and on delay-jitter, where delay-jitter is defined as the difference between the maximum and minimum delay. If an upper bound is provided on delay, delay-jitter will result only in early delivery, and can therefore be absorbed by buffering in the service provider if enough buffers are available. Therefore, delay-jitter is not listed as a requirement for the clients. However, it should be noted that a network which controls jitter would allow less buffer space to be allocated to the connection on the receiving end-system.

Typical continuous media applications include:

- *Compressed Video:* Variable rate; delays <= 300 ms, if interactive; DMA-like coding devices; loss tolerant; synchronization with audio or text; variable period (slow-motion or fast-forward) or suspend transmission (freeze frame).

- *Uncompressed CD quality digital audio:* Constant rate; loss sensitive.

- *Multiplexed, interactive digital voice:* Constant rate; delay <= 300 ms; loss tolerant; silent periods.

- *Multimedia distributed classroom:* Phases of instruction (e.g. lecture, movie); interactive.

From these examples we have determined the following list representing data transport requirements of *most* CM clients:

(R1)   Periodic delivery of data without gaps.

(R2)   Bounded delay of stream between sender and receiver.

(R3)   Logical stream abstraction: used to communicate redefinition of stream parameters to receiving client and data transport service provider, and for synchronization of data from separate connections.

(R4)   Delivery of all correctly received and possibly of corrupted data.

(R5)   Notification of undelivered or corrupted data.

(R6)   No requirement for explicit interactions with system for each data transfer (i.e. delivery of data is periodic, not event-driven).

4. Definition of the Continuous Media Transport Service

We now briefly describe a continuous media transport service (CMTS) designed to meet the needs of CM clients (a considerably more detailed specification of this service and its underlying protocol is given in [MoW91]). This service provides unreliable, in-sequence transfer (simplex, periodic) of *stream data units* (as defined below) between a sending client (C_S) and a receiving client (C_R), with performance guarantees on loss, delay, and throughput. The CMTS service satisfies all of the requirements (R1),...,(R6) as listed above. Data is passed from C_S to the CMTS entity on its end-system (CM_S) via a shared circular buffer. Synchronization between C_S and CM_S is provided via traffic and performance parameters, and through explicit synchronization variables.

All traffic and performance parameters are defined in relation to two basic units: the *stream data unit (STDU)* and the *periodicity* of the conversation (T). An STDU is a data unit whose boundaries must be preserved by the CMTS and indicated to C_R. It is C_S, which decides how the stream to be transferred to C_R is mapped onto a sequence of STDUs as illustrated in Fig. 1. Typically, for a CM application, the information to be transmitted (e.g. voice signal, sequence of images) is digitalized prior to its transmission by a coding process (or possibly a sequence of such processes). The coding process maps a continuous signal function (in the sense of coding theory, e.g. voice, moving scene) onto a sequence of code words (cf. coding theory again). A CM application then has several options in mapping these code words into a sequence of STDUs:

a)   Sequence of code words mapped onto sequence of bytes (byte stream), 1 byte corresponding to 1 STDU;

b)   one-to-one mapping of code words onto STDUs;

c)   concatenation of several code words to build one STDU, e.g. in the case, where different sub-streams are multiplexed (time-multiplex) by an application to form one overall stream;

d)   combination of a code word and a time-stamp into one STDU (where a "time-stamp" is a reference to the period the code word represents); e.g. relevant in transmitting a stream, which had been stored, along with the timing information required to reconstruct its original timing.

The periodicity, T, of a stream can also be specified by an application. The periodicity characterizes the frequency of coding events (typical values chosen for periods would be $T=k\times0.125$ ms in PCM-voice coding or $T=k\times33$ ms in transfer of a video stream, where k is an integer). The data corresponding to a period maps into an integral number of STDUs. The CMTS service recreates on the receiver, the stream that had been seen on the sender at the granularity (in time) of a period, T. This implies, in particular, that data associated with period Δt_i at the sender must arrive before the beginning of the corresponding period $\Delta \tau_i$ at the receiver. Of course $|\Delta t_i|=|\Delta \tau_i|=T$, for all i, if $|I|$ denotes the length of interval I. Fig. 2 illustrates the basic timing within transfer of a sample stream (e.g. of voice or video data). It should be noted, however, that this figure does not reflect the fact, that we allow some "work-ahead" to the sender of the stream and some time "behind-schedule" to the receiver, as described below.

The conversation between C_S and C_R consists of a sequence of logical streams, with intervals of (medium- or long-term) silence between streams. A logical stream consists of a periodic sequence of STDUs which may be fragmented for actual transmission by the network. C_S must inform CM_S of the beginning and end of each logical stream. The model we are using for interactions between C_S and C_R is that data generated during an interval Δt_i by C_S is needed by C_R during the corresponding interval $\Delta \tau_i$, where $t-t_0$ represents the elapsed time of the stream at the sender and $\tau-\tau_0$ represents the elapsed time of the stream at the receiver (t_0 and τ_0 denoting beginning of stream as observed by sender and receiver respectively). Therefore, after indicating the beginning of a logical stream, C_S is *obligated* to ensure that all the STDUs corresponding to a given period are in the shared buffer before the end of that period, where periods are defined at regular intervals (of length T) from the beginning of stream transmission (t_0 in this case). Because a sender may have difficulty providing data corresponding to an interval exactly within that interval (e.g. due to contention for CPU or memory bus), another parameter, N_{Sslack}, is defined as the number of bytes, C_S is allowed to provide to CM_S ahead of schedule (i.e. before t_{i-1} for Δt_i). The solution of prefilling buffers has also been suggested in [AHS90]. C_S is also obligated to obey the traffic characteristics specified for the conversation and for the current stream.

At the beginning of a logical stream, C_S may redefine some traffic and performance parameters for the duration of the stream. These must be no more strict than the parameters of the actual connection (e.g. a decrease in the data rate is allowed). Such information may be used to assist cooperation between C_S and C_R because the characterization of a stream can more accurately represent current traffic and performance needs than the long-lasting connection parameters, which must cover all possible streams to be sent on the connection. The stream characterization may also allow system and network resources to be conserved. In addition, C_S may indicate a value for $d_{Sstartup}$, the time between the time of the request, t_{start}, and the start of the stream, t_0, at the sender. This time is used by C_S to pre-load the buffer. An upper bound on the *duration* of the stream may also be specified, to be used in allocating resources at the receiver.

After being informed of the start of a new logical stream, CM_S is obligated to cooperate with CM_R to transfer all STDUs for each period to C_R before the beginning of the corresponding period on the receiver, provided that C_S met its contractual obligations.

The CMTS entity at the receiving end-system (CM_R) and C_R also interact via a shared circular buffer, stream parameters, and shared synchronization variables. CM_R must inform C_R of the beginning of a new logical stream. After that, CM_R is obligated to put data in the shared buffer before the beginning of the period in which it will be needed (i.e. before time $\tau=\tau_{i-1}$ for period $\Delta \tau_i$), and C_R is obligated to remove from the shared buffer all the STDUs corresponding to a period before the end of that period (i.e. before time $\tau=\tau_i$ for period $\Delta \tau_i$). Since a receiver may have difficulty removing data exactly during its corresponding period, the value N_{Rslack} is defined at the receiver to indicate how far C_R can fall behind without data being lost. More precisely, C_R can leave up to N_{Rslack} bytes in the buffer after the end of their corresponding period. As at the sender, the first period is defined to begin after a delay of length $d_{Rstartup}$, where $d_{Rstartup}$ includes $d_{Sstartup}$, as well as the delays introduced for smoothing and for tolerating delay-jitter in the network. In addition, CM_R must inform C_R of data loss (which includes late data) and corrupted data. In the implementation, STDU descriptors are used to maintain STDU boundaries and to indicate errors in the data.

In order to meet guarantees regarding buffer overflow and starvation avoidance at the receiving client, all four entities must be involved in a handshake at the beginning of the conversation so that each may approve traffic and performance parameters. At this time some of the entities may reserve system and network resources to ensure that they will be able to fulfill their contractual obligations. This handshake is accomplished as follows: C_S presents a proposed set of parameters to CM_S. If CM_S accepts these, it passes them on to CM_R, with some possible modifications and additions. If CM_R accepts the (revised) parameters, it passes a (possibly) revised version of the original set to C_R. If C_R accepts the conversation request, it informs CM_R, who informs CM_S, who informs C_S. The parameters of the handshake before a conversation are described below for the C_S/CM_S interface. Parameters for the handshake between other entities are analogous to those described here.

The parameters of the C_S/CM_S interface were chosen to capture those traffic characteristics of continuous media traffic that have the greatest impact upon resource requirements and to specify performance requirements applicable to CM applications. The most significant of these parameters are described below. (A full description can be found in [MoW91].)

Traffic characterization parameters

The first three traffic parameters allow for flexible definition of a CM connection. They cannot be changed for individual streams. $STDU_{max}$ specifies the maximum size of an STDU (in bytes), i.e. maximum size of a logical unit for which boundaries must be maintained. ($STDU_{max} = 1$ is allowed as a special case, leading to a transparent byte stream data transfer). $CONST\_SIZE$ and $CONST\_NUM$ are booleans used to further describe the type of service requested (i.e. byte stream, constant-size STDUs or variable-size STDUs).

The next group of parameters characterize the traffic pattern. All traffic and performance parameters are defined in terms of T, the basic period of the stream. S_{max} is the maximum amount of data (in bytes) which C_S may put in the shared buffer during *any* period. S_{avg} specifies an upper limit on the mean number of bytes presented by C_S for transport within a single period, calculated over any averaging interval consisting of I_{avg} consecutive periods. S_{min} provides a limit on the burstiness of the stream missing from other traffic specifications we have seen. Our model of CM streams is that even variable-rate streams will have *something* to transmit each period. S_{min} allows a user to specify the minimum amount of data which is *expected* to be transmitted each period. One common example of such a stream is a compressed video stream which transmits an image compressed with only intra-frame encoding followed by several frames that achieve higher compression using inter-frame encoding ([e.g. Leg91]). Without this parameter, we would have to make the worst-case assumption that all the data allowed in an averaging interval could be sent in a minimal number of periods with no data being transmitted during the rest of the averaging interval. (Note: C_S is *not required* to transmit S_{min} bytes each period, but if it does not, it cannot be sure of sending its entire allotment for the averaging interval without possible loss of data due to buffer overflow.)

We would like to briefly discuss some of the advantages of the traffic characterization described above for CM clients. As far as the authors are aware, present transport services allow clients to describe their burstiness by specifying a variability in the inter-arrival times of fixed-size messages (e.g. [AHS90], [FeV90]). While this is the proper characterization for network packets and for message-oriented clients, it is not a convenient manner for describing the burstiness of CM traffic (e.g. compressed video), which is better described as a variable-size message sent at fixed intervals. The characterization we have presented allows for variable-size "messages" (corresponding to the data produced in a period) at a fixed interval T. This characterization, along with the inclusion of S_{min} will allow for a better characterization of CM traffic and, therefore, more efficient utilization of network and end-system resources.

N_{Sslack} is the maximum workahead allowed to C_S as previously mentioned. It specifies the maximum number of bytes which C_S is allowed to put in the shared buffer early, i.e. before the beginning of the period to which the data corresponds.

Quality of service parameters

The parameters for indicating the quality of service (QOS) desired were chosen to sufficiently communicate the needs of most CM clients. As stated previously, the basic model of the CMTS service is that the stream on the sending end-system will be recreated on the receiving end-system at the granularity of a period. The service handles delay-jitter for the client (where delay-jitter is defined as the variability in delay) by ensuring the shared buffer at the receiving end-system is large enough to tolerate maximum possible jitter without overflow. Because there are no explicit interactions between the client and the service, the implications of delay-jitter on timing of the stream do not apply.

D_{stream} is the maximum acceptable delay of the stream, where the delay of a stream is defined as the time between the start of a stream at the C_S/CM_S interface and the start of the stream at the C_R/CM_R interface. Since D_{stream} must also be maintained for each period of the stream, it implies a deadline for data arrival at the receiver. (All data associated with the ith interval on the sender, Δt_i, must arrive at the receiver before the beginning of the same period at the receiver, i.e. before τ_{i-1}.) S_{err} allows the client to specify the maximum *granularity* (in bytes) of a data loss caused by data corruption or buffer overflow. This parameter is interpreted as an upper bound on the packet size used for this stream. W_{err} specifies a lower bound on the probability that a unit of data transfer (of size $\leq S_{err}$) is correctly delivered to the receiving interface. *REPLACE* is a boolean that indicates whether corrupted data should be replaced with dummy data (supplied by the user in a dummy data unit of size S_{err}) instead of being delivered as it is received or discarded. This service is useful for in-band signalling of data loss and for filling in holes in the data stream.

5. Basic Underlying Services and a Transport Protocol to Support CMTS

We presently implement, within the XUNET II project, the service described above as part of a real-time protocol suite. Connection establishment and teardown (including resource allocation) are provided by a connection administration service (RCAP, cf. [BaM91]), which will also handle the connection establishment and teardown functions for CM connections. A network service (RTIP, cf.[VeZ91]) which implements the schemes described in [FeV90] will provide network connections with real-time guarantees for delay, delay-jitter, throughput, and loss. RTIP will allow the transmission of packets via connections established within (possibly a hierarchy of) interconnected subnetworks with FDDI- and ATM-components.

In addition to these underlying services, the CMTS service requires relatively large buffers on the receiving end-system, as one of the main concepts underlying this service is to use buffers to smooth fluctuations in the arrival process of data during a stream as well as delay jitter introduced by the network and/or the end-systems. (Calculations with respect to buffer requirements will be given in section 6). A real-time clock with a high precision timer, and real-time scheduling of the CPU and network driver are also required.

To realize the CMTS, we defined the *Continuous Media Transport Protocol (CMTP)* which supports communication between CMTS peers at the sender and the receiver. The first version of the CMTP protocol could be kept relatively simple. This results primarily from the fact that several of the communication functions needed in conventional data communication (in particular, retransmissions for error correction, flow control, etc.) are not required in order to provide the CMTS service. Regarding retransmissions, we take the position (stated in [FeV90]) that most real-time applications will not be able to wait for retransmissions, and even if they could, the amount of data which would need to be stored to perform retransmissions on a high bandwidth-delay product network could not be justified for CM clients, which do not require perfectly reliable service. Similarly, resetting a data stream to an earlier status (period) is not possible as the resource requirements needed to set check-points in general are prohibitive for storing an intermediate status of a stream. Therefore in the case of a serious error, tear-down of a connection with successive re-establishment of a connection and initialization of a new stream seem to be the most appropriate measures. The simplex nature of the real-time connections used also represents an obstacle to a dialog between sender and receiver directly within such a connection. So, our assumption in the design of the CMTP protocol has been that a data stream between C_S and C_R is transmitted via exactly one (uni-directional) connection between C_S and C_R. This connection thus represents a data connection. As no multiplexing takes place in the CM transport layer, a one-to-one mapping of the addresses of communicating clients to the address of the network connection (connection-id provided by RCAP in the case of XUNET II) can be used to solve the addressing problem.

As an extension to the current design, we assume that CM_S and CM_R are able to exchange (reliably) control information concerning the state of the data connections presently established between them. In a similar way, we assume that the CMTS service reliably transfers C_S- / C_R-control information between clients to support an application-oriented protocol between them (separate communication, in addition to the exchange of a data stream). This solution can be viewed as an "out-of-band-signaling" between the communicating CM clients.

Until now, the CMTP protocol has only been specified for covering communication within the data connection. Experiences of the CMTS implementation are considered to be indispensible prior to a protocol extension and will be taken into account in the completion of the CMTP protocol.

The first version of the CMTP protocol is based on three types of protocol data units (PDUs):

- ON_PDU: Signal start of stream; redefine parameters, facilitate synchronization
- OFF_PDU: Signal end of stream
- DATA_PDU: Transmit one STDU, one fragment of an STDU or a number of bytes (in case of a byte stream).

The beginning of a stream must first be signaled by the sending client via an ON_PDU. Reaction after receipt of an ON_PDU is according to the service specification (cf. section 4). To increase reliability, two copies of an ON_PDU are transmitted at the beginning of a stream (separated by an interval, which is considered to be large enough in order to make errors in both ON_PDU transmissions sufficiently independent of each other). The receipt of at least one of these copies by CM_R is sufficient for the correct initialization of the stream. Loss of both ON_PDUs will be considered as a serious error situation, requiring connection tear-down and re-establishment.

After stream initialization (if successful), DATA_PDUs can be sent to transfer data of the stream. Data of the stream (i.e. the STDUs) is mapped onto DATA_PDUs either 1:1, by concatenation of STDUs (in particular in byte streams) or by fragmentation of STDUs (respecting maximum packet-size of the underlying network service as well as maximum size of acceptable loss, specified by C_S). Each DATA_PDU is transported in exactly one packet.

The current stream ends when an OFF_PDU arrives, indicating a silent period will follow. An alternating 1-bit stream identifier is used to cleanly separate successive streams from each other, even in the case of error situations (e.g. loss of successive OFF and both ON_PDUs).

Additional control information exchanged between CM_S and CM_R (protocol extension) could refer, e.g., to

- acknowledgement (ACK/NAK) for ON_PDU;

- acknowledgement (ACK) for OFF_PDU;
- indication of buffer overflow at receiving interface to CM_S;
- PING to check whether sender is still alive, when neither data nor an OFF_PDU has been received for a given interval; and
- some control signal, used by the sender to check whether the receiver is still alive.

6. Some Implementation Considerations

Because of the desire to eliminate most client/system interactions, data is transferred between the service provider and clients via a shared circular buffer at both the sending and the receiving site (C_S/CM_S- and C_R/CM_R- interfaces). This implementation also lowers the number of data copy operations. Fig. 3 depicts some essential types of interactions between components used in the implementation.

Shared synchronization variables are used to inform the producers of data (C_S at the sender and CM_R at the receiver side) as well as the consumers (CM_S at sender and C_R at receiver) of the present state of the buffer (e.g. amount of data in the buffer). STDU descriptors are used to delineate (variable-size) STDUs and to provide an indication of data errors without explicit interactions between clients and the service provider. Fig. 4 depicts the use of descriptors for the buffer (B_R) shared between CM_R and C_R.

In the prototype implementation, CM_S will check the buffer (B_S) it shares with the sending client once per period, packetize the data found there (up to a maximum controlled by the maximum burst and average rates as required by the underlying network service), and schedule its transmission.

[MoW91] contains the derivation of the following conservative estimate for b_S, the minimum size of buffer B_S guaranteed to prevent data loss due to overflow of this buffer:

$$b_S = 2 \times S_{max} \qquad \text{(data for current and next periods)}$$
$$+ N_{Sslack} \qquad \text{(workahead for } C_S\text{)}$$
$$+ b_{sm} \qquad \text{(for smoothing)}$$
$$+ b_{align} \qquad \text{(for aligning STDUs in buffer)}$$

Similarly, for the size, b_R, of the buffer B_R, we obtain:

$$b_R = b_s$$
$$+ N_{Rslack} \qquad \text{(amount of data } C_R \text{ can be late in consuming)}$$
$$+ 2 \times S_{max} \times \left\lceil (d_j / T) \right\rceil \qquad \text{(delay jitter in ON\_PDU and stream transmission)}$$

The buffer requirements are a direct result of the fact that enough buffer space must be provided to absorb the maximum possible delay jitter and variation in the data rate being observed at the sending and receiving end-systems. Buffer requirements at the receiver can be reduced if CM_S uses additional information with respect to the timing of the stream in order to delay the transfer of STDUs at the sending end-system, thereby absorbing workahead at the sending end-system. For a connection being used to transport fixed sized STDUs (in particular for a byte stream), b_align = 0, as alignment problems do not exist in these cases.

7. Summary

Most important properties of the transport service and protocol introduced in this paper are a consequence of our effort to tailor the service and protocol design to the data transport requirements of continuous media applications. Therefore, the service introduced supports the continuous delivery of a data stream to a receiving CM client, error-handling mechanisms that can be adapted flexibly to the typical demands of CM applications, the use of a priori knowledge with respect to the future arrival pattern to be expected for a given stream, etc. The basic concepts introduced, such as the notion of stream data unit, as well as the large variety of parameters offered at the service interface can be used by communicating CM applications for a relatively flexible characterization of (e.g. voice or video) streams. This flexibility is also provided for the mapping of STDUs onto packets of an underlying network service (1:1, fragmentation or concatenation as options), where the mapping may even be controlled by the transport service users (e.g. by specifying the maximum granularity of a data loss). The solution chosen can easily support the possibility of allowing a (de-) coding process to react to the state of the communication system, as suggested e.g. in [GiG91]. The communication system's state might be considered to be reflected by the actual occupancy of the shared buffers (B_S and B_R) on sending and receiving end-systems, which could lead to a variation of the (de-) coding rate.

To complete the present design it will still be necessary to integrate the experiences gained in the prototype implementation of CMTS in an extended service/protocol design. The extensions will have to specify, in particular, additional possibilities of reacting to protocol errors as well as the exchange of different types of control information.

Limitations of the solution primarily concern the buffer requirements in the end-systems, which may become significant in those cases when delay jitter within the network and within the end-systems will become large and additionally large traffic fluctuations exist within the arrival process of the stream. However we believe that in future computer systems (even in workstations and personal computers) we can expect provision of communication buffers in the range of (a few) MByte at least for CM applications, if this leads to significant simplifications and performance improvements. For some dialog-oriented applications the stream delay resulting from our approach (of typically > 100 ms) may become disadvantageous as well. Realization of multi-point connections (e.g. required in video-conferencing) by means of (a possibly large number of) point-to-point connections, which would be the solution based on the CMTS service (within XUNET II architecture), may also lead to some inefficiencies. These inefficiencies could, of course be eliminated if the underlying network service would support multipoint communication.

In parallel to the CMTS prototype implementation, presently a modeling study is being carried out in order to gain some insight into the impact that configuration parameters of end-systems (such as buffer sizes, run-times of communication software, etc), properties of the underlying network service (such as packet delay jitter, packet loss rate, etc.), and the local load of the end-systems may have on the quality of the CMTS service as observed by CM clients (e.g. expressed by the probability of a buffer overflow with resulting loss of data and/or by the probability of late arrival of data in the B_R buffer).

8. Acknowledgements

The authors would like to express their particular gratitude to Amit Gupta and Francesco Maiorana for their engagement in the implementation of the CMTS prototype, and to Eckhardt Holz for his detailed simulation study to analyse the behaviour of the CMTS service under various boundary conditions.

A large number of in-depth discussions with a lot of resulting stimuli have taken place during the CMTS design within Tenet research team at International Computer Science Institute and University of California at Berkeley. In particular, Prof. Domenico Ferrari as head of Tenet team and the group members Riccardo Gusella, Bruce Mah, Dinesh Verma, and Hui Zhang have provided very valuable suggestions during the preparation of this paper. This support is sincerely acknowledged by the authors.

Special thanks also go to Prof. David Anderson and Ramesh Govindan for their comments which helped to improve an earlier version of this paper.

9. References

[AHS90] D. Anderson, R. Herrtwich, C. Schaefer, "SRP: A Resource Reservation Protocol for Guaranteed-Performance Communication in the Internet", Int. Comp. Sci. Inst., Technical Report No. ICSI TR-90-006 (1990).

[BaM91] A. Banerjea, B. Mah, "The Real-Time Channel Administration Protocol", Proc. 2nd Int. Workshop on Network and Operating System Support for Digital Audio and Video, Heidelberg (November, 1991).

[Che88] G. Chesson, "XTP/PE Overview", 13th Conf. on Local Computer Networks, IEEE Comp. Soc. (October, 1988), 292-296.

[ChW89] D. R. Cheriton, C. L. Williamson, "VMTP as the Transport Layer for High-Performance Distributed Systems", IEEE Commun. Magazine, Vol. 27, No. 6 (1989), 37-44.

[CLZ87] D. D. Clark, M. L. Lambert, L. Zhang, "NETBLT: A High-Throughput Transport Protocol", ACM SIGCOMM Workshop on Frontiers in Comp. Netw.(1987).

[DDK90] W.A.Doeringer, D. Dykeman, M. Kaiserswerth, B.W. Meister, H. Rudin, R. Williamson, "A Survey of Light-Weight Transport Protocols for High-Speed Networks", IEEE Trans. on Commun., Vol. 38, No. 11 (1990), 2025-2039.

[FeV90] D. Ferrari and D. Verma, "A Scheme for Real-Time Channel Establishment in Wide-Area Networks", IEEE J. Sel. Areas in Comm. SAC-8 (April, 1990).

[GiG91] M. Gilge and R. Gusella, "Motion Video Coding for Packet Switching Networks: An Integrated Approach," SPIE Conf. on Visual Commun. and Image Processing, Boston (November, 1991).

P. Haskell, K. H. Tzou and T. R. Hsing, "A Lapped-Orthogonal-Transform Based Variable Bit-Rate Video Coder for Packet Networks," Int. Conf. on Acoustics, Speech and Signal Proc., Glasgow, Scotland, May 23-26, 1989.

[HSS90] D. Hehmann, M. Salmony, H.J. Stuettgen, "Transport Services for Multi-Media Applications on Broadband Networks", Computer Commun., Vol. 13, No. 4 (1990), 197-203.

[ITC91] Proc. Workshop on "Continuous Time Media", Information Technology Center, Carnegie Mellon University, Pittsburgh (June, 1991).

[Leg91] D. Le Gall, "MPEG: A Video Compression Standard for Multimedia Applications," Commun. of the ACM, Vol. 34, No. 4, (1991).

[LiH91] M. Liebhold, E. M. Hoffert, "Toward an Open Environment for Digital Video", Commun. ACM, Vol. 34, No. 4 (1991), 104-112.

[LPS91] T. F. La Porta, M. Schwartz, "Architectures, Features, and Implementation of High-Speed Transport Protocols", IEEE Network Magazine, Vol. 5, No. 3 (1991), 14-22.

[MoW91] M. Moran, B. E. Wolfinger, "Design of a Continuous Media Data Transport Service and Protocol", unpublished (1991).

[NRS90] A.N. Netravali, W.D. Roome, K. Sabnani, "Design and Implementation of a High Speed Transport Protocol", IEEE Trans. on Commun., Vol. 38, No.11 (1990), 2010-2024.

[VeZ91] D. Verma, H. Zhang, "Design Documents for RTIP/RMTP", unpublished (1991).

[Wat89] R. W. Watson, "The Delta-t Transport Protocol: Features and Experience", Proc. IFIP Workshop on Protocols for High-Speed Networks, North-Holland (1989), 3-18.

[WrT90] D. J. Wright, M. To, "Telecommunication Applications of the 1990s and their Transport Requirements", IEEE Network Magazine, Vol. 4, No. 2 (1990), 34-40.

[Zit91] M. Zitterbart, "High-Speed Transport Components", IEEE Network Magazine, Vol. 5, No. 1 (1991), 54-63.

[Zha91] L. Zhang, "Virtual Clock: A New Traffic Control Algorithm for Packet-Switched Networks," ACM Trans. on Computer Systems, Vol. 9, No. 2 (1991), 101-124.

10. Figures

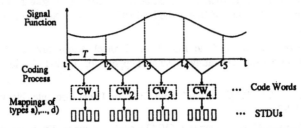

Figure 1: Coding of a stream and its mapping onto a sequence of STDUs

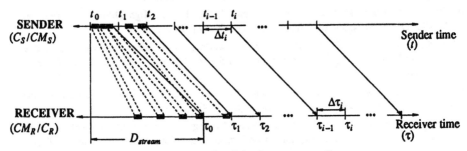

Figure 2: Basic timing during the transfer of the data corresponding to a stream

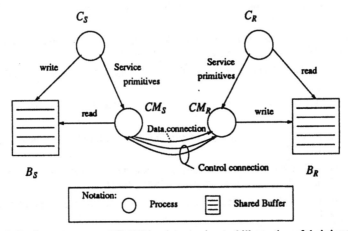

Figure 3: Basic components of CMTS implementation and illustration of their interactions

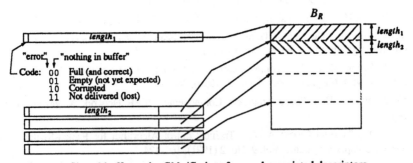

Figure 4: Shared buffer at the CM_R/C_R-interface and associated descriptors

Session VI: Projects I

Chair: Andy Hopper, Olivetti Research Center and University of Cambridge

The papers presented in Session VI dealt with four very different multimedia projects which are summarized in chronological order.

The first paper was "Design Considerations for a Multimedia Network Distribution Center" by Riccardo Gusella of Hewlett-Packard Laboratories and Massimo Maresca from the University of Genoa. In his talk, Massimo examined the issues involved in the design of a central facility called the Multimedia Network Distribution Center, which manages a number of different multimedia applications. The design revolved around issues of asymmetric communication, support for heterogeneous networks, adaptability to changes in networks or host loads as well as system integration.

Experiments were done on an Ethernet network with SUN hosts running UNIX, including the TCP/IP protocol set. Video frames were in the order of 100 Kbytes. These frames were stored in compressed form on a mass storage system or can be taken from an uncompressed source and then compressed by special hardware before being sent out onto the network. A compression factor of 15 was used. This produced a required throughput of 200 Kbytes/s for video conferencing applications. Performance results were presented for memory copying, TCP processing, UDP processing, decompression, visualization directly to the frame buffer and visualization through the X Server.

"Next Generation Network and Operating System Requirements for Continuous Time Media" was presented by Scott Stevens from Carnegie-Mellon University. Scott highlighted the requirements for building complex multimedia applications in which different elements of the application can be combined in various ways. The proposed technique involves using fine-grain elements and a rule-based system with a facility to specify how these elements should be combined and displayed.

Other mechanisms are used to deal with the scaling of images, synchronization, etc. It was pointed out that in order to build such systems a more abstract method of defining the multimedia elements is needed which allows different elements to be readily connected together. Synchronization at the frame level was also proposed to allow users to specify additional responses based on a given action.

The paper "Dynamicity Issues in Broadband Network Computing," joint work with Jose Diaz-Gonzalez, Russell Sasnett and Vincent Phuah, was presented by Steven Gutfreund from GTE Laboratories. It looks at dynamicity issues for multimedia applications in a distributed environment. These issues include the movement of applications from one network to another, the migration of active applications, dynamic changes in network topology and the ability to assemble applications from a number of different sources.

The SHOWME application is a multimedia environment which deals with these issues by defining a uniform homogeneous term called an Element which can be used to describe different multimedia entities. The binding of Elements is handled by a Resource Dispatcher which provides a tuple space, similar to Linda. Tuples are used to control the multimedia facilities specified by different elements. The system supports pattern-directed binding in which the binding between elements occurs at runtime.

The last paper of the session came from Ralf Cordes of Telenorma ("Managing Multimedia Sessions on a Private Broadband Communication System," Ralf Cordes, Dieter Wybranietz, Rolf Vautz). This paper addresses the problems of multimedia applications in which connections to different servers are dynamically changing depending on the interactions of one or more clients. To provide these facilities, software systems must be geared to support a number of new features, including in-call bandwidth modification, quality-of-service negotiation, integration of servers, transaction-oriented protocols and object-oriented structuring of generic applications of service elements.

Object classes are used to define composite objects called Pages and monomedial objects called Particles as well as ways for linking and anchoring objects and specifying telecommunication service elements. Transaction support is provided via the use of an ATM network, support for multipoint communications and the use of fine-grained set-up and roll-back facilities. An advanced distributed communication system for the customer premises market was also proposed. This network will also facilitate the use of several network interfaces and is based on FDDI-II.

DESIGN CONSIDERATIONS FOR A MULTIMEDIA NETWORK DISTRIBUTION CENTER

Riccardo Gusella
Hewlett-Packard Laboratories
1501 Page Mill Rd.
Palo Alto, CA 94303 - USA

Massimo Maresca
DIST - University of Genova
Via Opera Pia 11A
16145 Genova - Italy

ABSTRACT

In this paper we consider a distributed system in which a central facility, called a Multimedia Network Distribution Center, serves a number of clients and handles simultaneously different multimedia applications. These applications include digital TV distribution, interactive TV (consisting of hypermedia techniques applied to TV), as well as collaborative applications such as virtual distributed classrooms. Clients are allowed to request services independently of each other, or groups of clients can make joint service requests.

In order to satisfy a variety of applications and to support the largest possible number of clients, a server must be able to deal with a heterogeneous environment, in which clients range from small PCs to powerful workstations with or without special hardware for compression/decompression and with different visualization throughput. Moreover, the server must be able to handle shifting network load conditions and to offer concurrent synchronous and asynchronous access.

The server is connected to a set of devices (either live or playback) that generate digital video and audio streams under its control and to a set of networks and links used to transport the streams to the clients. We define the functionality and the performance requirements for an MNDC. To verify the validity of the MNDC functionality, we have built and measured a prototype system and taken performance measurements using applications that moved video and audio data across a local-area network.

1. INTRODUCTION

A Multimedia Network Distribution Center (MNDC) is a facility in a distributed system that can be used to deliver multimedia information to a number of clients simultaneously. Because communication of multimedia information is much more demanding than data communication–not only are there real-time and synchronization issues, but there are also new error modes, and more demanding quality-of-service requirements–the systems at the server and client ends must be closely coordinated. We can think of two classes of applications that can be categorized as MNDCs.

The first class is characterized by the *one-to-one* communication paradigm. Several users of this class of applications (clients) can be active simultaneously, but they are independent of each other and each one of them opens a point-to-point, bidirectional connection with the MNDC. An example of an application in this class is interactive video, in which client viewers of stored documentaries can choose to follow one of several different paths of the documentary at a number of predetermined points during playback. The means of making the selection–a typed command, a button click, or a voice command–is a user interface question, which we will not address here; we will simply assume that the user's choice is signaled, using the return path of the connection, to the server, which will then begin to deliver the new documentary segment to the user. A second example of an application that requires a one-to-one communication paradigm is an MNDC that provides multimedia database access. In this type of database, queries can return text, images, sounds, video clips, and other types of timed data [3]. It is different from the previous application in that the focus is on queries, making the interaction between client and server much more frequent.

The second class of applications that can be classified as MNDC is characterized by the *one-to-many* communication paradigm. In this case, several clients are concurrently active and the server, possibly on the basis of input provided by the clients, decides autonomously on the sequence of the information delivered to the clients. A simple example of this class of applications is video distribution of the type available today over CATV networks; in this application the return communication channels from the clients to the server are never used: to change video channel clients listen to different port numbers. A second example is represented by a teaching system in which the instructor is in control of the server, which in turn sends clients (students) lecture information. Occasionally, one of the clients may interrupt the lecture flow by asking a question. Rather than providing full physical connectivity and connections to all other clients, it is the server that relays the question to the other clients. The slight additional delay is a small price to pay in return for the great simplification in system design. (This is similar to what happens in large conferences when a question from the audience is repeated by the speaker before he or she provides the answer.) A third example of an application that fits this communication structure is a medical application in which the server handles the instrumentation in an operating room. This instrumentation may comprise a set of cameras directed to an ongoing surgical operation as well as a number of sensors monitoring blood pressure, heart and brain activity and other vital signs. The server directs this multimedia information to a number of medical students and a few specialists, whose task is to supervise the operation and help the surgeon during its most critical phases. The specialists may volunteer a comment, or they may be asked questions from the operating room. In either case, the server will relay the data to the various output lines as we have discussed in the previous example.

An MNDC may transmit live data, recorded data or a combination of the two within the same application. The video distribution and database applications are examples of systems that transmit recorded data. The medical application is an example of live data transmission;

the teaching system may transmit a combination of both types of data. This distinction is important as we will see that live and recorded data will be handled differently by the network.

2. GENERAL DESIGN PRINCIPLES

Before discussing specific design requirements for an MNDC, it is important to consider several fundamental characteristics of multimedia data transmission that make the design of an MNDC different from that of traditional data communication systems. First of all, the MNDC transports mainly continuous media data, which has two key properties: it consists of a sequence of messages sent at a fixed rate, and the quality of its presentation can be adapted to the varying load conditions of the network and of the hosts. In addition, the MNDC is based on an asymmetric connection between one server (or in some circumstances a limited number of servers) and a set of clients; in contrast, in other multimedia distributed systems connections among the hosts may be more uniformly distributed.

The asymmetric structure of the MNDC creates certain requirements for hosts and network. The server host must have enough power to handle a large number of continuous media channels simultaneously. One means of obtaining this power is to use special hardware devices and file systems for storing/retrieving continuous media information in real time and special hardware devices for compression/decompression. Client hosts, instead, need not be high-performance, as they will be handling a smaller number of channels: clients may be simple PCs or workstations with no special hardware. The network should be structured in a hierarchical fashion so that links closer to the server have greater bandwidth than links closer to the clients.

One common issue in the design of distributed multimedia applications, which arises in the MNDC context as well, is how the network can provide real-time communication. One approach is based on the reservation of the network and computing resources during the connection establishment in order to guarantee the performance [2], while another approach is based on the development of flexible communication systems, in which the emphasis is on the adaptability of the quality of service to the varying load conditions of the network and of the hosts [4]. The MNDC follows this latter philosophy; the only assumption about communication management is that the network is able to separate data belonging to different media. This ability to distinguish and separate various media is critical because each medium has different performance parameters and different quality-of-service demands (e.g., audio requires, in general, more stringent error handling than video).

The asymmetric structure allows identifying some specific design goals, which have to be met by distributed systems that support an MNDC. Such design goals are listed below.

1) *Heterogeneity*: the MNDC must support networks and hosts having different characteristics, performances and costs. For example, clients of different performance classes must be able to receive and present the same video stream simultaneously. The quality of the presentation in each client will depend on its capabilities.

2) *Adaptability*: the MNDC must be able to adapt the quality of the presentation to load variations in the network and the hosts. For example, when the load in a client host

grows to the point at which the client no longer can present the frames received at the rate they are produced, the client must reduce the quality of the presentation, in a manner that will be the least disruptive or perceptible to the viewer.

3) *Integrated approach*: the MNDC must handle multimedia data no differently from the way it handles regular computer data. Neither special devices for continuous media presentation nor special communication channels for continuous media transmission should be used.

3. DESIGN REQUIREMENTS

The system design requirements of an MNDC involve two sets of issues: first network architecture and communication protocols, and second, computer organization and operating system architecture to support the multimedia data traffic. Much work is in progress in these areas [7], but in this paper we will only address the issues that are relevant to the design of an MNDC, and we will start with networking issues.

While current bus bandwidths are on the order of 100 Mbytes/s and newer, higher-parallelism busses promise to be much faster, the networking community is working hard to develop large scale networks in the Gbit/s range (equivalent to 128 Mbytes/s) within the next several years. Despite the increased capacity possible with such large scale networks, however, network bandwidth will still be insufficient for the aggregate traffic that can easily be produced by several workstations generating multimedia data such as high-resolution motion video. Thus, image compression is necessary to support these applications.

The most prominent image compression standards, JPEG, MPEG, and H.261, involve transform coding techniques [12]. But these algorithms are computationally very intensive, precluding their implementation by software means if real-time performance is required. However, in the case of an MNDC, since it is conceivable that the server will be considerably more powerful than client workstations, one could exploit asymmetric compression schemes that require a considerable amount of work during compression but could use quick table lookup methods for decompression. Alternatively, especially in the case of one-to-many communication serving heterogeneous clients, hierarchical or pyramidal compression schemes appear very promising. Displaying a rough picture would be quite cheap, and more powerful receivers could obtain better images by decoding more and more subbands.

In our view, multimedia traffic will be transported by general purpose, integrated networks. Because current internet packet switching nodes do not distinguish between various types of traffic, temporary congestion in a network segment affects all traffic across the segment. Provided that network access of multimedia users is controlled so that the total amount of multimedia traffic is always below a certain bound, we claim that it is possible to satisfy the real-time properties of multimedia traffic without changing the network in any drastic way. We suggest that all that is needed is an appropriate queuing algorithm in the internetwork gateways and buffer management in the receiving host. Although a detailed description of our network architecture is outside the scope of this paper, one issue–how to map the quality-of-service requirements to the services provided by the transport protocols–is of major importance to the design of an MNDC.

Each gateway will have separate queues for high-priority traffic and low-priority traffic. The difference between the two types of queues is primarily in the jitter they introduce in the packet delivery process, which could be quite high for low-priority queues. We also assume that the network bandwidth will be much higher than the bandwidth required by a single conversation, so that transmission can proceed at speeds faster than real time even on lower priority queues.

We have classified the traffic produced by the various applications listed in the Introduction as live and playback. Since playback traffic can be transmitted ahead of the time is it required by a client, we assign playback traffic to lower-priority queues and use large buffers in the receiving clients to correct the jitter introduced by the communication system. This arrangement will allow us to reserve the high-priority queues exclusively for live traffic, and, assuming that the proportion of the total live traffic is small, we can design a queue service discipline that in most cases will produce small delay and small jitter. We believe that this kind of network architecture and buffering scheme would produce the performance required by multimedia traffic under appropriate traffic conditions.

The second set of requirements for the design of an MNDC is concerned with computer organization and operating system architecture. Since cache memories do not help much when the flow of data is from the network interface to the frame buffer memory, a first fundamental problem is to try and avoid data paths that include main memory—by far the slowest component in today's workstation architecture: the DECStation 5000/200, the HP 9000/720, and the Sun Sparc 2 machines have respectively main memory chips with clock cycles of 100 ns, 80 ns, and 70 ns, respectively. A related computer organization problem is the position of the decompression engine with respect to the frame buffer. We claim that, in order to reduce main memory traffic and achieve the highest performance, the two should be next to each other, connected through a separate bus.

Another important issue is how to deal with the skews between the clocks of the server and those of the clients. The problem arises because the server and a client produce and consume data at the same rate, but the respective rates are determined by their own clocks, which, running at even slightly different speeds, may in the long run cause queue under- or overflows. To quantify the amount of skew, let us assume that clocks diverge no more than four seconds over 24 hours and that we transmit 30 frames per second. Then, in 60 minutes we may be off, in either direction, of up to five frames.

One simple way to deal with this problem is to have the operating system synchronize the rate of a client's clock with the rate of the clock of the server using algorithms analogous to those presented in [6]. The alternative method of letting an application do the resynchronization is not optimal because a client may have several application programs running simultaneously, all of which would have to apply the clock transformation on their own. However, since a machine may be part of an administrative domain whose clocks are synchronized independently of the clock of the MNDC, if the rate of a client's clock is changed, then the client must have two clock sources, one for supporting multimedia timed operations, the other for the regular OS time services that needs to be synchronized locally. Notice that popular time synchronization programs such as `ntp` [9] and `timed` [5] affect an operating system software variable and not the hardware device that UNIX uses to produce interval timer interrupts for user processes requesting them.

In terms of operating system support other issues are include media synchronization, real-time scheduling support, performance. We believe that for these issues the solutions that have been proposed for general multimedia systems [1] apply as well to MNDCs.

4. EXPERIMENTS

To verify the validity of the MNDC design principles stated in Section 2 and to evaluate the feasibility and the cost of design solutions meeting the specific MNDC goals and requirements outlined in Sections 2 and 3, we ran certain video and audio distribution experiments on an MNDC testbed. In this section we concentrate on describing the video experiment, which is the most demanding one in terms of computing and communication throughput.

The testbed was an Ethernet network connected to a number of different hosts; for our experiments, we confined our analysis to Sun hosts. These hosts all run the Unix operating system including the TCP/IP protocol set. The experiment with full motion video consisted of the distribution of sequences of frames of CIF size [8] (352×288 8-bit pixels, about 100 Kbytes). In the reminder of the paper we use the term "frame" to refer to a CIF video frame.

The MNDC server reads the frames of a video stream from its mass storage or from an input device, compresses them and sends them to one client (one-to-one communication) or to more clients simultaneously (one-to-many communication) at a speed of 30 f/s (frames per second). Considering that a compression factor of 15 produces little degradation (in terms of user perception) in video-conference type video sequences, the resulting required throughput is about 200 Kbytes/s. The clients receive the frames, absorb the jitter introduced by the network by synchronizing the stream with a local timer, decompress the frames, and display them, either writing them directly to the frame buffer or using the local X window server as a virtual display.

In order to meet the design goals introduced in Section 2, namely heterogeneity and adaptability, each client must have control over the quality of the presentation at its site, while the server must structure the transmission in order to support a presentation of the best possible quality. Because of the need to be flexible and adaptive, the structure of the client subsystem is a critical part of an MNDC, and was the main focus of our study, the remainder of this section is devoted to the analysis of this issue.

4.1 BASIC ASSUMPTIONS ON THE EXISTENCE OF A TRANSPORT SERVICE

We began with the general assumption that there was "good-quality transport service". With good-quality transport service we mean that a lightweight connection [11] can be established between the server and the clients. We define lightweight connection to mean that routes are chosen at establishment time and kept fixed during the session. Once a lightweight connection is established, no error checking need be done on the packet's data segment (we accepted some corruption), only rate-based flow control techniques are used (i.e. the receiver cannot delay the sender), and no message buffering and reordering is performed. Frames or fragments of frames are transmitted in datagrams following the same routes, but neither their delivery nor their correctness is guaranteed.

As mentioned before, in a multimedia system the network should be able to separate different media, and schedule the messages of each of them independently. In the experiment, however, we waived this requirement and relied on the FCFS-based scheduling techniques of the TCP/IP protocol suite, in the current Internet. In particular, we chose to use the UDP/IP protocol, which provides the service closest to the one we desired, offering unreliable datagram delivery, as we needed, and in practice also offering ordered datagram delivery (for the datagrams that are delivered) in local or low-complexity environments, in which the routing is trivial and fixed.

4.2 MULTI-PROCESS ORGANIZATION

An MNDC client program could be logically split into three processes, running concurrently and asynchronously. The first process, called the *receiver process*, queues the received messages in a FIFO buffer called the receiver queue. The second process, called the *transformation process*, extracts the messages from the receiver queue, does the necessary transformations (e.g., decompression, recombination and possibly error correction) and copies them to another FIFO buffer, called the presentation queue. The third process, called the *presentation process*, extracts the processed messages from the presentation queue and either displays (if video frames) or plays (if audio) the data contained in such messages using the proper output drivers.

Because of the dynamically varying load in the client host, it may happen that at the time a new frame is to be presented some of the frames received must be dropped, because it is discovered that they are late. Since these frames have been already decompressed, the computing power expended on their processing is wasted.

A two-process structure reduces the chances of such wasted processing; a *receiver process* receives the incoming frames from the network interface in the receiving buffer, and a *presentation process* checks that the frame is on time and, if it is, does the decompression. Some jitter, in the form of delay variation, may be introduced locally at each client, because the decompression may not require exactly the same time for each frame and because spending a large amount of time in processing the data (the decompression of a frame using the scheme described in this section takes about 30 ms in a Sun Sparc 2) increases the probability that other Unix processes will be served during such period of time. However, our experiments show that, in terms of user perception, the delay jitter introduced by decompression at display time is negligible.

4.2.1 PROCESS ACTIVATION MECHANISM

Assuming that the MNDC client subsystem is composed of the concurrent processes described above, we must decide how these processes are to be activated. The primary alternatives are the `signal` mechanism, the lightweight process library, and different Unix processes.

We chose to use the `signal` mechanism, which makes it possible to manage asynchronous events (at a certain minimum granularity). As soon as the client program starts, it sets up two different `signals`, one to be delivered by the I/O handler at each new frame received (`SIGIO`) and another to be delivered by a local timer at fixed intervals (`SIGALRM`).

The receiver process is activated upon receiving a message, while the presentation process is activated upon receiving an interrupt from the timer.

The choice of having the presentation process activated at regular time intervals rather than by an I/O signal originated by the presentation device whenever the driver is ready to accept new data offers the advantages of uniformity in the treatment of video and audio and explicit control, at the user level, over the length of the output buffer.

4.2.2 SYNCHRONIZATION

Video frames and/or audio messages must be synchronized in order to be presented at the right time. Each message leaving the server (we use the term message to refer to both video frames and audio packets) is assigned an increasing *sequence number*, which determines univocally the relative time (with respect to the beginning of the sequence) at which the message is supposed to be presented, according to the expression:

$$presentation\_time = sequence\_number \ / \ frame\_rate + start\_time$$

5. BASIC OPERATIONS

The experimental client subsystem presented in the previous section, as well as the analysis of the type of processing that needs to be performed in the acquisition, decompression and presentation of integrated continuous media data, shows that there are a number of basic operations that must be performed on each frame by each client, and that depending on the performance of these operations in each client host, a different quality of presentation is achieved.

These basic operations are *memory copy, protocol processing, decompression* and *visualization*. We have analyzed the performance of a specific system, a Sun Sparc 2, as a representative of the class of the current-generation high-performance workstations, to understand to which extent such a class of machines is suitable for use as MNDC clients without additional hardware. Table 1 shows the results obtained; the performance results are given in bytes per second, frames per second (f/s) and compressed frames per second (cf/s) assuming a compression factor equal to 15.

TABLE 1. – PERFORMANCE OF BASIC OPERATIONS IN A MNDC CLIENT

OPERATION	MBYTES/S	FRAMES/S	COMPRESSED FRAMES/S
Memory Copy	9.7	100	-
TCP Processing	4.4	-	675
UDP Processing	6.7	-	1035
Decompression	2.8	29	-
Visualization (directly to the Frame Buffer)	8.2	85	-
Visualization (through the X server)	3.2	33	-

The performance of the *Memory Copy* basic operation was studied by trying a number of different techniques (e.g., execution of integer and double precision assignments and use of the memcpy() library routine) to copy a memory buffer from one area to another. The experiments were done with a buffer of very large size (4 Mbytes) in order to eliminate the effect of the cache memory. In fact, continuous-media data processing does not exhibit data locality, as each message is processed at most one time (for decompression) and then consumed by its visualization. Many researchers have recognized this fact and, as a consequence, there is a wide consensus that the architecture of workstations oriented to multimedia data processing must be improved to permit workstation to bypass the cache memory when it is not needed. The timing results are given in terms of uncompressed frames per second, because obviously this type of data movement accounts for much of the processing time.

The performance of the *Protocol Processing* basic operation was studied both for connection oriented (TCP) and connectionless (UDP) communication. In order to evaluate the throughput of TCP/IP and UDP/IP, a communication session was established between two processes in the same machine through the loopback interface, so to avoid generating any data-link layer traffic. The performance figures are only given in terms of compressed CIF frames per second, considering that it is expected that only compressed images travel in the network.

The performance of the *Decompression* basic operation was studied by adopting a space-domain intraframe compression algorithm. up analysis of each frame (only intra-frame compression). The reason of this choice is that in a heterogeneous system there may be client hosts which are not equipped with special hardware for decompression; these hosts are likely to use space domain techniques to decompress in real-time. Our algorithm processes the image sequentially in blocks of size 16x16 pixels, according to a quadtree coding scheme [10]. For each block, the variance and the mean are computed. If the variance is smaller than a prespecified threshold, which controls the compression factor, the entire block is encoded using its mean value. Otherwise, the original block is subdivided into four square subblocks and the procedure is repeated. The smallest block size of 2x2 pixels is not further subdivided. This method does not require floating point computation and can be implemented using a recursive program.

The performance of the *Visualization* basic operation was studied both considering the case in which the frames are copied directly from the main memory to the frame buffer, using the library routines made available by SunOS (pixrect()), and the case in which the frames are copied to the frame buffer through the X server. In this second case, the X server and the client program exchange data using shared memory, in such a way to avoid the overhead of interprocess communication.

6. DISCUSSION

Using the figures presented in the previous section, it is now possible to discuss the performance requirements of a distributed system to support the implementation of the MNDC. We consider the case of Sun Sparc 2 workstations and the case of Ethernet and FDDI networks.

In Table 2, we indicate the physical-layer throughput of Ethernet and of FDDI. It is evident that the transmission of uncompressed frames, which would require 24 Mbit/s, is not supported by Ethernet. Compressed video, instead, assuming a compression ratio of 15 to 1, can be transmitted in real time and only takes 16% of the network bandwidth. Supposing that 50% of the network bandwidth is available for multimedia traffic, up to 94 cf/s can be transmitted, corresponding to three video channels of 30 cf/s each. In contrast, again assuming that 50% of the bandwidth is available, FDDI supports the simultaneous transmission of up to two uncompressed video channels or as many as 31 compressed ones.

TABLE 2. – THROUGHPUT OF ETHERNET AND FDDI

| Ethernet | 1.25 Mbytes/s | 12 f/s | 180 cf/s | 6 video channels |
| FDDI | 12.5 Mbytes/s | 120 f/s | 1800 cf/s | 60 video channels |

Let us now examine the minimal number of operations that each video frame must undergo from the network interface to the visualization device. The first step is *reception*, which involve one memory copy of a compressed frame from the network interface to the workstation main memory; the second step is *protocol processing*, which may or may not involve the calculation of a checksum (in our experiments it does because we have used UDP); the third step is *decompression*, which may be performed either by specialized hardware or by software (in our experiments we have used software decompression); the fourth step is *visualization*, which includes the communication between the client program and the X server.

TABLE 3. –REQUIRED OPERATIONS INSIDE AN MNDC CLIENT

OPERATION	TIME IN MS
Memory Copy	0.66
UDP Processing	0.96
Decompression	34.48
Visualization through X	30.22

Table 3 shows the times required to carry out each of the steps above, as measured in our experiments. As expected, the most time-consuming operations inside the client workstation are those requiring the generation and/or processing of uncompressed frames, namely decompression and visualization. The decompression time can be reduced by using special hardware for decompression, while the visualization time is bounded by the speed of the frame buffer (30 ms per frame; see Table 1) and can be reduced by optimizing the X server path from the client program to visualization (notice from Table 1 that visualization directly to the frame buffer is almost three times as fast as visualization through the X server) and/or by adopting a faster frame buffer.

7. SUMMARY

We have presented the architecture of a distributed system in which a central facility, called a Media Network Distribution Center, serves a number of clients and handles simultaneously different multimedia applications. We have described a set of possible applications of such a system, such as interactive video, database access, video distribution and distributed classroom. We have outlined the design principles upon which a MNDC is based and we have introduced a set of specific requirements to be met.

We have then focused on continuous media data and in particular on video, taking the case of CIF video (352 × 288 pixels) as a case study and performing a set of experiments on video sequences of CIF frames to verify the validity of the design principles and the feasibility of proposed design solutions meeting the specific system requirements. Our experiments concerned the implementation of an MNDC in a local environment based on Ethernet networks. We focused on the client program and partitioned the client part of the MNDC system into a set of basic processing steps that must be carried out in sequence on each video frame as it moves from the network to the frame buffer. We have measured the performance of each of these processing steps, by running a set of specific experiments, and have reported the results. Although building a successful MNDC certainly requires additional basic research, on the basis of our experiments, we conclude that the MNDC architecture that we have outlined in Sections 1, 2 and 3 is a valid one.

REFERENCES

1. Anderson, D. P. and G. Homsy, A Continuous Media I/O Server and Its Synchronization Mechanism, *IEEE Computer 24*, 10 (1991), 51-57.

2. Ferrari, D. and D. Verma, A Scheme for Real-Time Channel Establishment in Wide-Area Networks, *IEEE Journal of Selected Areas in Communications 8*, 3 (1990), 368-379.

3. Fox, I. A., Advances in Interactive Digital Multimedia Systems, *IEEE Computer 24*, 10 (1991), 9-21.

4. Gilge, M. and R. Gusella, Motion Video Coding for Packet-Switching Networks – an Integrated Approach, *SPIE Conference on Visual Communications and Image Processing*, Boston, 10-13 November, 1991.

5. Gusella, R. and S. Zatti, The Berkeley UNIX 4.3BSD Time Synchronization Protocol, Computer Science Technical Report, UCB/Comp. Sci. Dept. 85/250, University of California, Berkeley, June 1985.

6. Gusella, R. and S. Zatti, The Accuracy of the Clock Synchronization Achieved by TEMPO in Berkeley UNIX 4.3BSD, *IEEE Transactions on Software Engineering 15*, 7 (July 1989.), 847-853.

7. Multimedia Information Systems, *Special Issue of IEEE Computer Magazine*, October 1991.

8. Liou, M., Overview of the px64 Kbit/s Video Coding Standard, *Communications of the ACM 34*, 4 (1991), 59-63.

9. Mills, D. L., Internet Time Synchronization: The Network Time Protocol, *RFC 1129*, October 1989.

10. Tanimoto, S. and T. Pavlidis, A Hierarchical Data Structure for Picture Processing, *Computer Graphics and Image Processing*, April 1989, 104-119.

11. Zhang, L., A New Architecture for Packet Switching Network Protocols, Ph.D. Thesis, Dept. of EECS Massachusetts Institute of Technology, July 1989.

12. Digital Multimedia Systems, *Special Issue of Communications of the ACM 34*, 4 (April 1991).

Next Generation Network and Operating System Requirements for Continuous time Media

Scott M. Stevens

Software Engineering Institute
Carnegie Mellon University

Abstract

Accessing massive multimedia databases will require multiple representations of those databases. Initial access may be through visual representations of the database. However, traversing numerous levels of tree-like structures will quickly find the user lost. Simple database queries may overwhelm users with information.

To overcome these problems the Advanced Learning Technologies Project at Carnegie Mellon University's Software Engineering Institute embeds in multimedia objects the knowledge of the content of those objects over several dimensions. With this model, variable granularity knowledge about the domain, content, image structure, and the appropriate use of content and image is embedded with the object. In ALT, a rule base acts as a visual director, making a judgement on what image to display and how to manipulate it. This provides the ability to present disparate text, audio, images, and video, intelligently in response to users needs.

It is difficult to move through information that has an intrinsic and essentially fixed temporal element such as video. While detailed indexing of video can help, users often wish to peruse video much as they flip through the pages of a book. Two techniques developed for this project will facilitate such searches. First, detailed, embedded knowledge of the video information will allow for scans by various views, such as by content area or depth of information. Second, partitioning multimedia data into smaller objects reducing bandwidth problems associated with accessing central data in large video files. Concatenation of logically contiguous files allows for seamless, continuous play of long sequences

A. Introduction

The volume of information becoming available to the user through continuous time media is such that it can no longer be assumed that the human-computer system is composed of an intelligent user accessing tractable amounts of static information. The new model must be one of intelligent-dynamic information aiding intelligent users in entertainment, learning, and working tasks.

The Object Lens and Athena Muse projects at the Massachusetts Institute of Technology, the Information Visualizer project at Xerox Palo Alto Research Center, the Course Processor project at Carnegie Mellon University, and the User Interface Designer's Assistant and the Advanced Learning Technologies projects at CMU's Software Engineering Institute have investigated solutions to the problems of storing, accessing, and visualizing multimedia information. Tasks that involve huge information spaces overwhelm both users and today's electronic workspaces. Accessing massive multimedia databases will require multiple representations of those databases. Initial access may be through visual representations of the database. However, traversing numerous levels of tree-like structures will quickly find the user lost. Simple database queries may overwhelm users with information.

To overcome these problems we embed in multimedia objects the knowledge of the content of those objects over several dimensions. Much current work treats multimedia objects, especially video, more as text with a temporal dimension [1,2]. The Advanced Learning Technologies (ALT) project at CMU's Software Engineering Institute has developed a multidimensional model of multimedia objects. With this model, variable granularity knowledge about the domain, content, image structure, and the appropriate use of content and image is embedded with the object. These often orthogonal descriptions of an information database promote both usability and accessibility. For example, in ALT an expert system acts as a visual director, behaving intelligently in the presentation of images. Based on a history of current interactions (input and output) the system makes a judgement on what image to display and how to manipulate it. This provides, for the first time, the ability to present disparate text, audio, images, and video, intelligently in response to users' needs.

B. Constant Rate Continuous Time (CoReCT) Media

Many researchers have noted the unique nature of motion video. Differences between motion video and other media, such as text, are typically attributed to the temporal nature of video. Every medium has a temporal nature. It takes time to read (process) a text document or a still image. However, each user does this at his or her own rate. Often it is possible to assimilate visual information holistically, that is to come to an understanding of complex information all at once. The creative process is similar, and need not be restricted to the visual domain. Mozart said that he conceived of his compositions not successively but in their entirety [3]. Clearly time is an intrinsic part of music. While Mozart may have "seen" his whole composition at once, the temporal aspect of a piece must have been present in that same instant. Subjective and real time may be similarly contracted or expanded, by the user, while reading text or viewing an image.

If one likens the scrolling of text to viewing a motion video sequence we see the real difference between video and audio and other media. For example, if a system fixed the scroll rate of text for a mythical average reader, say at 400 words per minute, a reader that read at even 401 words per minute would soon be out of synch with the text. Even a 400 word per minute reader would undoubtedly find passages that, because of complexity or interest, require a much slower rate. Obviously, no one would argue that the scroll rate should be fixed. But video and audio must be played at a constant rate, the rate at which they were recorded, to have almost any meaning at all. A user might accept video and audio played back at 1.5 times normal speed for one ore two minutes. Even if digital signal processing is used to maintain a near normal audio tone, it is unlikely that users would accept long period of such playback rates. Moreover, the information transfer rate would still be principally controlled by the system.

The real difference is that video has a constant rate output that cannot be changed without negatively impacting the user's ability to extract information from the stream. Video is a Constant Rate Continuous Time (CoReCT) medium. Its temporal nature is constant due to the requirements of the viewer/listener. Text is a Variable Pace Continuous Time Medium (VaPid). Its temporal nature only comes to life in the hands of the users.

Documents are, hyperlinks aside, also continuous. The difference between a text document and motion video or audio is that the constraints on scheduling and synchrony are much more severe for video and audio. User's performance can be adversely affected when text is presented at less than fifteen characters per second [4]. With video, users will not accept a wait of over one thirtieth of a second (or, bowing to international pressures, one-twenty-fifth of a second). A user may accept a few second delay between the appearance of a page of text and an associated but separately stored table or image. If the synchrony between video and audio is violated by more than about a fifteenth of a second users will both note this difference and find it unacceptable. But the continuity and synchrony differences between video plus audio and text are ones of scale, not of kind.

The differences between CoReCT and VaPid media become critical when a user is searching for information. The human visual system has an aptitude for quickly looking at an image or a page of text and finding a desired piece of information while ignoring unwanted information (noise). This is what makes flipping through the pages of a book a relatively efficient process. Even when the location of a piece of information is known a priori from an index, the final search of a page is aided by this ability. But the motion video analog of this process, fast forward, is not nearly so efficient.

It is difficult to move through information that has an intrinsic constant rate temporal element such as video. While detailed indexing of video can help, users often wish to peruse video much as they flip through the pages of a book. Current methods make this a difficult process. Analog videodisc scanning, jumping a set number of frames, may skip the target information completely. To be comprehensible, a simple scan such as a VCR's fast forward often takes too much time. Network and even bus bandwidth problems aside, displaying motion video at twenty times normal rate presents the information at an incomprehensible speed. And it would still take three minutes to scan through a one hour video!

Even if the visual system could make sense of such accelerated motion, a short two second shot would be presented in one tenth of a second. With human and system reaction times potentially adding to a second or more, significant overshoots will occur.

Audio is of no help. Beyond 1.5 or 2 times normal speed audio becomes incomprehensible. At faster playback rates frequency shifts make it inaudible. Even digital signal processing techniques to reduce the frequency shifts associated with high speed playback of audio fail at three or four times normal rate. With about 150 spoken words per minute one hour of video contains 9,000 words, which is about 15 pages of text. If the information being sought is from "talking head" video, such as a lecture, or worse yet from audio only, a comprehensible high playback rate of 3 to 4 times normal speed is a totally unacceptable search mechanism. Assuming the target information is on average half way through a one hour video file it would take 7.5 to 10 minutes to find. No user today would accept a system that took 10 minutes to find a word in 15 pages of text.

Three techniques are proposed for facilitating such searches. First, detailed, embedded knowledge of the video information will allow for scans by various views, such as by content area or depth of information. Second, video is partitioned into many separate small files. First pass searches retrieve a small segment of one to two minutes of video. Bandwidth problems associated with the transfer of large video segments are eliminated. Continuous play of a long sequence is accomplished by seamlessly concatenating logically contiguous files. Third, scans can be performed by changes in visual information, such as by scene change. Most compression schemes will permit easy analysis of scene change since major image transformations affect the compression and decompression algorithms. In cases where this is not efficient, embedded knowledge about the content of individual frames can substitute. Much like noticing chapter and section headings in a book while flipping pages, all three of these types of scans will permit information based scanning of CoReCT material.

C. ALT Paradigm for Use of CoReCT Media

It is obvious that both the type and the volume of information accessible by computers is growing at an exponential rate. What is less obvious is that much of this information should be available from different viewpoints (from different information types). A reader of a Shakespearean play will glean a perspective unavailable to a viewer of a video of that same play. Likewise the viewer gains a different perspective than the reader, especially if the viewer can compare more than one director's and actor's interpretation of that same play. What is even less obvious is that different descriptions of a multimedia database are necessary to promote usability and accessibility.

Current multimedia operating system descriptors are little different than for a graphics file. The information necessary for use of the data includes traditional "file control block" information such as the name (usually of the scene or story), the size of the file in bytes, the creation date, and the last modification date. The type of media should be available (e.g. audio, video, or video with audio). Lastly, media-specific descriptors must be recorded. For audio these include, sampling rate, filtering, attenuation, and compression algorithm and for video these include frame rate, position, hue, saturation, and compression algorithm used.

With this information an application that knows which video or audio segment is required can perform basic functions. For video with audio these would include, play, setting the position, clipping or scaling, if need be performing color corrections, and adjusting volume and tone. These functions form the complete set used by the typical multimedia application of today and also is the limit of most system support for digital multimedia.

Beginning in 1987 the Advanced Learning Technologies Project (ALT) at the Software Engineering Institute of Carnegie Mellon University began developing a radically new paradigm for interactive multimedia [5]. For more sophisticated applications such as ALT, additional descriptor requirements include, abstract representations of the data (as opposed to simply the physical representations of the data) and information identifying any procedural actions to take following access of the data (i.e., control information).

The goal of the ALT Project was to create a virtual workplace where a user could learn about and experience a software engineering technique called an inspection. Inspections are structured meeting where development workproducts are systematically examined to identify defects. Users are placed in a virtual building where they may walk to, amongst other rooms, a multimedia library, their own office, and a conference room where they may participate in a simulated inspection.

In the conference room the user sees three members of the inspection team enter the room, sit down, and begin the discussion. The user may talk to the video personae at any time through a natural language interface, or the user may simply sit back, look and listen. Of course, as in a real meeting, the personae will sooner or later ask the user to participate. This is a high fidelity simulation of a group discussion. The traditional multimedia/interactive video paradigm is play a video sequence, stop, wait for user input, branch. A more appropriate term for this in "interrupted video."

Consider a hypothetical simulation of a conflict between two parties, used to teach mediating skills. In the canonical interrupted video version the output data may consist of two conversations, one for each party. The student can see the two conversations played in sequence, or may see only one. This results in four possible permutations, as shown in figure 1. If on the other hand each of these conversations is broken up into smaller, meaningful pieces, then there are numerous possibilities to alter the information presented by the simulation according to the student's mediating actions (input). The conversations might be partitioned into expressions of emotion, position statements, monologues supporting a position, and conclusions. Depending on the student's effectiveness as a mediator, conflicts may be resolved and compromises made. The finer granularity provides for the ability to create a much richer, more highly interactive simulation than with a simple interrupted video paradigm.

Partitioning video into small files can facilitate the creation of high fidelity simulations. However, as the complexity of the simulation increases, there exists a greater need to augment audio and video building blocks with descriptive data to make them usable by the simulation. The ALT system uses this paradigm to implement a code inspection simulation.

The ALT system simulates four participants and the interactions between them in an inspection. ALT provides the ability for the user to take any role in the process with the system simulating the other three participants. This is achieved, in part through a rule-based expert system that was developed to model the participants. The expert system defines the "personalities" and controls the dialogue between the simulated members of the inspection team and the user. Throughout the simulation the expert system continuously composes the video and the audio.

The ALT system uses Intel's Digital Video Interactive (DVI). DVI permits up to seventy-two minutes of full screen, full motion video to be stored on one compact disc. The simulation of the inspection requires approximately ten hours of audio and two hours of motion video with audio. We are able to increase the apparent storage by saving fractional screen images and composing the full image from still images plus motion sequences. In analog video, images are fixed during production and post-production. For practical purposes, the image on a videodisc or video tape cannot be altered significantly during playback. If the information of interest takes up only a fraction of the screen one full frame of analog video storage must still be allocated. Since images are stored digitally in DVI neither of these limitations is true. We are able to compose the visual image during playback (see figure 2). Since we wished to store the images of the actors separately we save significant storage space by saving only the section of the images that is of interest.

The rule base consists of over 8,000 lines of OPS/83 code implementing approximately 200 complex rules and associated procedures. It makes decisions in areas such as who should speak, the tone they should speak with, the content of what is to be spoken (context space search), and who is the persona speaking to. The rule base also controls the conversation and models the personalities of the participants.

The rule base uses the ten hours of audio, two hours of video with audio, and several thousand still images to dynamically compose the scene. The audio and video alone consist of over five thousand objects. All of these objects are organized in a multi-dimensional structure.

The rule base composes the visual and spoken dialogue from a four dimensional multimedia database (see figure 3). The first dimension is the topic under discussion (context space). The second dimension is the speaker (persona). The third dimension is the specificity within the topic. This is related to the temporal aspect being modeled in the conversation. (As people speak on a topic, they tend to get more specific and build on what was said previously, much as a text builds on previously chapters.) The fourth dimension is affect (emotion).

Abstract information about objects in ALT varies in granularity from scene headers describing information that is globally constant for a set of frames, to frame headers, information that is local to a single frame. Figure 4 suggests one digital file representation to implement both scene and frame headers for multimedia objects. In the current implementation, descriptions cannot be tagged to individual frames of motion video. This limitation was imposed by prior operating system limitations of the DVI environment. However, for still images and files of multiple still images that are animated to create a limited motion video, information at the frame level is available.

Figure 5 depicts the scene header information for the each of the video objects available to the rule base. Along with traditional "file control block" information, embedded in each object is information on camera angle, field of view (both objective points of view), character, topic, specificity within topic, tone, who the scene is addressed to, opinion, emotional subject, discussion resolution status, pointers to other topics, and gesticulation (all subjective points of view). The objects under the subjective points for both scene and frame (in the image database) header define the dialogue element's location in the four dimensional space. The rule base determines the points in this space which are needed to compose the appropriate dialogue.

An interesting part of the rule base is called Hitchcock, the visual director of the system. With digital video, we can manipulate the images when we play them back. In fact, for the ALT system we must in order to compose scenes with different personae in them. We want the expert system to behave intelligently in presenting these images, much like directors do today when shooting a scene, or when they edit it. But in ALT, the system does it during playback. In the simplest case, Hitchcock must determine who the speaker is addressing and display appropriate images (i.e. the actor must be looking at who he is talking to). More interestingly, if the user is dominating a conversation, the system may present a camera angle of the participants on the screen which is slightly lower. Years of experience and many studies have shown that images such as this tend to portray the viewer as more dominant [6]. Figure 6 three shows the scene and header information for each image.

In ALT there is a surrogate travel interface which allows the user to walk around the workplace and move into and out of a variety of rooms (see figure 7). The image sequences for this are contained in one scene, along with the header information shown in figure 8. With the addition of subjective information it becomes easier for the application to know what room the user is near and thus what rooms are accessible. This paradigm gives the application detailed information about the virtual world by associating that information with the data. Updates to the data, such as newly accessible space, are automatically passed to the controlling application.

Since the ALT Project began, other researchers have discussed related schemes, defining cinematic primitives for use by multimedia systems [7]. Unfortunately their current implementations use analog videodiscs as video sources [8]. In these systems, areas of interest are, at least in part, identified by frame number, even to the point where the user may need to use frame numbers for video access. With 54,000 frames on a single half hour videodisc it is hard to see how users can effectively interact with the video unless a form more abstract than frame number is used.

Having the video and audio in digital form is crucial to the ALT paradigm. The descriptions of an information database, being closely tied with the information, should remain with the information data rather than be implemented in an application working on the information. Abstract descriptions of multimedia greatly enhance usability for both applications and users.

D. Future Directions and Conclusions

The current implementation of ALT could be significantly improved with additional digital video operating system functionality. The principle added functionality needed is the ability to access multiple data streams from single or multiple sources. In our system, this capability would allow more than one reviewer to move at once and more than one audio stream to be played at the same time. With the addition of fully integrated scene and header information image databases could be stored as a series of motion video files. This header information should not be overly specified, e.g., a restrictive maximum size should not be enforced. These capabilities would provide better visual flow, improve flexibility, and save storage via video compression. The overall objective of these needs is to be able to manipulate multimedia data dynamically in different ways.

An improved simulation scenario might be the following:

Suppose that in the middle of a comment one of the simulated personae, Andy, directs a caustic remark toward one of the other personae, Michelle. Andy finishes neutrally by asking for further comments from the other people in the meeting.

The rule base would like to:

- Play the audio for this dialogue.
- Show video images of the Andy looking frequently at the code but occasionally at the other reviewers.
- At the appropriate time (noted by audio scene's frame headers), show video images of the Andy looking at the Michelle and being aggressive.
- Simultaneously, show reactions of the Michelle.
- Conclude with video images of the Andy looking at each reviewer

Even without these added capabilities, the ALT paradigm provides for:

- The ability to associate common information to a set of frames via a scene header, and to differentiate frames according to information in a frame header
- The ability to access this header information quickly without having to process the actual scene or frame data associated with this information
- The ability to output selected frames from a given scene without any display artifacts from the unselected frames, and with minimal timing constraints

With ALT paradigm, variable granularity knowledge about a domain, content, image structure, and the appropriate use of content and image is embedded with the object. These often orthogonal descriptions of an information database are shown to promote both usability and accessibility. This provides, for the first time, the ability to present disparate text, audio, images, and video, intelligently in response to users' needs.

Acknowledgements: I would especially like to thank Michael Christel who is the software engineer on the Advanced Learning Technologies project and contributed significantly to every phase of the project. His review of this article and help in preparations of its figures was invaluable. This work was sponsored by the U.S. Department of Defense

References

1. "A Construction Set for Multimedia Applications," Matthew E. Hodges, Russell M. Sasnett, and Mark S. Ackerman, *IEEE Software*, January 1989
2. "Intermedia: The Concept and the Construction of a Seamless Information Environment," Nicole Yankelovich, Bernard J. Haan, Norman K. Meyrowitz, and Steven M. Drucker, Brown University, *IEE Computer*, January 1988
3. Hadamard, J., *The Psychology of Invention in the Mathematical Field.* Princeton University Press, 1945
4. "The Effect of VDU Text-Presentation Rate on Reading Comprehension and Reading Speed," Jo. W. Tombaugh, Michael D. Arkin, and Richard F. Dillon, Proceedings of ACM CHI '85 Conference of Human Factors in Computing Systems, 1985
5. "Intelligent Interactive Video Simulation of a Code Inspection," Scott M. Stevens, *Communications of the ACM*, July 1989 Volume 32 Number 7
6. Kraft, R. Mind and media: The psychological reality of cinematic principles. In *Images, Information & Interfaces: Directions for the 1990's*, D. Schultz and C.W. Moody, Eds. Human Factors Society, New York, 1988
7. "Cinematic Primitives for Multimedia," Glorianna Davenport, Thomas Aguierre Smith, and Natalio Pincever, *IEEE Computer Graphics & Applications*, July 1991
8. "Parsing Movies in Context," Thomas G. Aguierre Smith and Natalio C. Pincever, USENIX-Summer '91, Nashville, TN

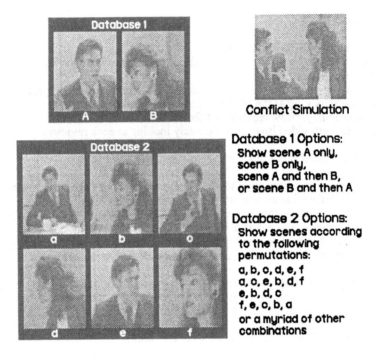

Figure 1. Effects of Granularity on Simulation Fidelity

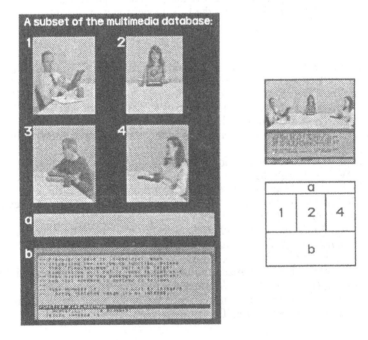

Figure 2. ALT Compositional Components

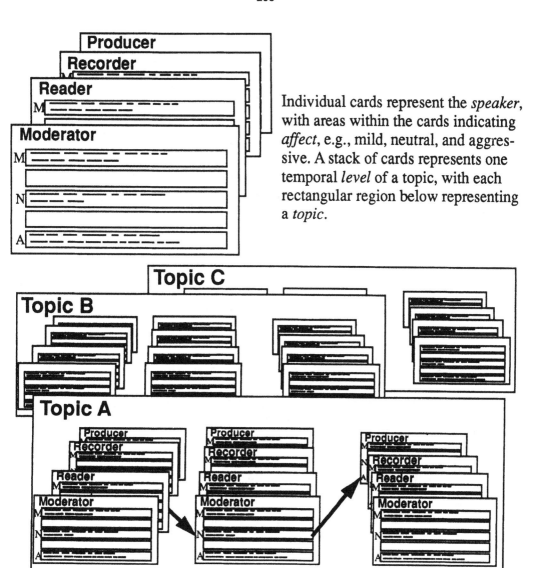

Individual cards represent the *speaker*, with areas within the cards indicating *affect*, e.g., mild, neutral, and aggressive. A stack of cards represents one temporal *level* of a topic, with each rectangular region below representing a *topic*.

Figure 3. ALT's Four Dimensional Dialogue Structure.

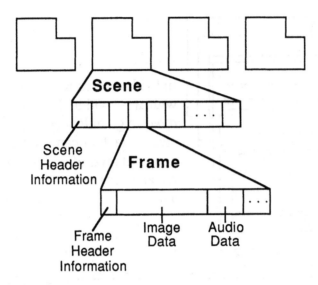

Figure 4. ALT file representations for scene and frame headers.

Scene Header for Inspection Video Dialogue:

- Traditional "file control block" information
    - Name of scene
    - Size in bytes of scene
    - Creation date of scene

- Media type (audio, video, audio and video)

- Media-specific descriptors
    - Audio (sampling rate, filtering, attenuation,...)
    - Video (frame rate, position, compression algorithm,...)

- Objective point of view
    - camera angle
    - field of view

- Subjective point of view
    - character
    - topic and specificity within the topic
    - tone
    - who the scene is addressed to
    - opinion
    - the personal focus (emotional subject)
    - resolved status
    - bridge to another topic
    - gesticulation

Figure 5. ALT video scene header information.

Scene and Frame Headers for Inspection Images

- **Scene header:**
  - Objective point of view
    - camera angle
    - absolute location
    - field of view
  - Subjective point of view
    - character
    - emotional level

- **Frame header:**
  - Objective point of view
    - hue and saturation
  - Subjective point of view
    - direction of gaze
    - gesticulation

Figure 6. ALT image scene and frame header information.

Figure 7. One still image from ALT surrogate travel interface.

```
┌─────────────────────────────────────────────────────────────────┐
│ Scene and Frame Headers for Surrogate Travel:                    │
│                                                                   │
│ • Scene header:                                                   │
│     • Objective point of view                                     │
│         • Field of view and depth of field (constant throughout   │
│           the scene)                                              │
│     • Subjective point of view                                    │
│         • Geographic boundary information                         │
│                                                                   │
│ • Frame header:                                                   │
│     • Objective point of view                                     │
│         • Camera angle                                            │
│         • Absolute location (x, y, z)                             │
│     • Subjective point of view                                    │
│         • Near room(s)                                            │
│         • Accessible room(s)                                      │
│         • Indication of frame as a room entrance                 │
└─────────────────────────────────────────────────────────────────┘
```

Figure 8. Surrogate travel scene and frame headers.

Dynamicity Issues in Broadband Network Computing

Steven Gutfreund
Jose Diaz-Gonzalez
Russell Sasnett
Vincent Phuah

Distributed Multi-Media Applications Project
GTE Laboratories, Inc.
40 Sylvan Road
Waltham, MA 02254 USA
jdiaz@gte.com, sgutfreund@gte.com
vphuah@gte.com, rsasnett@gte.com

1.0 Abstract

A compound multi-media document consists of a mix of continuous media elements (audio, video, instrument sensors, etc.) and computational elements that display, chart, record, and process the media elements. In a network environment, where these elements exist at different locations on the communication fabric, there are obvious synchronization problems. However, there is also a class of problems we call dynamicity issues, which are concerned with the need to: reconfigure the network, add and remove connections that bind the elements, deal with mobile elements. In this paper we characterize the different types of dynamicity that occur in networked compound documents and present a top-level architecture for managing dynamicity.

2.0 Background

For the last two years, our group has been involved in the design of multi-media environments. This work has consisted of two parts:

1. We have been an active participant and contributor to the design and development of the Athena Muse multi-media authoring system by the Visual Computing Group at MIT Project Athena [Hodge89]. Our collaboration also consisted of the loan of one of our group members, one of the original designers of Muse, to the Muse project.

2. The construction, in cooperation with other research groups at GTE Laboratories, of a hybrid, digital broadband network capable of supporting multi-media connectivity among our offices. This hybrid network utilizes a (GTE proprietary) SONET-compatible digital switch that was used to provide video connectivity to/from multiple sources, a computer controllable ISDN switch for voice connections, and an Ethernet network for data connections. We have ported and made modifications to Muse to make use of this network and created prototypes with Muse of cooperative documents for multi-user, multi-media applications.

3.0 The Next Effort

Our next effort involves the migration from a three network interconnect (ethernet, SONET, ISDN) to a single unified ATM-based [Miner89] gigabit LAN. In this configuration we will be able to provide a much tighter degree of integration in our compound multi-media applications. For example, we will be able to provide real-time synchronization between video clips and computer simulations, or to have a computer visualization driven by real-time instrument sensors connected to a laboratory experiment.

In order to create such applications one may have to combine many different continuous (e.g. audio, video), discrete (e.g. data bases), and computational (e.g. visualization programs such as AVS [AVS91]) elements. Each of these elements may reside on a different node. The bandwidth, latency, and quality-of-service (e.g. error rate) between nodes can have a significant impact on the design and performance of an application. Furthermore, various *dynamic* changes in the network, software topology, loss/addition of video elements, and movement of the user dictate the need for a flexible element interconnection architecture. However, before creating an interconnection architecture for multi-media applications we need to examine and understand the different aspects of these dynamic network changes.

4.0 Dynamicity

Dynamicity in distributed multi-media applications exist at four levels:

1. Transportability

2. Mobility

3. Reconfigurability

4. Plasticity

4.1 Transportability

Transportability involves the need to move applications from one network to another. When an application moves one has to re-bind it to the various media, data, and compute elements that participate in the application. This re-binding can be more complicated than what can be accomplished with a simple directory service. Because of the different topology, throughput, and line quality at the new location, different choices may have to be made concerning which compute and media elements will be part of the application.

4.2 Mobility

Mobility is an issue concerned with the migration of active applications. One would like to be able to download a running application from an office workstation to a multi-media notebook connected via a cellular phone, and then move to a new location (e.g. home) where the application would be transferred to the home system. If the application is shutdown then this is a transportability issue. However, if the application needs to be active during the migration phase, then we have a mobility issue. Mobility issues are more difficult than transportability issues in that the binding of the multi-media document needs to be constantly re-checked. In a transportable document, binding only needs to be performed at start-up.

4.3 Reconfigurability

The location in the network where the user is running the system is only one element of dynamics. Another involves the dynamic changes to the topology of the network and the addition and loss of compute and media resources. A reconfigurable multi-media application can dynamically re-organize its granularity of distribution to make use of more compute resources. It is also capable of discovering better links to equivalent classes of media and data resources.

4.4 Plasticity

Compound multi-media applications are usually not created from scratch. They involve the assembling of various pre-built elements such as video, audio, visualization programs and databases. Pre-built elements come with their own set of user interface widgets. When creating the user interface for the compound application, one does not want to merely concatenate the underlying user interfaces. Instead one would like to create one unified interface. Such an interface could synchronize related controls on several different underlying components. Additionally, a user should be able to customize an interface to make use of more sophisticated interaction devices that are available at different user stations on the network.[1] We call such customizable interfaces plastic interfaces.

5.0 SHOWME

SHOWME (SHared Object-oriented Workbench for Media Elements) is a multi-media environment that attempts to address the dynamicity issues that arise in distributed applications.

SHOWME applications consist of a set of *elements* that are distributed over a broadband network. An element is a single uniform homogenous term to describe the items that are composited to form a SHOWME application. Elements can be continuous data resources such as video or audio sources. Alternatively, an element can be a discrete data source such as a data base. Elements can act as sources or sinks of information in SHOWME composite documents. For example, a video element can be the source of an application and the sink can be a database (figure 1).

Elements form the fundamental unit of distribution of SHOWME applications. In other words, a SHOWME application is constructed by assembling a set of elements, where the elements reside at different locations in a distributed environment. Elements provide a single uniform component for describing and specifying an application.

Elements can be computational. That is, while many elements in a SHOWME application are mere data sources or sinks, some are computational elements that transform data. Data is piped through a computational element from its source to its sink. These sources or sinks can in turn be computational elements. Thus, one can create applications which con-

---

1. With the acceptance of the X window environment, it has become much easier to detach an application from its user interface and attach a new set of virtual widgets.

sist of a stream from a video source, computational transformational elements that extract events from the video stream, and windows that act as sinks to display the video and the extracted event stream.

Figure 2 shows the basic SHOWME architecture.The binding of elements is handled by the Resource Dispatcher. The Dispatcher is a hybrid of Linda [Carri89], the x-kernel [Peter90], and the Message Backplane [Reiss90]. It provides a distributed shared memory called tuple space. Objects in tuple space can be located by pattern matching, range queries, or user procedures. Elements of a multi-media application register themselves in the shared memory. Dynamic binding of elements is performed by pattern matching and queries to the Dispatcher (figure 3). The actual data flow between continuous media and compute elements can either be through the shared memory or over out-of-band channels (e.g., ATM virtual circuits [Miner89]) depending on the bandwidth needs of the connection.

Connected to the Dispatcher are the Dispatch Analyzer and Dispatch Controller. The Analyzer is used to conduct performance tests for different Resource Dispatcher implementations. It collects throughput, latency, and error rate statistics and can chart performance for different packet and burst sizes. The Dispatch Controller is used for tuple space maintenance. It can browse, add, and remove tuple spaces in an individual Dispatcher. It can also be used to change hashing, layout, and storage parameters associated with a tuple space. In the future, it will be used to manage data segmentation among multiple Dispatchers.

Since elements are frequently pre-built, and sometimes are commercial off-the-shelf software packages, a wrapper must be applied to elements to re-direct their data sources and sinks to tuple space. This is performed by the Bonds. There are four different types of information that a Bond will forward to an element from tuple space: control, data, widget, and script.

Control data are tuples that are used for starting and stopping elements. A tuple can be placed in tuple space to start the execution of an element. Other control tuples in tuple space will specify where to obtain the executable of the element, which node will run it, what the command parameters are, and what the X-resource defaults are.

Data tuples are used to specify the input/output files that the element uses (e.g. standard input/output). The data tuple can either directly contain the entire contents of the file, or specify a handle for an out-of-band ATM circuit (typically a direct channel to a persistent data base).

Widget tuples provide values for the widgets that the element needs. Each action that the user performs on his interface (e.g. a button press) will result in a widget tuple being placed in the Dispatcher. The Bond-wrapper looks for these tuples and forwards them to the element disguised as a normal X-event.

Lastly, the user can write scripts in a high level language (a derivative of the Muse script language [Hodge89]). Scripts are typically used to automate widget actions. For example, a script can produce a stream of widget tuples that simulate user actions. Scripts can also be used to produce control and data tuples, however there are limits to the scope of the scripting language.

6.0 Conclusions

Pattern-directed binding is very flexible. Binding between elements occurs at run-time allowing applications to be both mobile and transportable. Furthermore, compute tasks can be registered in shared memory and dynamically dispatched to compute servers - allowing a degree of reconfigurabily of an application based on the presence of more compute resources. Media can be bound by type, i.e. if one is on a network link that supports non-compressed HDTV video one can bind directly to an HDTV feed. Otherwise, the pattern-based query may result in a binding to a lower grade of service such as compressed NTSC. Plastic interfaces can be created by either substituting widgets that produce equivalent widget tuples or by using scripts to simulate the invocation of widget controls.

We believe that the SHOWME architecture shows great promise in being able to meet the dynamicity goals outlined in the previous sections. We have currently implemented the central parts of Dispatcher system and preliminary performance studies have been very promising.

7.0 Acknowledgments

We wish to thank Dean Casey and Harry Mussman for their contributions in setting up our SONET video network. We would also like to thank Dave Decker, Bill Griffin, and Jerry Jones for providing the management support and climate conducive to this research.

References

[AVS91]   AVS, Scientific visualization package, Stardent Computer Corp, Newton, MA.

[Carri89] Carriero, N., and Gelernter, D., "How to Write Parallel Programs: A Guide to the Perplexed," *ACM Computing Surveys*, Vol. 21, No. 3, pp. 323-357. September 1989.

[Gutfr91] Gutfreund, S., "Integrating Distributed Visualization Systems with Multi-Media," *Workshop on Computer Graphics in the Network Environment, Proceedings of SIGGRAPH '91*. July 1991.

[Hodge89] Hodges, M., Sasnett, R.M., and Ackerman, M.S., "A Construction Set for Multimedia Applications," *IEEE Software*, Vol. 6, No. 1, pp. 37-43. January 1989.

[Miner89] Miner, S., "Broadband ISDN and Asynchronous Transfer Mode (ATM)," *IEEE Communications Magazine*, Vol. 27, No. 9, pp. 17-24. September 1989.

[Peter90] Peterson, L., Hutchinson, N., O'Malley, S., and Rao, H., "The x-kernel: A Platform for Accessing Internet Resources," *IEEE Computer*, Vol. 23, No. 5, pp. 23-33. May 1990.

[Reiss90] Reiss, S., "Integration Mechanisms in the FIELD Environment," *IEEE Software*, Vol. 7, No. 4, pp. 57-66. July 1990.

214

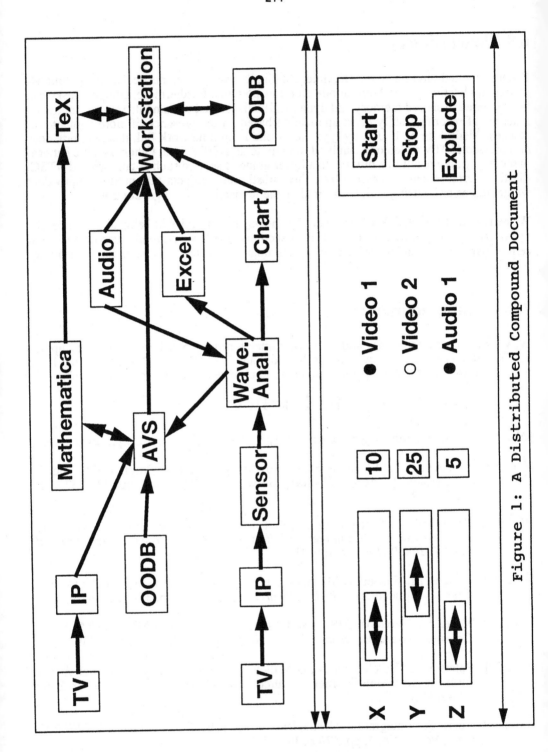

Figure 1: A Distributed Compound Document

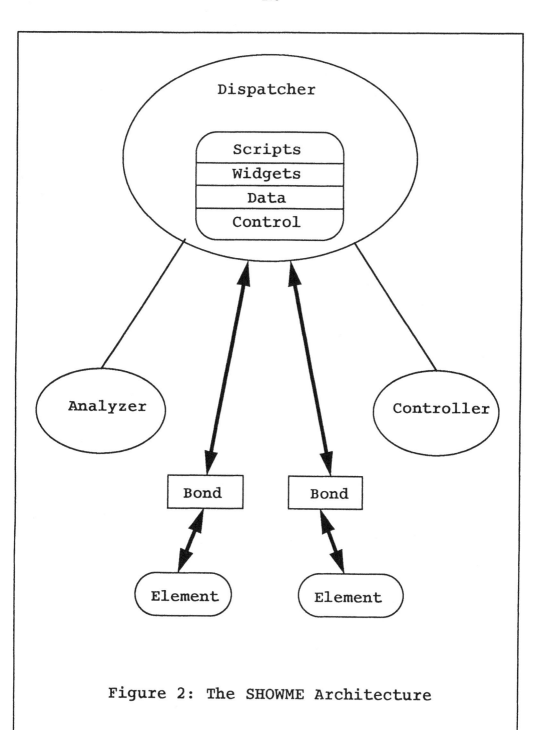

Figure 2: The SHOWME Architecture

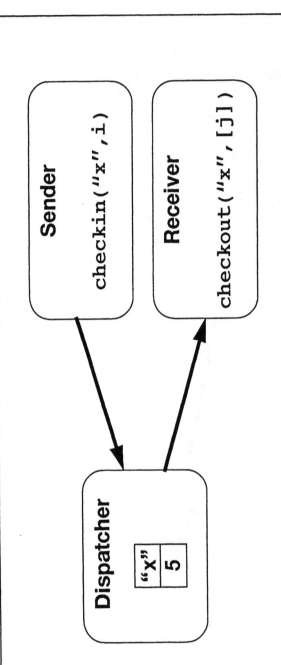

Figure 3: Pattern Matching in Tuple Space

MANAGING MULTIMEDIA SESSIONS ON A PRIVATE BROADBAND COMMUNICATION SYSTEM

An object-oriented transaction-oriented approach including signalling aspects

Ralf Cordes   Dieter Wybranietz   Rolf Vautz

Telenorma GmbH
Bosch Telecom
Advanced Systems Development Division
Mainzer Landstraße 128 - 146
D-6000 Frankfurt/Main 1

1 Scope

Multimedia communications comprising the exchange of audio, video and data information will play an increasingly important role in the telecommunications market (2,4,8,12,15). These new applications set up new requirements for the communication infrastructure ranging from the physical interconnection to the management and control of these services (1,3,5,7,11,16). Different classes of multimedia communication services are discussed as a case study which is based on a prototype of a distributed broadband communication system. By an advanced distributed multimedia application we show how timing and structuring requirements, session management and synchronization are supported by distributed control software including signalling and call control.

2 Scenario and Outline

Assume an enterprise-wide communication network of a publishing company. Using a multimedia terminal a reporter initiates a video conference from site T2 with his partner, who is the editor of a journal for which the reporter had written an article (see Figure 1).

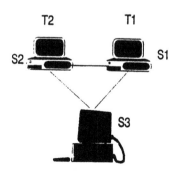

Figure 1: The Publishing Scenario

Before, the reporter had received his reviewed and annotated document (voice and text annotations are possible) from a multimedia document server (site S2). The editor sitting at site T1 accepts the video conference request and the communication starts. During this session the editor mentions that he has made some additional comments on the paper which he would like to discuss with the reporter. This version of the article is sent via a file-transfer to site T1 from site S1 where it was stored. By comparing the different versions using a joint pointing device which is visible on both screens they work cooperatively on the document. Assume further that during this conference the editor proposes to look up information stored in a database to clarify controversely discussed items. The editor (T1) initiates a query against a distributed database management system (DDBMS) located at sites S1, S2, and S3; whereas the global management resides at site S2. The request and the response from the DDBMS are displayed simultaneously on both screens. During the whole conference, both partners can see each other in a dedicated window on the screen.

This is one example of a typical future application scenario employing multimedia information exchange. There are a variety of other examples including distributed language laboratories, travel agencies, hospitals, education etc. as compared to todays communications facilities (16). All of these applications have in common that during one "logical" call several basic connections exist. For example, in the scenario presented above one-to-one audio/video and data connections have to be established or to be released. Thus, a multimedia call does not just consist of several connections, but these connections are subject to change dynamically. Take the database query of the sample application given, the signalling and call control software of a future communication system therefore has to offer primitives and mechanisms to handle these new requirements.

In particular, there are some main problems to be solved. Take as one example the establishment of the videophone connection. If the two videophone terminals offer different qualities, both sites have to agree on a minimum context with a special kind of Q.o.S. Therefore, we have to use a special type of negotiation by setting up the connection provided by signalling primitives. On the other hand the multimedia retrieval call against a distributed database has to be supported by special protocol primitives offering a multilevel transaction protocol. These protocol primitives should provide the facility of setting up connections from a third site, in our case the DDBMS has set up connection to its subsystems which are located somewhere across the network. To avoid a bottleneck at the DDBMS site, every subsystem should be able to set up connections to the requestor sites. With these functionalities we are able to cascade connections in a multicast- and broadcast-oriented manner providing different levels of end-to-end control.

To summarize, we have to use new techniques for system software supporting multimedia communication sessions on private communication systems. Examples are:

- in-call bandwidth modification
- negotiation of communication contexts with suitable Q.o.S.
- integration of servers providing cascaded connections with end-to-end control
- transaction-oriented protocols providing logging, roll-back, and security aspects
- object oriented structuring of generic application and service elements.

After giving a rough impression about our object oriented approach, we show the timing diagramme of the possible transactions supporting the scenario showing above. After discussing the architecure of our underlying communication system we present the appropriate primitives for the multimedia signalling protocol.

3. Structuring of generic service and application components

To gain more flexibility by integrating services on a multimedia terminal we have to structure the multimedia service components in an appropriate manner (4). This approach should be independent from a dedicated application and should provide generic components for different applications which are easily configurable to complex services and/or applications. These components are called generic application and service components. For us the concept of hypertext in combination with an object oriented approach of structuring the multimedia information units offers the suitable features meeting the requirements mentioned above. These generic multimedia application components are running on multimedia terminals (5) connected to a basic communication system.

In particular, we distinguish between object classes for composite (complex) objects (so called *PAGES)*, for monomedial objects (called *PARTICLES*), for linking and anchoring objects, and for telecommunication service elements. Excerpts from our class structuring is given in the Figure 2. The conceptual basis is given by the DEXTER model (6,10) serving as a framework for a network wide information information service (open link server) and the object oriented structuring of multimedia material given by the MHEG proposals (9). According to the DEXTER model we distinguish between the

- presentation,
- storage, and
- physical

layer for the generic components.

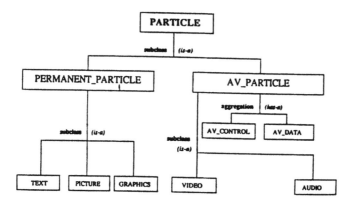

Figure 2: Excerpts from the Class Structure

The instantiated objects in the storage layer of our system have a dedicated slot providing the default presentation of this object. The presentation manager decides whether the default presentation should used or to present it according to the calling application or service context. Each *PARTICLE* contains the information about connections (links) to other particles and/or pages. These *LINKS* are also first class objects and can be subdivided into two classes. One class is dedicated to the traversal of the network wide information pool (*NAVIGATIONAL LINKS*); the other one supports the synchronization of media streams, especially for connecting time independent particles with time dependent ones (*SYNCHRONIZATIONAL LINKS*). In more detail, we subdivide the *NAVIGATIONAL LINKS* according to the source and destination of the link like *PAGE, PARTICLE*, and *TELECOMMUNICATION ELEMENT*. Using this extra structuring of linking information we are able to encapsulate and hide typical telecommunication related information for services and components to provide dialing, setting up connections, and accessing protocols like FTAM or MHS.

4 Transaction oriented Support for Multimedia Services

The object oriented structuring of service and application components is embedded in a multilevel transaction system (13,14) which offers:

-- *Usage of the flexibility of an ATM network*: To provide flexible allocation and release of bandwidth for parallel or modified virtual channels during one physical multimedia session.

-- *Support of parallel processing* according ot the object oriented structuring of the generic service components based on message passing.

-- *Avoiding of bottlenecks* during one multimedia session by flexible routing and flexible allocation of network pathes using multipoint protocol primitives.

-- *Usage of fine grained set-up and roll-back functionalities* according to the touched objects by appertaining of subtransactions on one special level.

We will demonstrate our approach by analyzing the scenario described in Section 2 and presenting a diagramme (see Figure 3) of possible transactions and their levels.

The begin of the multilevel transaction is the request from T1 at the global management of the DDBMS at site S2: *req(S2)*. S2 has to invoke the subsystems at S1, S2, and S3 with the dedicated query for each site: *req(Si)*. In parallel on each site the search process runs. These processes produce the result of the query which will be delivered in parallel to the sites T1 and T2: *send(Ti)*. If the information has arrived, an acknowledgement is send to the sites Sj from every Ti: *ack(Sj)*. If the processes running at sites Sj dealing with the query are finished without an error, an acknowldegement is send the global management at S2 from every site: *ack(S2)*. At the end if there are no errors the calling site T1 will receive an *ack(T1)* and the whole distributed query has been managed by the system.

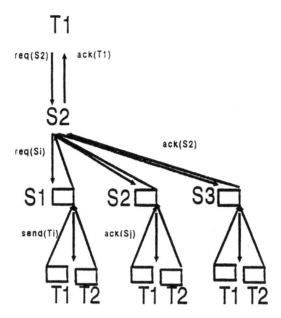

Figure 3: Timing Diagramme of Transactions in Scenario

5 Architecture of the communication system

Our considerations are based on an advanced distributed communication system for the customer premises market. Such a system is currently being developed in the course of the DAMS project (Dynamically Adaptable Multiservice System). The main objective of DAMS is to address the problems associated with the integration of services which are based on either circuit-switched or packet-switched techniques within the business environment (see Fig. 4). DAMS will provide for both delay-sensitive telephony services and time-indepented data services a fully integrated and cost-effective system which will fulfill future customer requirements. Further goals are flexible use of bandwidth, adaptability to changing user demands, increased realibility and availability.

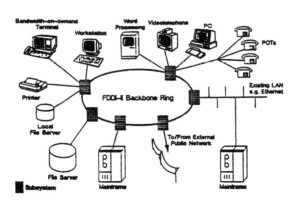

Figure 4: The DAMS Architecture

In order to cater for evolving customer requirements, flexibility has been a major concern in the specification of the system. It consists of autonomous subsystems, which can operate in a stand-alone mode or can be connected by a backbone system to build large networks or extend the system to greater distances. So-called DAMS nodes which can adapt to various interface types (e. g. ISDN basic access, ISDN Primary Rate Access, Ethernet, Token Ring, IEEE 802.9 IVDTE Terminals etc.) offer a local switching capacity and can be interconnected by a backbone system. The DAMS backbone unit is based on the FDDI-II standard defined by ANSI-subcommittee X3T9.5. In principle, the DAMS nodes can be interconnected by different topologies and transfer technologies. The bandwidth for the interconnection of the subsystems can be extended by the use of parallel backbone systems or a backbone with higher transmission rate.

The DAMS node internal switching is realized by a special hybrid mechanism operating on a bus system. In order to be prepared for the new standard in telecommunication, the evolution of the DAMS system concept towards ATM has already been considered. The switching principle of DAMS allows the integration of isochronous as well as non-isochronous data switching. The hybrid ATM/STM local switch will be achieved by introducing a connection-oriented switching technique based on the ATM-header structure defined by CCITT. An ATM-header is added to each incoming packet which is then switched based on a contents of the virtual channel identifier (VCI). The DAMS node internal signalling and control procedures are supported by signalling protocols based on the ATM-cell structure where these cells are routed using the VPI/VCI contents. The allocation of non-isochronous slots is controlled by a distributed media access protocol. Packets gain access to the bus by using non-isochronous slots. The isochronous traffic is circuit-switched between the related units allowing fixed and variable bandwidths (in terms of multiples of the basic access rate 64 kbit/s).

The distributed hardware is controlled by a distributed operating system called PordOS which fulfils the above mentioned requirements. In this context, support of real-time processing, migration of system modules during operation as well as load sharing capabilities are provided. The operating system is composed of the kernel and a set of services running on top of the kernel. The kernel provides process management, communication facilities, exception handling, I/O-support, memory management as well as test and debug aids. The services on one hand implement the environment needed to run the application programs and, on the other hand, mechanisms and strategies to support advanced systems features. PordOS also provides a good basis for the implementation of high level structuring facilities as, for example described in (17).

6 Signalling Aspects

The system described above provides basic mechanisms to reserve isochronous bandwidth for arbitrary bitrates and the transfer of data. The hybrid approach allows the software-controlled allocation of isochronous and non-isochronous bandwidth. The multimedia scenario given above requires more advanced facilities. Several of these basic connections have to be combined into one multimedia call. We will discuss possible modifications to the Q.931 protocol to make it compatible with the new requirements of broadband communications. Although, at this moment, it is not likely that an enhanced Q.931 protocol will be the final solution for B-ISDN, the identification of the parameters to be inserted to handle ATM connections and complex multimedia services will make easier the development of a new protocol. In the following discussion, the message structure as defined in Q.931 is maintained. Therefore, every message may comprise a protocol discriminator, call reference, message type, and other information elements as required. While most of the Q.931 messages may remain such as "call information phase" messages (resume, suspend, user information etc.), the "call clearing" messages (disconnect, release etc.) and the "miscellaneous" messages (facility, information, notify etc.) changes are needed in the "call establishment" messages (alerting, call proceeding, connect, set up etc.). In this paper, we will focus on two new protocol procedures: Context negotiation and the request for allocation of new channels during an established call.

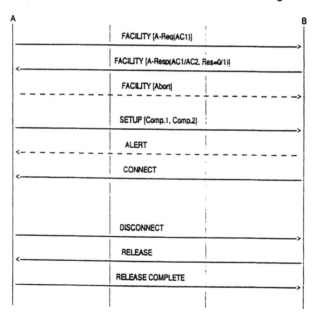

AC = Application Context
A-Req = Associate-Request
A-Resp = Associate-Response
Res = Result = 0 (positve) / 1 (negative)
Comp. = Service Component

Figure 5: Context Negotiation before Call Establishment

The "context negotiation" procedure is described as a supplementary service which can be invoked before the request to establish a call (see Fig. 5). The introduction of "context negotiation" as a supplementary service will have the following advantages: Invoking this supplementary service before call set up will provide the user with the possibility to negotiate with the remote user terminals all necessary parameters required for the call with the distinguished service. If there are terminals at the called user site which are fully compatible to the requested service the call will be established only to one of these terminals. If the distinguished service cannot be supported at the called user site but if there are terminals which provide an alternate compatible service, the calling user will have the opportunity to decide if a call with this alternate service shall be established. Without initiation of the "context negotiation" there is no guarantee that the call will be established with a remote terminal with the highest level of compatibility. In this case, if the requested service is not mandatory, the call would also be established with a terminal with a lower level of compatibility that responses faster. In Figure 6 we show the structuring of the Facility elements for context negotiation.

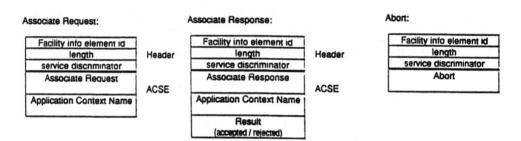

Figure 6: Facility information elements for Context Negotiations

The "allocate" message is sent from the user or the network to request the allocation of new channels during an established call. In the allocate message more than one connection can be requested at the same time and the information elements needed to characterize them are the same as those presented in the set-up message. As an enhancement of the allocate message, a "modify" message is introduced which can request the modification of active channels. Both, the modifiy and allocate messages, are especially needed for multimedia calls as in the application scenario described above. In Figure 7 the modification during an active call is shown in detail.

7 Current Status of Work and Outlook

A first demonstrator of a distributed multimedia application based on a private broadband switching system has been presented at the TELECOM 91 in Geneva in October. The described prototype of the distributed communication system will be available in 1992 within in ESPRIT and RACE programme for supporting multimedia communication sessions.

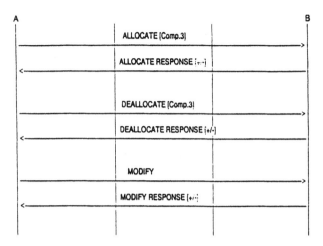

Figure 7: Modification during active Call

Acknowledgement

Most of the work presented has been supported by the European Community within the projects DAMS (ESPRIT 2146), MCPR (RACE 1038) and BUNI (RACE 1044). We would like to thank all the partners and institutions for fruitful discussions and their cooperation.

References

(1) Berra, P.B. et al.; Architecture for Distributed multimedia Database systems, Computer Communications, vol.13, no.4, May 1990

(2) Bulick, S. et al.; The US West Advanced Technology Prototype multimedia communication System, GLOBECOM89, Texas Nov. 1989

(3) CACM Special Issue: Standards and the Emergence of Digital Multimedia Systems, April 1991

(4) Cordes, R.; Kummerow, T.; Multimedia Communication and Information Management Based On Available and Emerging Standards; Proc. IEEE Workshop on Telematics, Cheju Island, Korea, 1991

(5) Cordes, R.; On the Way to Hypermedia and Multimedia Services and Terminals; Proc. Interactive Communication Tools, Paris, May 1990

(6) Halasz F., Schwartz M; The Dexter Hypertext Reference Model; in (10)

(7) Little T.D., Ghafoor A.; Synchronization and Storage Models for Multimedia Objects. IEEE Journal on Selected Areas in Communication, Vol.3, No.3, April 1990

(8) Ludwig, L., Dunn D.F; Laboratory for the Emulation and Study of Integrated and Coordinated Media Communications, Proc. SIGCOMM Conf., Aug. 1987, Stowe Vermont

(9) MHEG Document S.3, Summer 1991

(10) NIST, Hypertext Standardization Workshop, Jan. 1990

(11) Sheperd W.D., Salmony M.; Extending OSI to support Synchronization Required by Multimedia Applications; Computer Communications Vol.13, No.7, Sept. 1990

(12) Watanabe K. et al.; Distributed Multiparty Desktop Conferencing System: MERMAID, Proc. Conf. on CSCW, Los Angeles, 1990

(13) Weikum G., Principles and Realization Strategies of Multilevel Transaction Management, ACM TODS Vol.16, No.1, 1991

(14) Weikum G, Haase C.; Multi-Level Transaction Management for Complex Objects: Implementation, Performance, Parallelism, ETH Zürich, No. 162, July 1991

(15) Wybranietz, D.; Cordes, R.; Stamen, F.-J. ; Support for Multimedia Communication in Future Private Networks; 1st Workshop on Network and Operating System support for Digital Audio and Video, Berkeley, Nov. 1990

(16) Wybranietz, D.; Strategies for Future Private Communication Systems, to appear: 3rd International Symposium on Systems Research, Informatics and Cybernetics, Baden-Baden, Aug. 1991

(17) Wybranietz, D.; Buhler, P.; The LADY Programming Environment for distributed operating systems; Future Generation Computer Systems, Vol.6, North Holland, 1990, pp 209-223

Session VII: Multimedia Abstractions I

Chair: Jonathan Rosenberg, Bellcore

The session "Multimedia Abstractions I" comprised three presentations about abstractions that provide a framework for supporting software that manipulates digital audio and video. These abstractions serve the same purpose as existing system abstractions, such as I/O libraries and windowing systems. They hide the complexity of underlying hardware and software and provide organized mechanisms for performing common manipulations. These abstractions also ease the design, implementation and debugging of software and promote portability. As we begin to see progress on the fundamental systems issues for supporting digital media (such as operating systems and networks), we can expect such multimedia abstractions to become more important.

The first presentation was by David Anderson from the University of California at Berkeley ("Toolkit Support for Multiuser Audio/Video Applications," David Anderson and Pamela Chan). David discussed COMET, an extensible toolkit providing a set of abstractions designed for distributed, multi-user applications that use continuous digital media. An example application is a teleconferencing system that provides multiple users with video, audio and graphics communications.

The set of abstractions provided by COMET is implemented by a set of object-oriented classes. There are classes for abstract I/O devices and audio mixers and a mechanism for programs to link objects of these classes to form the nodes of a graph. It is the specific linking of objects that defines the architecture of an application. Once an application has created and linked objects as desired, the application calls a COMET-supplied setup routine. This routine realizes the graph by creating processes and establishing communications as necessary.

In COMET, nodes of the graph must know about their neighbors to allow the calculation of delay requirements and the determination of data conversions. The nodes communicate this information among themselves by obeying a set of conventions defined by COMET. The extensibility of COMET comes from allowing programmers to define new classes and include objects of these classes within the graph. As long as the objects obey the communications conventions, they participate as full-fledged COMET objects.

Data conversions along a path are needed to match the source data format to the capabilities of processing and presentation devices. For example, an input source might provide a stereo audio data stream, while the system on which it will be played supports only mono audio. In this case, the stream must be converted from stereo to mono before it enters the speaker. The places at which data conversions are performed is important because they determine the delays along paths and the amount of network traffic.

The second presentation was by Duane Northcutt of SUN Laboratories ("System Support for Time-Critical Applications," J. Duane Northcutt and Eugene M. Kuerner). Duane began his presentation by giving his personal views about the state and future of multimedia. In his view, multimedia is currently a lot of hype with little substance. He suggested that multimedia was the application area that would evolve to support what people will do with the next generation of workstations. What will people be doing with these workstations? Duane believes that we will use the machines primarily to support communications, not to support computation as is done today.

The bulk of Duane's presentation was about abstractions for supporting digital media within a workstation. The major change required is the introduction of time as a first-class notion within workstation operating systems. To this end, Duane and his colleagues are working on a high-level operating system model for supporting time.

An operating system for supporting multimedia is composed of two parts. The "kernel" of the operating system provides fundamental resource management for the machine (such as bus access scheduling and process allocation). Sitting on top of this is the "system," which augments the kernel's capabilities to provide appropriate multimedia support to user processes. It is, however, important to note that the system level is not expected to be exported directly to application programs. Rather, libraries will be built to present a more appealing interface to the machine's capabilities.

A central notion for the system level is resource management, which must be based on time constraints to support multimedia. This will require new operating system abstractions. Duane made an analogy to network support in operating systems, which necessitated new abstractions, such as ports and sockets.

In supporting time, the researchers feel it is necessary for the definitions of time constraints to be modular. This is necessary to isolate the effects of constraints to make their implementation more manageable. In addition, this modularization aids software reuse. Furthermore, they decided to concentrate all notions of time within the new programming abstractions as opposed to distributing the mechanisms throughout the operating system interface. This avoids the need to augment existing operating system functions with additional parameters to express time constraints.

Duane closed by stating that a user-level implementation of the system exists and a kernel-level implementation has begun.

The third presentation was by Daniel Ingold of ETH in Zurich ("An Application Framework for Multimedia Communications," Stefan Frey and Daniel Ingold). Daniel discussed the design of an object-oriented framework to support multimedia applications. This work is part of the ETHMICS project, which is building a testbed for integrated multimedia communications. Their design provides abstractions for controlling multimedia devices, signaling among such devices, and controlling the presentation of media.

The design provides an object hierarchy rooted at the class Stage, which defines a means for signaling and event propagation among objects. A tree of Stage objects controls the multicasting of a single media stream. The branches of the tree are made up of alternating objects of class Device and Channel.

The Device class specifies common properties of physical resources such as input and output actions. A Device instance connects an incoming Channel to an outgoing Channel. The Device class provides methods for specifying bandwidth demands and jitter requirements.

Device objects, used to connect Channel objects, support 1-to-many transmission of a single medium stream. Device objects are responsible for requesting bandwidth and propagating information concerning synchronization requirements.

The mapping from an object to its presentation is specified by MediaView objects. For example, the VideoView subclass defines the destination area for video on a display and allows the video to be scaled and moved as any window. As another subclass of MediaView, the AudioControl object provides an interface for audio manipulations as well as for mixing multiple audio channels.

Daniel concluded his presentation by presenting an example object hierarchy that implements a simple picturephone application. This application has been partially implemented.

At the end of the presentations, there was an open discussion among the workshop attendees. The discussion was sparked by Ralf Guido Herrtwich's comment that dataflow models appear to be the appropriate basis for multimedia abstractions, but it was not clear how to turn such models into efficient low-level scheduling decisions.

One group of participants suggested that it was more important to define proper abstractions than to worry upfront about performance. Someone commented that this was an instance of a common, but mistaken view: if one gets the abstractions correct, then appropriate performance "magically" follows. The response to this was that there is no substitute for basic engineering competence. If an abstraction cannot be implemented efficiently, it is due to poor design engineering.

This led to a heated discussion about the importance of obtaining experience from building applications as opposed to pushing on fundamental technology. This is the classic "technology push" versus "demand pull" argument. On the side of demand pull, it was argued that proceeding without applications experience was likely to produce "solutions looking for a problem," technology with no apparent practical application. The other side, technology push, stated that multimedia was a field that was fundamentally limited by technology. Therefore, it was virtually impossible today to build many of the promising multimedia applications, or even to conceive of them given today's technology.

TOOLKIT SUPPORT FOR MULTIUSER AUDIO/VIDEO APPLICATIONS

David P. Anderson
Pamela Chan

Computer Science Division, EECS Department
University of California at Berkeley
Berkeley, California 94720

October 25, 1991

ABSTRACT

Comet is a UNIX/C++ toolkit for writing programs that involve multiple users and that use digital audio and video. Comet provides a simple programming interface: the application builds a graph of objects representing speakers and microphones, mixers, files, and so on. Comet then realizes the graph by creating processes to handle mixing and file I/O if needed, and linking them by network connections to audio/video I/O servers. In addition, Comet addresses the interrelated issues of client requirements and resource management. It determines delay and throughput requirements, process placement, and data type conversion; it deals with resource managers on the application's behalf. These mechanisms are based on a negotiation protocol among the components of the object graph.

1. INTRODUCTION

"Multiuser CM applications" are those that 1) use audio/video media (*continuous media*, or CM), and 2) are run concurrently by multiple users at different locations, allowing real-time interaction and collaboration. Such applications typically provide conferencing (each participant hears and perhaps sees the other participants) and may also involve storage and playback of CM material. However, applications may differ in their system-level requirements and preferences: For example, distributed music rehearsal needs low delay while distributed music recording needs high data quality and low loss.

Integrating CM data in the standard framework (operating system, networks, file system, user programs) provides many advantages, but it makes multiuser CM application difficult to develop. First, they are highly distributed: a typical application might involve a graphical user interface program running on each user's workstation, servers on each workstation for discrete- and continuous-media I/O, processes for audio mixing and file I/O, and a central process to manage conference membership. Second, the application must determine its performance requirements and inform the underlying system, perhaps to reserve resources. Finally, since many implementation details depend on hardware and network properties (such as the availability of multicast), it is hard to write portable programs.

We believe that many of the implementation issues of multiuser CM applications should be handled by a software layer that we call a *multiuser CM toolkit*. Applications based on such a toolkit specify the desired CM functionality in high-level terms, and the toolkit manages the details. In a distributed system whose components provides real-time semantics (network channels, CPU scheduler, real-time file systems, *etc.*) the CM toolkit layer acts as an "overseer" that interacts with these system components to determine an application's implementation structure and the corresponding performance requirements.

As a proof of concept and to explore the issues, we are developing a multiuser CM toolkit called *Comet*. Comet supplies a set of C++ classes representing abstract CM components: microphones, speakers, files, audio mixers, and so on. The application creates instances of these classes and links the objects to form a dataflow-style graph. Comet then implements this abstract graph by 1) creating processes, if needed, to handle CM data, and 2) connecting existing server processes to each other and to the new processes. It determines the delay and throughput requirements of each component, and informs resource managers accordingly. Hardware and network properties are encapsulated within Comet classes, so they are transparent to the programmer and can be changed easily.

A prototype version of Comet has been completed, and has been used as the basis for a transcribed conferencing application. The implementation of Comet uses ACME [2], a server that provides shared network-transparent access to workstation CM I/O devices. Comet also uses the IPC and I/O multiplexing features of InterViews [5], a C++ toolkit for X11. However, the principles of Comet do not depend on ACME or Inter-Views.

2. COMET FEATURES

2.1. Master/Slave Architecture

Comet applications have a master/slave structure. Each participant runs a separate instance of the slave program, and there is a single instance of the master program. The master program can run as a daemon at a well-known host, or can be run on demand by the slave program. The hosts must run some version of UNIX, but need not share a common file system; Internet TCP connections are used for all communication.

The slave program provides the user interface (handling mouse/keyboard events and generating window system requests) while the master program maintains global state and issues CM-related commands to Comet. In a basic conferencing application, for example, the slave provides mouse-based controls for joining and leaving conferences, while the master maintains conference membership and tells Comet how CM I/O devices and files are to be interconnected. In a collaborative editing program, editing may be local to the slave, but any globally visible changes must be propagated through the master.

Communication between the master and the slaves is done using remote procedure calls (RPCs). RPCs may be initiated at either end, and include application-defined RPCs as well as Comet's internal RPCs. The master and slave programs are single-threaded. Comet uses the InterViews Dispatch library to multiplex among I/O sources (window system events and RPCs on the slave, RPCs from multiple slaves on the master).

2.2. Continuous Media Objects

The master program manages the continuous media (audio/video) components of the application. The Comet toolkit on the master side provides a set of C++ classes, derived from a base class CM_NODE, that represent these components. The current set of CM_NODE types is as follows:

- MICROPHONE and SPEAKER represent CM I/O devices. These devices are "abstract" in the sense that several of them (perhaps from different Comet applications) may be mapped simultaneously to the same physical I/O device. Constructor arguments specify the slave where the device is located.

- INPUT_FILE and OUTPUT_FILE represent disk files storing CM data. Constructor arguments specify the file name, the slave whose file system stores the file, and (for OUTPUT_FILE) CM data type preferences.

- AUDIO_MIXER represents a component that takes in N audio streams and generates as output the sums of each $N-1$ of these streams, as well as the sum of all N. Each input and output has a corresponding PORT (see below).

- SOURCE_PROCESS, SINK_PROCESS, and FILTER_PROCESS represent processes executing user-defined programs that act as a CM data source, sink and filter respectively. Constructor arguments specify the program name and the slave where the program is stored.

2.3. The CM_NODE Graph

A PORT object represents a source or sink of CM data. A CM_NODE has one or more associated PORTs. The operation

```
p->join(PORT* q)
```

creates an (abstract) connection from output port p to input port q. Each port can have at most one such connection.

Using join(), the application builds a directed graph of CM_NODEs. When the CM_NODE graph has been constructed, the application calls setup() on any node of the graph. This call causes Comet to realize the graph (see Section 3) and starts the flow of data between the devices, processes, and files represented by the graph.

The CM_NODE graph may be modified (by adding or removing nodes or links) during program execution. For example, as new participants join a conference, new MICROPHONE and SPEAKER objects are created and connected to the AUDIO_MIXER. After an addition is complete, the application must call setup() on a CM_NODE in each modified subgraph.

3. IMPLEMENTATION

Comet must convert a graph of CM_NODEs (which are simply C++ objects in the master program) to a set of processes and network connections that realize the graph. Our approach is modular: each CM_NODE manages its own implementation (this facilitates adding new CM_NODE types). To implement itself, a CM_NODE needs information about its neighbors. For this purpose, Comet defines a C++ interface, which we call the "CM_NODE Protocol"[1], between CM_NODEs.

The CM_NODE Protocol addresses four issues: 1) whether data streams require low delay, 2) the data representation on each stream, 3) the message size on each stream, and 4) connection establishment. The protocol defines a set of functions that each CM_NODE must provide. These functions, enumerated in the following sections, are implemented differently in each derived class of CM_NODE.

3.1. Delay Bound Determination

If end-to-end delay exceeds 200 milliseconds or so in a conference application, conversation becomes difficult. When CM data is being written to a file, however, the end-to-end delay is unimportant. We define a CM data stream to be *low-delay* if it involves data being sent between human users (*e.g.*, from a microphone to a speaker) and *high-delay* otherwise. For many CM components, the optimal handling of CM data depends on whether it is low- or high-delay: Low-delay streams must be handled by high-priority processes and cannot tolerate the delay of buffering, while high-delay data streams can be buffered and can tolerate high processing delays.

The CM_NODE Protocol allows each CM_NODE to learn whether the data streams it handles are low- or high-delay. The function from_rt_source(PORT*) returns True if the data stream from the given output PORT includes data originating from a "real-time" source (a CM input device). Similarly, to_rt_sink(PORT*) returns True if the data stream entering the given input PORT is destined for a real-time sink (a CM output device).

The implementation of these operations is type-specific. For example, AUDIO_MIXER implements to_rt_sink() by examining the outputs to which the given input stream contributes, calling to_rt_sink() on each of the input ports connected to these outputs, and returns True if any of these calls returns True. The result is then cached in the PORT object for subsequent calls (this is necessary to avoid infinite recursion).

3.2. Data Type Negotiation

The "type" of an audio stream is determined by the number of samples per second, the number of bits per sample, and possibly a logarithmic compression of samples. For video, the type includes the image size, the number of frames per second, and so on. We assume that

- There is a fixed finite set of CM data types.

- There is a partial order $<$ on the set of data types. $S < T$ means that S has less information, and typically a lower data rate, than T.

- For any types S and T there is a least type $U = sup(S, T)$ such that $U > S$ and $U > T$.

Some pairs of types, such as 44 KHz mono and 22 KHz stereo, may be incomparable; the preference of one over the other is then application- or user-specific (see Figure 1).

---

[1] We call it a "protocol" because function calls must be made in a certain order; it is not a network protocol.

The `CM_NODE` Protocol allows `CM_NODEs` to negotiate the data types of the CM data streams that connect them. The goals of the negotiation is to find an assignment of types to streams that 1) is feasible (the conversions are implementable by the `CM_NODEs`); 2) provides the maximum possible quality at the outputs; and 3) minimizes network traffic by doing conversions as far "upstream" as possible.

A `CM_NODE` provides functions

```
TYPE_SET feasible_types(PORT* p);
TYPE_LIST prime_types(PORT* p);
```

where p is an input `PORT` of the `CM_NODE`. `feasible_types()` returns the set of types the `CM_NODE` is able to accept on the port. `prime_types()` returns the list of feasible types that have no redundant information; *i.e.*, for which no lesser type will produce identical output. The list is sorted by decreasing preference, allowing the `CM_NODE` to rank incomparable types.

For an output device such as `SPEAKER`, `prime_types()` depends on the speed and width of the DAC. For example, the prime types for a 22 KHz monoaural DAC might be 22 KHz mono and 8 KHz mono (in that order). 44 KHz mono is not prime because it would produce the same output as 22 KHz mono. Incomparable types can be ranked, if desired, by the application. `feasible_types()` depends on the conversion capabilities of the I/O server. If the server can convert 44 KHz stereo to 22 KHz mono in real time (along with its other tasks) it would list 44 KHz stereo as a feasible type.

The determination of prime and feasible types for `AUDIO_MIXER` is more complex. Suppose I is an input stream of the mixer, and let S be the set of output streams to which I contributes. A type T is feasible for I if, for each $s \in S$, the mixing agent[2] is able to mix input in format T with the other inputs of s, and convert the result to a type that is feasible for the destination port of s. The prime type list for I might be determined as follows. Fix a type T and a stream $s \in S$. Let $<P_1 \cdots P_n>$ be the prime types of the destination port of s. Let $V(s)$ be the least i such that $Pi < T$ ($V(S)$ is the "value" of type T for stream s). Define $R(T) = \sum_{s \in S} V(S)$; $R(T)$ is the aggregate value over all output streams to which I contributes. The `prime_types` list is then formed as follows: enumerate all types by increasing value of $R(T)$; delete from this list types T that are not feasible or for which there is a type U such that $R(T) = R(U)$ and $U < T$. An example is shown in Figure 2.

Each `CM_NODE` supplies a function

```
TYPE actual_type(PORT* p);
```

where p is an output `PORT` of the `CM_NODE`. This returns the actual type to be output on the given `PORT`. The policy is type-specific. For example, an `AUDIO_MIXER` might use the following policy to determine the actual type for an output stream R. Let $S = s_1 \cdots s_n$ be the input streams that contribute to R. Call `actual_type()` on the nodes that generate these streams to learn their types $T_1 \cdots T_n$, and let $\overline{T} = sup(T_1 \cdots T_n)$. Let U be the set of types S in the feasible set of the destination

---

[2] This agent may be an I/O server or a separate mixing process; see Section 3.6.

such that the mixer can convert the output stream to S. Let $T_0 \cdots T_n$ be the prime types of the port to which R is connected. Let i be the least such that $T_i < \overline{T}$ and there is a type $T \in U$ with $T_i \leq T$. T is then the actual output type of R. An example is shown in Figure 3.

3.3. Message Length Negotiation

The length of messages (the units in which data is written on connections) is an important issue for low-delay connections. There is a tradeoff between packetization delay and per-message overhead. To minimize delay and overhead simultaneously, the temporal message size on data stream should approximate the maximum of the I/O interrupt periods of the input and output devices (see Figure 4).

To allow negotiation of message size, each `CM_NODE` provides a function `message_length(PORT* p)`. If p is an input `PORT`, this returns the largest message length (measured in milliseconds) that will minimize packetization delay. For a `SPEAKER`, this is determined by the device interrupt period. For an input stream S of an `AUDIO_MIXER`, it is the minimum of the `message_length()` values of the input ports to which S contributes.

If p is an output `PORT`, `message_length()` returns the actual message length to be sent on p. In general, this is computed as $max(N, M)$, where N is the message length of the connected input `PORT` and M is the minimum of the message lengths of data streams that contribute to p.

3.4. Connection Establishment

Comet uses TCP connections to convey CM data. Each TCP connection is used as a simplex channel. By convention, the sending end plays the active role in connection establishment (in BSD UNIX terminology, the sender does the `connect()` and the receiver does the `listen()` and `accept()`). Each `PORT` object includes a network address (host Internet address and port number) for the corresponding TCP socket. Each `CM_NODE` must provide the following operations:

```
bind();
connect();
accept();
```

`bind()` creates listening sockets for all input `PORT`s, and stores their addresses in the `PORT` objects. `connect()` sets up outgoing connections for the object's output ports. `accept()` accepts incoming connections, finishes setup in a class-specific way. Each function then performs the same operation on all neighboring objects. The `setup()` function (called by the application to "activate" a `CM_NODE` graph) simply calls `bind()`, `connect()` and `accept()` on the target `CM_NODE`; these calls eventually propagate throughout the graph.

3.5. Collapsing Subgraphs

It is sometimes useful to "collapse" portions of the `CM_NODE` graph, and have one `CM_NODE` assume the responsibility for implementing some of its neighbors. For example, an `AUDIO_MIXER` object handles the implementation of any `MICROPHONE` or `SPEAKER` objects to which it is connected (see Section 3.6). This is accomplished by having a type field in the `CM_NODE` base class. The `setup()` function of `MICRO-PHONE` checks if it is connected to an `AUDIO_MIXER`, and if so it simply forwards the call there.

3.6. Implementation of Some `CM_NODE` Types

Abstract CM I/O devices (`MICROPHONE`, *etc.*) are implemented using ACME. When a slave arrives, the master sets up an RPC connection to the ACME server on the user's workstation. If a CM I/O device is connected to an `AUDIO_MIXER`, then the graph is collapsed and the implementation is left up to the `AUDIO_MIXER` (see below).

File I/O (`INPUT_FILE` and `OUTPUT_FILE`) is implemented using the Comet slave library. The `bind()` and `accept()` operations for an `OUTPUT_FILE`, for example, makes RPCs to the Comet library in the appropriate slave. The `accept()` handler accepts a CM connection, then creates an I/O activity (using the InterViews I/O multiplexer) that reads data from the socket and writes it to a disk file.

An `AUDIO_MIXER` object can implement itself in either of two ways (see Figure 5). Both implementations use an *audio mixer program* that runs on the master host. The mixer program takes an arbitrary number N of digital audio input streams, and produces output streams for the total sum and for each $N - 1$ sum. It performs conversions between different data types, and it distinguishes between low- and high-delay streams, dealing appropriately with each one.

The constructor for `AUDIO_MIXER` takes as an argument the estimated number K of `MICROPHONE`s and `SPEAKER`s to be directly connected. If K is below a system-dependent threshold, a *distributed* implementation is used: `AUDIO_MIXER` examines its neighboring nodes and, for `MICROPHONE`s and `SPEAKER`s, connects the corresponding ACME servers directly. Each ACME server receives audio streams from the other ACME servers, and mixes them itself. If other objects (*e.g.*, files) are also connected to the mixer, then an audio mixer process is created to handle all such objects.

The above approach (direct interconnection of ACME servers) minimizes delay, but does not scale well: it generates $O(N^2)$ network traffic, and causes each ACME server to do $O(N)$ work for mixing. Therefore, if K exceeds the threshold, `AUDIO_MIXER` implements itself using a central mixing process only; this reduces network traffic to $O(N)$ and ACME workload to $O(1)$, at the cost of roughly doubling the delay.

Other implementations are possible. Network multicast capabilities could be exploited. If a mixer joins devices clustered in two LANs connected by a WAN, it could by implemented by a central mixer processes in each LAN, with the two processes linked by a single WAN connection.

4. RELATED WORK

Comet was inspired by graphical user interface toolkits such as InterViews [5]. These toolkits provide abstractions (menus, editors, *etc.*) for displaying and interacting with discrete data such as graphics and text. Comet is concerned purely with continuous media.

A *centralized CM toolkit* provides access to CM I/O at a single workstation. Such a toolkit, in combination with an existing window-system toolkit, makes it easy to define objects for acquiring and playing sound; these objects can then be added to mail or document systems. The Andrew Toolkit [7] takes this approach; we also developed a centralized toolkit for ACME that provides similar functionality. While this approach is useful for some applications, it has two fundamental limitations. First, it does not support multiuser applications well because management of global state, such conference membership and floor control, requires complex protocols for synchronization and failure handling. Second, handling CM data in the client program is nonoptimal in some situations: the telephony application directly connects the I/O servers, and file playback may be done more efficiently by sending data directly from a file server to an I/O server rather than having data pass through the client program.

A system described by Bates and Segal [3] offers object-oriented software layers for writing multi-user CM applications. The abstractions are similar to those of Comet: I/O, storage, mixing and multicasting. Ludwig [6] describes an analogous system based on a hierarchical dataflow model. These systems differ from Comet in that they are grounded in the telecommunications world. Because the underlying hardware is a dedicated network or crossbar switch, they have no mechanisms corresponding to the CM_NODE protocol. They are targeted at developing "services" rather than user-defined applications, so they do not emphasize integration with existing programming environments.

Other distributed CM toolkits are based on a "conference server", reflecting the viewpoint that the CM portion of most multiuser CM applications is simply generic audio/video conferencing. Examples include MMConf [4] and CoLab [8]. Comet takes a different viewpoint: we think of conferencing as itself being a class of applications with a broad range of requirements (scalability of the number of speakers and listeners, delay bounds, synchronization of recorded material playback, floor control policies, security, *etc.*).

5. CONCLUSION

Like other systems for multiuser audio/video applications, Comet provides a set of classes representing CM resource types; low-level details and hardware dependencies are hidden within the implementation of these classes. Comet makes two additional contributions. First, the CM_NODE Protocol, embodied in the C++ interface between CM_NODE objects, addresses issues that are crucial for handling CM data in general-purpose distributed systems: delay bounds, data type, message size, and connection establishment. Because these issues are dealt with in a uniform way, new CM_NODE types can be added easily. Second, Comet's master/slave model simplifies the management of global state (handled by the master) while providing fast response to local GUI interactions (handled by the slaves).

The Comet prototype lacks some key features. Comet provides audio capabilities only; we expect that many of the ideas will extend to video. Since the CM data graph is closed, cross-linking between discrete and continuous media (as would be needed for speed recognition or synthesis) is not possible. Comet should be combined with a system for CM data storage, linkage and indexing more sophisticated than the UNIX file system. Finally, it would be convenient to have classes that combine GUI and CM (*e.g.*, a control panel for playing a sound file).

Similarly, the Comet implementation might be extended in several ways. Currently, mixer processes are run on the master host, and user-defined processes and file I/O on the slave host. More generally, these processes might be placed on faster or less loaded machines, perhaps taking network communication costs into account as well. Beyond load-balancing, one might ask that resources (CPU time, network and disk bandwidth) be reserved so that the application will receive a guaranteed performance level. Issues of distributed resource reservation are discussed in [1].

REFERENCES

1. D. P. Anderson, "Meta-Scheduling for Distributed Continuous Media", UC Berkeley, EECS Dept., Technical Report No. UCB/CSD 90/599, Oct. 1990.

2. D. P. Anderson and G. Homsy, "A Continuous Media I/O server and its Synchronization Mechanism", *IEEE Computer*, Oct. 1991.

3. P. C. Bates and M. E. Segal, "Touring Machine: A Video Telecommunications Software Testbed", *First International Workshop on Network and Operating System Support for Digital Audio and Video*, Berkeley, CA, November 8-9, 1990.

4. T. Crowley, P. Milazzo, E. Baker, H. Forsdick and R. Tomlinson, "MMConf: An Infrastructure for Building Shared Multimedia Applications", *Proc. 1990 CSCW Conference*, Oct. 1990, 329-342.

5. M. Linton, J. Vlissides and P. Calder, "Composing User Interfaces with InterViews", *IEEE Computer 22*, 2 (Feb. 1989), 8-22.

6. L. Ludwig, "A Threaded/Flow Approach to Reconfigurable Distributed Systems and Service Primitives Architectures", *Proc. of ACM SIGCOMM 87*, Stowe, Vermont, Aug. 1987, 306-316.

7. A. J. Palay, "The Andrew Toolkit: An Overview", *Proceedings of the 1988 Winter USENIX Conference*, Dallas, February 9-12, 1988, 9-21.

8. M. Stefik, G. Foster, D. Bobrow, K. Kahn, S. Lanning and L. Suchman, "Beyond the Chalkboard: Computer Support for Collaboration and Problem Solving in Meetings", *Comm. of the ACM 30*, 1 (Jan. 1987), 32-47.

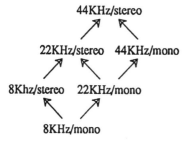

Figure 1: Comet assumes that the data types for a particular medium (audio or video) are partially ordered by $<$; $S < T$ means that S contains less information than T. Audio types might be ordered as above (an arrow from S to T means $S < T$).

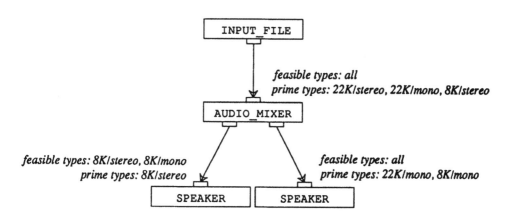

Figure 2: Each input port is assigned a set of *feasible* types and a list of *prime* types. For output devices, these are determined by the hardware of the I/O server. For other CM_NODE types, they are determined by the corresponding lists from the ports to which the data is destined, and by the conversion capabilities of the CM_NODE.

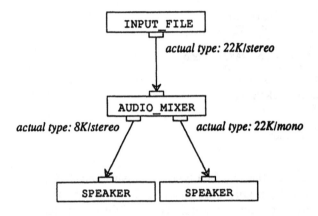

Figure 3: A `CM_NODE` selects the actual data type to be sent on each output `PORT` based on the feasible and prime types of the corresponding input `PORT`. In this example (continued from Figure 2), the `INPUT_FILE` stores 44KHz stereo data. Because this high quality is not usable, it converts the data to 22 KHz stereo before sending it to the `AUDIO_MIXER`.

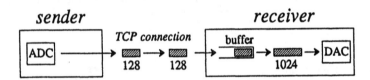

Figure 4: In this example, 128-sample messages are being sent to an output device with a 1024-sample interrupt period. Since the device must receive 8 messages before it can output, it would be more efficient to use 1024-sample messages. Increasing the message size beyond 1024, however, would increase packetization delay.

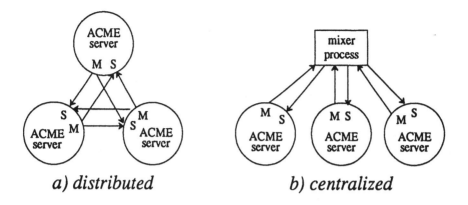

a) distributed   b) centralized

Figure 5: An `AUDIO_MIXER` object can implement itself in either of two ways. If the number N of participants is small (a), it directly interconnects the microphone (M) of each ACME server to the speakers (S) of the other servers, and has each server do its own mixing. If N is large (b), it creates a separate mixer process; each ACME server sends its input to the mixer process and receives a mixture of the other $N - 1$ inputs.

System Support for Time-Critical Applications

J. Duane Northcutt and Eugene M. Kuerner
Sun Microsystems Laboratories, Inc.

A number of interesting application areas have time constraints associated with them — most notably in the area of multimedia. Today's workstations do not provide adequate support for applications whose definition of correctness is a function of time. This paper presents a set of system-level mechanisms that permit the incorporation of multiple streams of sustained, high-bandwidth, time-critical information as first-class data types within a distributed workstation computing environment. This work is being done as a part of a broader research effort being conducted at Sun Microsystems Laboratories to develop fundamental system technology in support of multimedia applications.

Introduction

The primary objective of this work is to develop system-level mechanisms which support the integration into the workstation of applications that manipulate time-critical information (e.g., audio and video [Hopper 90]). The definition of correctness for this class of applications is a function of time. It is not sufficient to simply perform a computation, it must be completed by a specific point in time for it to be of maximum value to the application [Northcutt 87].

The management of application timeliness requirements is one key concern. The ability to express the timeliness constraints of various activities and to temporally coordinate (i.e., synchronize) activities within, and among, application programs are the major issues that must be addressed. Furthermore, whenever a system is called upon to support activities with associated time-constraints, it is necessary to address the cases where the system lacks sufficient resources to meet all of the given time-constraints.

This paper describes a portion of a larger effort that is attempting to introduce the notion of time-criticality into the distributed workstation programming environment. The emphasis of the work described here is on the creation, and validation through use within a larger system context, of a programming interface that supports the needs of time-critical applications within the workstation. This interface will provide the means for expressing an application's timeliness requirements to the system (to permit the system to manage its resources so that application time constraints can be met) as well as the application's overload handling policies (so that the system can properly deal with the case where the application's time constraints cannot be met).

What is called for is the active management of resources by the system in order to meet the expressed time-constraints of its instantaneous application mix. In effect, the system must close the control loop of activities that have timeliness constraints associated with them.

This perspective implies that time-critical applications must interact with the system in a fundamentally different way than they do today. In particular, different information (in both kind and amount) must be exchanged between the system and its users, and the control relationship between applications and the underlying system must be significantly different (e.g., involving asynchronous notifications, up-calls, etc.).

Problem Statement

There is an entire class of problems to which today's workstations cannot be applied [HRV 90]. This application area is characterized by the inclusion of information or activities whose correctness (or value to the user) is a function of time. Multimedia audio and video, visualization, virtual reality, transaction processing, and data acquisition are examples of this class of applications. Current workstation system architectures fail to support time-critical applications for three main reasons. The system does not have the proper resources, the resources are not correctly organized, and the system resources are not managed in the proper fashion. This work is aimed at augmenting workstation functionality in order to allow applications to integrate time-critical information into the workstation.

Time Regulation of Application-Level Activities

In this paper, time-critical activities are defined to be operations that involve information that has time constraints associated with its manipulation. The manipulation of time-critical information is defined to include the following types of activities: *acquisition* (i.e., getting information into the system); *processing* (i.e., performing computations on the information); *transport* (i.e., moving the information between sets of sources and destinations); *storage* (i.e., buffering the information for later access); *coordination* (i.e., regulating the time when other forms of manipulation are performed on the information); and *presentation* (i.e., the delivery of the information to its end-user).

For the purposes of this work, individual units of time-critical information are referred to as *samples*, while sequences of such units are known here as *streams*. Each sample has a time constraint associated with it, and each manipulation of the data in the end-to-end sequence inherits its time constraints from the samples currently being dealt with. Thus, some activities are inherently time-critical (i.e., they are involved in the manipulation of time-critical information), while other activities acquire time-criticality through their association with other time-critical activities (i.e., if a function is performed on behalf of a time-critical activity, the function must inherit the client's time constraints).

In addition to meeting the timeliness needs of individual activities, it is necessary for the platform to support application-specified temporal alignments of separate time-critical activities. For example, the audio and video streams in a teleconferencing application require one type of time coordination, while the cueing of audio and video tracks in a compound multimedia document calls for another type of coordination. It is essential that the system support many different forms of temporal relationships [Steinmetz 90]. These range from strong forms of temporal alignment, where samples occur at specific times (e.g., generate an image when a word is clicked in a document), to weaker forms where samples of the constituent streams occur at given rates (e.g., generate audio, video, text according to a scripted presentation).

Key Technical Objectives

The most critical issues guiding the solution to the problem of time-critical application support are, high utilization of system resources, graceful degradation of functionality in the face of insufficient resources, and modularity in the expression of applications time constraints.

A high degree of system resource utilization is considered essential, because workstations traditionally provide best-effort services and applications areas such as multimedia are essentially open-ended in terms of resource demands. In effect, the system objectives should be to deliver the highest degree of value to the users that is possible, given an instantaneous mix of application demands and available system resources. This precludes the use of techniques for meeting timeliness needs that rely on excess assets [Liu 73].

For similar reasons, the notion of graceful degradation is considered quite important to solving timeliness problems. With respect to its time-critical behavior, the system will need to provide a range of different quality-of-service levels. In cases where there are insufficient resources to meet application time constraints, the system may shed some of its load to reclaim system resources or degrade the quality of service provided to an application. The specifics of such policy decisions should be left to the application to define, and the system should provide support for a wide range of such policies.

In addition, the notions of modularity and extensibility are of great value. Each application should be able to dynamically express its instantaneous time constraints to the system. In turn, the system should make every effort to deliver the greatest amount of value to the users as is possible. While global knowledge and cooperative programming will almost always result in greater overall effectiveness, the system should require that each application express only its own timeliness needs. Furthermore, timeliness mechanisms must also permit the dynamic interconnection of applications with arbitrary topologies and the transparent insertion of new functions into time-critical information manipulation paths.

Research Context

This paper describes the system interface component of a whole-system approach to bringing support for time-critical applications to the workstation. Given the strong degree of interaction frequently required between the various components of a multimedia system (and the relatively unknown nature of this application domain) it has been valuable to develop an overall system framework within which various components could be developed. The following is a brief description of this framework.

The *Time-Critical Applications Layer* exists at the same layer as ordinary workstation applications, where both time-critical and ordinary applications will coexist within a dynamic, multitasking, distributed environment. Time-critical applications will typically not generate explicit time constraints, nor usage policies, but will instead make use of programming packages that provide abstracted interfaces to the system's time-critical services.

The *Application-Specific Programming Packages* provide interfaces that are familiar to (or specialized for) the practitioners within a specific application area (e.g., audio and video editing, compound document authoring, etc.). The software packages at this level provide the top-level policies for the management of specific types of time-critical information (e.g., media), and do not expose the distinct entities that are involved in the management of timeliness, nor do these packages require the programmer to express its explicit time constraints to the system.

The *Device Virtualization Layer* is responsible for abstracting out many of the limitations of physical devices and providing the higher layers of software with the desired logical view of the underlying system. This layer manages mapping of virtual to physical devices — i.e., it handles (perhaps transparently) the multiplexing of logical devices onto physical ones. These new abstractions are not necessarily specific to a particular type of media but provide generic abstractions for dealing with the issues related to the management of time-critical information. Time issues are much more concrete at this layer, and this is where much of the policy for time-critical activities resides. Users of this layer express desires for more abstract behaviors of virtual resources (e.g., "multi-way interactive" versus "stored unidirectional" channels of video), which are used in conjunction with this layer's policy decisions (e.g., permissible degree of jitter or delay, definition of exceptions, desired exception behavior, etc.) to generate the specific parameters required by the underlying software layers.

The *Operating System Layer* consists of two sub-layers: the kernel portion and the system portion. Both of these are logically part of the operating system. They form the basic functionality of the system as it is delivered and are separated from higher layers of software by some form of protection.

The *System Sub-Layer* is where the work described in this document is focussed. These are new system-level programming abstractions supporting time-critical applications. These abstractions are responsible for translating the more abstract requests for the manipulation of time-critical information into the specific parameters needed by the kernel-level mechanisms to ensure that system timeliness constraints are met. The programming abstractions also serve to direct manipulations of time-critical information into stylized forms that the operating system can manage. Furthermore, these programming abstractions concentrate timeliness-related concerns into a common mechanism, as opposed to requiring that concerns for timeliness be distributed throughout the entire operating system interface. This level implements the user-selected policies that provide the remaining information required by the kernel-level mechanisms to meet the more abstract timeliness needs expressed by this level's clients.

The *Kernel Sub-Layer* is where the management of the fundamental system resources (e.g., processor, memory, i/o, etc.) is performed. In order for the system to meet the demands of time-critical applications, it is critical for the resource management decisions to be performed in accordance with the time constraints of the computations making the individual resource requests. To accomplish this, a technique known as Time-Driven Resource Management (TDRM) is being used [Northcutt 88] [HRV 91a]. This approach requires that time constraints be associated with each time-critical activity, as well as directives as to what should be done in the event that the constraints cannot be met. This approach is quite different from traditional ones which attempt to encode the orthogonal attributes of importance and timeliness into a single value (typically, known as process priority) and involves quite different interactions between the operating system and its users.

Technical Approach

The following describes the major technical decisions that lie behind the definition of the system-level timeliness mechanisms and some of the major system-level implications of these decisions.

Key Design Decisions

First and foremost, the decision was made to develop a set of programming abstractions to encapsulate all notions of timeliness at the system level, and communicate specific timeliness requirements to the operating system kernel. In addition, the decision was made to view all issues related to the management of time-constraints as manipulations of time-critical information (i.e., a data-flow, as opposed to a control-flow perspective). This structures all manipulations of time-critical information into stylized activities that can be managed by the operating system. Finally, a model of time and synchronization was adopted to provide a framework for specifying the entire scope of temporal relationships that time-critical activities may assume (both individually, and with respect to one another) and the policies that might be followed in dealing with overload conditions.

Programming Abstractions for Timeliness

Two new programming abstractions are proposed here. First, a mechanism is provided to encapsulate the endpoints of time-critical information streams. This mechanism provides a uniform interface to a wide variety of physical, as well as logical, sources and sinks of time-critical information. These objects encapsulate the state associated with each endpoint entity includes such characteristics as: the maximum, minimum, and nominal sample rate (e.g., 44.1kHz, 30 Frames/second, etc.); the units of sampling granularity (e.g., video fields, video frames, blocks of audio samples, etc.); the amount of buffering provided; and the current state of the unit (e.g., running, stopped, buffer over-/under-flow, etc.). Secondly, active entities are provided to regulate the manipulation of the time-critical information streams flowing between the (passive) source and destination entities. This regulation involves the exercise of control over the time at which individual samples are acquired, stored, transported, processed, and presented.

The manipulation of time-critical information is regulated by a pair of logical activities that exist within the time-regulation entity. The first of these activities is involved with the acquisition of samples from the stream's source object and the buffering of these samples within the regulation object. The other logical activity involves the movement of samples from the regulation object's buffer area to the destination object. Providing that the internal buffer neither overflows nor underflows, the regulation object can control the time at which the samples are delivered in accordance with the timing constraints associated with the information.

The objects that encapsulate time-critical information stream sources and sinks are known as **Transducers,** and the objects that serve to connect sources to sinks and regulate the flow of time-critical information samples are known as **Conduits.** (The concepts embodied by these mechanisms are similar to those described in [Herrtwich 90] and [Escobar 91].) Both of these objects are described in the following chapter (and in greater detail in [HRV 91b]), while the remainder of this chapter provides descriptions of the major issues surrounding the definition of these entities.

Data-Flow-Based Approach

Integral to the foregoing definition of the timeliness programming abstractions is the point of view that all manipulation of time-critical information can be represented in terms of the timely transport of time-critical information samples.

This approach allows all user-level activities that have time constraints to be cast into a common form. This form is the timely transport of information from which all other manipulations (e.g., processing, storage, etc.) derive their time constraints. According to this model, all time-constraints stem from constraints on the time at which samples are delivered from the sources to the destinations.

The key technique in regulating time in this environment is the use of pre-fetching and buffering of samples in order to regulate their delivery times. Buffering can only smooth out transient timing disruptions (i.e., the phasing of sample delivery), so source and destination rates must be matched in the long-term. Therefore, higher level mechanisms are required to deal with long-term rate mismatches between connected sources and sinks of time-critical information.

The timely execution of processing activities is accomplished by the implicit propagation (or inheritance) of time constraints from the streams to the processing activities that operate on them. Computational elements are added to the delivery path of a time-critical information stream by encapsulating them within stream endpoint objects. The processing elements thereby acquire the necessary time constraint information.

Definition of Temporal Relationships

For the purposes of this work, there are two major forms of temporal relationships of interest. These are termed coordination and synchronization.

Event Coordination

Coordination is defined to be the general temporal alignment of events. This is a very general notion that involves only the act of ensuring that a set of events occur within a given interval of time with respect to each other. This alignment is not relative to any clock but is a function of the relative ordering of a set of events. In the case of time-critical information, coordination is viewed as part of the overall timeliness problem. Furthermore, all of the issues of time-critical information manipulation are combined into the single problem of ensuring that samples are delivered at the proper rates and phases with respect to each other.

Event Synchronization

Synchronization is defined here to be the temporal alignment of events relative to a physical clock. When time base is a real-time clock (or derived therefrom), then this conforms to the common notion of keeping an event stream flowing in real-time. In effect, synchronization is a special case of the more general notion of event coordination. As with coordination, there are both inter-stream and intra-stream synchronization issues. Aside from the addition of a real time base, all of the issues related to coordination apply to synchronization.

Generally speaking, a stream of time-critical samples can be considered synchronized if the samples are delivered to the destination at a specified rate. By defining synchronization criteria in terms of rates (as opposed to absolute time periods), it is possible to avoid the imposition of unnecessarily strict timing constraints on the underlying system. That is, rate-based synchronization does not require that each sample in a continuous stream be delivered at a specific time from the start of the stream. Rather, rate-based synchronization requires only that a given number of samples be delivered in some unit of time. This definition allows the users to express only that degree of synchronization required of their specific application. This provides the system with the least restrictive set of time constraints that are appropriate for the application mix.

Major Implications

In general, the support of time-critical applications requires new forms of interaction between applications and system software. These new interactions include the exchange between the application and the system of information concerning the timeliness constraints of individual computations (e.g, the amount of execution time required over the next real time interval), as well as policy direction as to what should be done in the event that the time constraints cannot be met.

Furthermore, overload (i.e., the case when resource requests exceed the system's current supply) will be a common case in typical multimedia systems (as opposed to an exceptional one). Therefore, the system must provide mechanisms for efficient interactions between the system and the applications. When resource requests cannot be satisfied, the requesting entities are to be notified. It is up to the client of these services to choose how and when notifications of such conditions are to be performed.

Time Regulation Mechanisms

This section describes the external view of the basic objects and the means by which they are combined into useful higher-level structures. In addition, a high level description is given of how time constraints are specified in practice and of the time-regulation mechanisms' internal structure and behavior.

External View

Basic Mechanisms

As mentioned earlier, the two new programming abstractions provide for the support of time-critical applications are known as *transducers* and *conduits*. These abstractions are combined in various ways to provide practical time regulation objects.

Transducers

Transducers are the endpoints in a time-regulated delivery chain. They are passive entities that provide uniform access to a diverse set of potential sources and sinks of time-critical information. Much like device drivers, transducers achieve their uniform appearance through the use of a common interface for supplying and consuming data. Transducers appear as system objects (i.e., abstract data types), have unique identifiers ("handles"), encapsulate state information, and have operations defined on them.

The operations defined on transducers include control-oriented (e.g., start, stop, pause, or resume), and data-oriented (e.g., get a sample from the source, or put a sample into the destination) ones.

In addition to encapsulating a logical source or sink of time-critical information, transducers include state information that reflect the timeliness characteristics of the enclosed entity. Conduits, the active time regulation objects, use transducer state information to implement time-regulated information transport.

Conduits

Conduits actively regulate sample delivery from source to destination. The conduit object encapsulates buffer space for samples, state information associated with the sample stream instance, and a regulation unit that oversees the desired time regulated information transport. As with transducers, conduits have a standard set of operations defined on them, which can be invoked by referencing the unique identifier (handle) associated with each conduit.

In addition to control oriented operations (e.g., start, stop, pause, resume) and state get/set operations, the conduit interface includes a number of connection points for asynchronous signals. Particularly, clients (users or other objects) may register interests or notifications for certain conditions that occur during the conduit's operation. These signals are used to notify clients of status, aberrations, and exceptions. Clients that temporally align (coordinate) multiple conduits make use of these signals. Additionally, conduits have asynchronous signal inputs that allow for dynamic adjustments to the conduit's (instantaneous) sample rate.

Compositions of Mechanisms

In order to create useful time regulation units from these basic abstractions, it is necessary that they be composed into a complete unit. Conduits and transducers may be combined both in simple and arbitrarily complex ways.

Basic Constructions

A simple composition is one in which a source transducer is connected to a destination transducer via a basic conduit. Such a construction permits a time-critical sample stream between the source and destination transducers to be regulated according to the intersection of user defined requirements, system limitations, and transducer requirements.

Complex Constructions

Objects which result from complex compositions of conduits and transducers project the same interface as basic compositions and are functionally indistinguishable to the programmer. However, the creation of complex compositions does involve the creation of additional regulation activities that implement independent stream coordination.

The system supports the two following composition types: "serial" compositions and "parallel" compositions. A serial composition is the connection of a set of complex or basic constructions together into a pipelined logical stream. A parallel composition is a hierarchical combination of concurrent time-regulated streams.

The ability to create pipeline-like compositions with time-critical information streams is useful for the same reasons that byte streams, pipes, and filters are so powerful in the UNIX operating system. Serial constructions allow for transformations on single data streams by embedding processing into the time-regulated information stream. Using serial constructions, embedded hardware or software processing elements inherit the stream's time-constraints, becoming time-critical processing elements that manipulate the information. Furthermore, as in the case of message passing in distributed systems, the serial composition of streams makes the physical node boundaries of machines transparent to clients.

Logically independent time-regulated streams that require coordination are realized with parallel constructions. A practical example of such a construction would involve the combining of a pair of (left and right) audio channels into a single logical (stereo) audio stream. This logical audio stream could then be combined with a video channel to create a movie entity. In this way, the logical streams encapsulate all of the necessary coordination information so that they can be used in the same way as primitive elements. Similarly two or more entities can be composed into a higher level one (e.g., four audio channels for quadraphonic audio). Finally, the compositions need not be at the same level of abstraction (e.g., stereo audio, a composition itself, is combined with a primitive video stream).

Time Management

Due to this work's data-flow orientation and emphasis on the multimedia application domain (specifically, continuous media [Anderson 91]), the time regulation mechanisms focus on managing the delivery times of samples within a stream of time-critical information. This model is used even when the activities that require time-regulation consist of aperiodic, asynchronous, or sporadic events.

Synchronization

Time-critical information is regulated according to the notion of minimally constraining (or "weak") synchronization. Weak synchronization recognizes that different applications have varying requirements with regard to the manipulation of time-critical information. For example, applications may be able to specify average rates over some interval, rather than demand hard time constraints. Therefore, the operating system has more flexibility in managing system resources and maximizing system utilization.

Basic constructions take advantage of weak synchronization, when possible, by defining a synchronization interval over which some number of samples are to be transported from the source to the destination transducers. If the number of samples is (N) and the synchronization interval is (T), then the regulation efforts of the basic construction focuses on transporting N samples every T units of time. The flexibility of this approach is derived from the fact that the system makes no guarantees regarding when during the interval T these N samples are delivered. The only presumed guarantee is that the average rate N/T samples/second is maintained across each time interval T.

It is worth noting that applications are not required to define time-regulated transport according to weak synchronization. The model permits application requests that result in strong (versus weak) synchronization. The transport of one sample during a given short interval of time represents such a hard limit.

Software Phase-Locked Loops

The time-regulation and subsequent transport adjustments basic objects make are done through a software phase-locked loop mechanism. The analogy to hardware phase-locked loops is very strong. The regulation activity controls the rate at which samples are emitted from the regulation mechanisms. This rate is adjusted based on information (an error signal) generated by comparing the actual output of the stream to the desired output (which can be varied by an external signal) over a given time interval. The integration period defined by the evaluation time interval acts as a low pass filter in the mechanism's feedback path. This helps ensure stability of the control mechanism and imposes more manageable demands on the underlying kernel mechanisms. The regulation mechanisms generally emit samples at a nominal rate and periodically adjust their phases in an attempt to follow a reference event stream.

Much as with hardware phase-locked loops, there is a continuous clock that is used as the free running reference (i.e. the time-base input), and there is another signal used to speed up or slow down the stream. This provides rate adjustments derived as some time delta from the base frequency. The regulated streams attempt to track changes in the reference event stream by speeding up or slowing down sample delivery. In the event that the reference rate departs from the defined "capture range" of a time regulated stream, the regulation mechanism enters a free-running state. The samples are transported at the stream's nominal rate (or "center frequency"). In this way, lost or delayed reference events do not disrupt the flow of samples from time-regulated streams.

Mechanism Internals

Basic and complex constructions encapsulate some amount of state and a set of logical management activities. Additionally, their active nature requires that they generate and accept asynchronous events and carry out specific policies when certain predefined conditions occur.

Local State Information

A major portion of the state information encapsulated by a time regulation object is comprised of the sample buffer (when one exists). The size of this buffer area is defined in terms of the maximum number of samples that can be stored at a given time. This value is specified at instantiation time and can be dynamically modified (within bounds) at run-time. The buffer area is managed as a FIFO queue, and the instantaneous number of samples in the queue is included as part of the object's state information. The occupancy of the buffer is critical to the maximum effectiveness of the time regulation mechanism.

Therefore, state information is kept which defines the minimum, maximum, and expected occupancy values for the buffer (i.e., the high-water, low-water, and nominal queue depths). Note that the latency imposed by the mechanism is proportional to the depth of the sample queue. In order to support the limit case of minimal latency (at a cost in variations in inter-arrival time or sample jitter), it is possible to specify no buffering for a basic construction.

In addition to the buffer area and its parameters, a number of other parameters are encapsulated by the regulation objects. Some of these are parameters that define the minimum, maximum, and expected rates of the sample producers and consumers, some are statistics accumulated by the system to reflect the current state of the regulated stream (e.g., number of samples to be transferred in this time interval, number of sample actually transferred, etc.), and others are the external adjustments to the rate control process.

Activities

Three logical activities manage time-regulated sample transport between source and destination transducers in basic constructions. Within complex constructions, additional activities are introduced that manage the coordination between the underlying constructs.

Source Activity

The source activity is responsible for managing the source transducer and maintaining the input portion of the buffer (and its associated state). The source activity attempts to keep the nominal number of samples in the buffer. If the queue depth goes above the high-water mark, or overflows, the source activity signals the entities that have registered an interest in such events. Alternatively, if the buffer state indicates that samples are being transported faster than the source is currently producing them, the source activity may, if permitted, try to speed up source sample production to keep the input portion of the buffer at its nominal level.

Destination Activity

The destination activity manages the destination transducer and the movement of samples from the input portion of the buffer to the destination transducer. The destination activity's behavior is essentially identical to that of the source activity. It attempts to deliver samples to the destination at a rate specified by the regulation activity and signals queue under-run events.

Regulation Activity

The heart of time-regulated sample transport is the actual regulation activity. It uses the notion of a software phase-locked loop, described earlier, to evaluate sample transport progress and make adjustments as necessary.

Regulation is activated following each specified integration interval and the current state of the sample stream (i.e., how many samples have actually been delivered), is evaluated against the current desired number (i.e., how many samples should have been delivered — the number given by the user, plus or minus the instantaneous adjustment amount given by other things that the regulator is coordinating with). Based on the evaluation of the current time status of the stream (i.e., ahead, behind, or on-time), the regulation activity can issue modifications to the execution rate of the source and destination activities.

Composite Stream Coordination Activities

The serial and parallel coordination activities extend the regulation concept to a slightly higher level of abstraction. However, the same basic regulatory and adjustment technique is used. The complex construction regulation activities use the expected transport requirements of their components and compares them to actual component sample transport rates during some defined interval. Depending on the comparison of the expected versus actual samples transported over a coordination interval (and the coordination policy), adjustments to speed up or slow down component transport rates are made.

Control Signals

During the course of normal execution, basic and complex regulation objects generate events to indicate interesting state changes to higher-level software and other regulated object in order to effect coordination among time-coordinated sample streams. Likewise, all objects are allowed to receive such signals and use them to adjust their sample transport timing. Current control signal categories include status conditions, aberration conditions, and exception conditions.

In addition, regulation objects receive an external reference signal that is used as a time-base. In the case of a single processing node, the time-base can be the same for all time regulation objects and may be derived from the system clock. In the physically distributed case, a logical common time-base can be provided (e.g., [Mills 90]). It should be noted here that the degree of time control that these mechanisms can exert on a stream can only be as accurate as the time-base that is provided.

It is possible to effectively "gate the clock" (i.e., selectively enable the delivery of time-base signals) in order to perform cueing operations or on demand sample delivery. This supports time-critical activities that depend on asynchronous events. For example, a button push event could be used as a time-base input to cause the regulation activity to request immediate resources for the short-term delivery of a sample. Likewise, these mechanisms can be used to deliver samples at specified queuing times.

Parameterized Policy

Run-time behavior is largely directed by user or client defined policy. The response of objects to status, aberration, and exception conditions is defined by default policies, or policies specified by the user.

Conclusion

The work presented here is a component of a larger system architecture. This layer is intended to provide an internal interface, that mediates interactions between applications and the operating system kernel. We know of no current alternative means of achieving the functionality provided by this facility. As such, it will be difficult to evaluate its effectiveness, but should be easy to find willing and eager test subjects for its use.

This paper describes work in progress, with an initial implementation currently underway. There will undoubtedly be many changes called for as these mechanisms are used in ongoing research and advanced development projects (both within and outside of Sun Microsystems). An effort is underway to identify and exploit opportunities for collaboration in this area in order to speed the development of an effective and broadly applicable facility.

References

[Anderson 91] Anderson, D. P., Govindan, R., and Homsy, G.
Abstractions for Continuous Media in a Network Window System.
International Conference on Multimedia Information Systems, Singapore, January 1991.

[Escobar 91] J. Escobar, D. Deutsch, and C. Partridge
A Multi-Service Flow Synchronization Protocol.
Technical Report, Bolt Beranek and Newman, March 1991.

[Herrtwich 90] R. G. Herrtwich
Time Capsules: An Abstraction for Access to Continuous-Media Data.
IEEE Real-Time Systems Symposium, December 1990.

[Hopper 90] A. Hopper
Pandora — An Experimental System for Multimedia Applications.
ACM Operating Systems Review, 1990.

[HRV 90] High Resolution Video (HRV) Workstation Project
System Support for Time-Critical Media Applications: Functional Requirements
HRV Project Technical Report #90101, November 1990.

[HRV 91a] High Resolution Video (HRV) Workstation Project
High Resolution Video Workstation: Executive Software Specification.
HRV Project Technical Report — in preparation, October 1991.

[HRV 91b] High Resolution Video (HRV) Workstation Project
High Resolution Video Workstation: Video Programming Abstractions.
HRV Project Technical Report — in preparation, October 1991.

[Liu 73] Liu, C. L. and Leyland, J. W.
Scheduling Algorithms for Multiprocessing in a Hard Real-Time Environment.
Journal of the ACM 20(1):46-61, 1973.

[Mills 90] Mills, D.
Network Time Protocol (Version 2) Specification and Implementation.
Network Working Group Technical Report, Delaware, July 1990.

[Northcutt 87] J. D. Northcutt
Mechanisms for Reliable Distributed Real-Time Operating Systems: The Alpha Kernel.
Academic Press, Boston, 1987.

[Northcutt 88] J. D. Northcutt
The Alpha Operating System: Requirements and Rationale.
Archons Project Technical Report #88011, Department of Computer Science, Carnegie-Mellon University, January 1988.

[Steinmetz 90] R. Steinmetz
Synchronization Properties in Multimedia Systems.
IEEE Journal on Selected Areas in Communications, 8(3), April 1990.

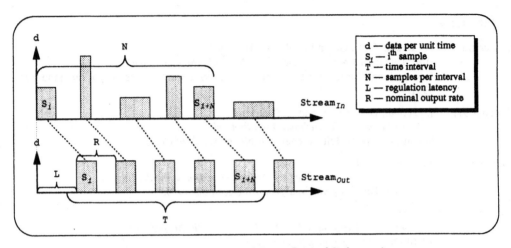

Figure 1 Regulation of Time-Critical Information

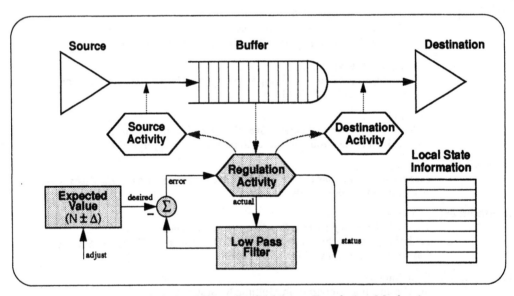

Figure 2 Software Phase-Locked Loop Regulation Mechanism

An Application Framework for Multimedia Communication

Stefan Frey and Daniel.P.Ingold

ETH Zürich

Abstract

Multimedia workstations will not find acceptance unless presented with attractive and easy-to-use applications which maintain a consistent user-interface. Since creating 'nice' applications is quite demanding and their lifetime in the rapidly evolving area of communications pretty short, a framework on which to base might be of some use:

This paper reports on the architecture and implementation of an object oriented framework with special emphasis on multimedia communication.
We introduce Device and Channel objects to control media-streams and to provide their signalling. Additionally we define a Service object as a base for multimedia communication between applications and we extend the existing view concept by new standard MediaViews to present and control audio and video streams.

The development of this framework is part of the ETHMICS project at the Computer Engineering and Networks Laboratory, ETH Zurich/Switzerland [1].

Introduction

The ETHMICS project will provide a workstation platform for integrated multipoint multimedia communication. Special hardware is under development to support broadband communication and integrated video views [2]. The network offers isochronous media-streams of guaranteed bandwidth and it delivers separate channels for each media, one of the channels supporting standard computer communication functions, which are used to control all other media.
Parallel to this hardware development, we work on an object oriented framework, featuring multimedia objects able to present any combination of audio, video and computer data via communication channels which can easily be set-up.
This Framework deals with problems discussed in [3] and [5]. While [4] uses communication objects on operating system level, we try to use them on a more general level within our framework.

Our work is based on the Macintosh™ operating system and MacApp™, a framework covering conventional computer media in an event driven paradigm.
In the following sections we introduce a hierarchy of dedicated objects which manage different kinds of data streams on different system levels. We also describe extensions to MacApp™ objects for displaying and manipulating full motion picture, audio and mixing of this data with conventional computer data.

To keep the system responsive, the workstation platform processes and forwards the media-streams (e.g. video) in hardware. The framework provides also objects that control these hardware or software devices and that direct the data streams within the system.

Framework Architecture

The proposed architecture addresses the following problems:
- Control Abstraction for Multimedia Devices
- Signalling among distributed Devices
- Multimedia Services Objects
- Media Presentation and Control Views.

We introduce a hierarchy of objects which controls the specific Multimedia Devices and which carries out the Signalling among them. These objects have a common origin, which we call 'Stage'. We manage each media stream by a tree of 'Stages' with the stream's source at its root and the data flowing out to the leaves. One tree controls the Multicast of a single media [Figure 1]. For bidirectional communications with multiple media, divers trees will be overlapping crosswise. This should provide enough flexibility to map most physical configurations on their controlling objects. Physical configurations may include custom hardware as well as QuickTime devices.

The 'Stage' defines means for hierarchical notifications and requests within the tree; e.g. they will allow for event delivery and delay estimation. Based on a 'media-type' field in each 'Stage', an application can decide which resources may be tied together to form a chain of stages, a so-called path.
Every path (from root to leaf) is composed of an alternating sequence of Devices and Channels (both descendants of the Stage).

A set of Devices objects may be assembled to form a multimedia 'Service' which is the building block for communication in distributed multimedia applications.
Presentation and Control of media, like full motion video, is implemented by descendants of the 'MediaView' object which, in turn, stems from the 'View' object (the basic visible entity of the underlying framework). Like any basic view of the framework these views can easily be incorporated in an application. Among other MediaViews we put a video-image with its user-handling (scalable, layered, grabbing) and an audio control (volume, mixing) at the programmer's disposal.

Devices and Channels

The Device object specifies some common properties of physical resources such as input, output or intermediary conversion units. A device basically connects an incoming channel to an outgoing one and inherits the stages' ability to forward events between the two.

It defines hooks for
- bandwidth demands/credits
- (restricted) blocking
- jitter measuring/compensation

Device objects are tied together by Channels to form the tree of Stages.
The Channel object encapsulates a 1-to-n propagation of a single media. It sets up the link, requests its bandwidth and forwards information concerning its synchronisation. Fine grain synchronisation (simultaneity) will be dealt with in the context of Services. Coarse grain synchronisation (of temporal relations) exploits the Stages' notification mechanism as a vehicle for annotations. The succession of events transmitted and received by the application may be recorded as a storyboard for presentations.

In the implementation, each Channel object, connecting multiple sites, is in fact distributed (instantiated on all sites involved) to implement the inherited Event propagation mechanism. To carry out the protocol used to open a specific physical channel is the task of a specialised descendant overwriting the methods involved.
We use persistent Stages for local resource management and configuration: Parts of the communication paths are predetermined by the resources which are physically available. These sequences of stages are allocated as persistent objects and their free end-points (devices) are hooked into the set of available local resources.

Distributed Devices

For us, an object is a state machine with fields holding state information and with methods performing state transitions. The object oriented approach provides a powerful tool for specifying extensible modules, but how can these be used to build a distributed application?
We are using events for connecting the objects among different applications and remote machines, since there is no other commonly used way of interaction. The event propagation is carried out by standard process-to-process communication.

We differentiate the following two basic kinds of event handling: We use lower level system events for communication set-up, and highlevel events to address the user (video annotations, conference management, pointing). Highlevel events are dispatched through the applications event queue and event handler, system events, however, asynchronously trigger a method (like a thread) in order to be efficient [Figure 2].
In an event driven system, the objects are extended to change their state in reaction to particular events.

Multimedia Service Object

Service objects are network visible access points for multimedia information exchange between applications. Depending on what media it is composed of, a Service administrates a set of different devices. Service objects are logic entities to build distributed multimedia applications. They are accessible in the network through their workstation and application names and contain directory methods which allow remote services to find out about functions which this service can provide. If common functions between services exists, these services can work together.

Channels attached to common source and destination Services (working in parallel) attempt to assure sufficient simultaneity of the various media belonging together (time continuous composition).

There are numerous ways to support complex interactions between applications - like majority voting, distributed queues or group scheduling. We propose to hide these interactions, usually called 'Groupware Support', within specialised services. Access to their devices would trigger the common channel's voting or queueing algorithm.

User Interface: MediaViews

The Device in the Service that receives a video stream will be a network socket at the head of a chain of Stages: Network socket -> Channel -> image decompressor -> Channel -> video scaling/overlay unit.

The last device in such a chain is controlled and presented by a specialised MediaView, in this case by a 'VideoView':

The VideoView defines the destination area for the overlay unit and allows the user to scale and move the motion image the way he is used to work with any other window on his screen.

The Video-View itself handles the local Mouse events used to drag it around, used to focus on a part of the image and to scale back to a full sight as well as pointing on an image location so that all other devices connected to the same source device get a proper indication.

A VideoView can print out a selected part of the image or 'copy'-'paste' it to any other application. The image size chosen (and the resolution required) is propagated back to the source. On its way, this request may be tuned down to the available bandwidth, such that the source is aware of what bandwidth to deliver.

An audio output device usually does not need a visual presentation of its data. The MediaView attached to it, the so-called 'AudioControl' object, allows for all usual audio manipulations (e.g. volume slider) as well as for mixing multiple channels in conference situations, thereby indicating the active speakers name. If the AudioControl view is attached side by side to a Video-View of the same Service, panorama control of the loudspeakers will reflect the position of the views on the screen.

If a service offers multiple sound channels, e.g. for a film with a separate commentary, both receiving audio Devices may be attached to the same (mixing) AudioControl MediaView or handled separately.

Compiling a Sample Application

A sample application like a picture phone instantiates a service object supervising video and audio devices, possibly also providing a shared pointing device and a blackboard device for synchronous exchange of textual and/or graphical information.

To do this, the application must search the set of available devices for the device-types needed and attach them to its service object.

Based on the user's needs, the application will then present the user with a standard 'Browser'. He may then select among remote applications with services that can exchange information with its own service object. That means that the service describer (list of device-types) of the local Service intersects with the service describer of the remote service.

For each device of the service chosen, the local service assigns a channel object to the corresponding local device. It is then the channel object's task to get the media stream delivered to the sink, i.e. to request the physical channel [Figure 3].

The interface of such a sample application would present the user with video-views and audio-controls. These MediaViews might be placed in a common window, which a user can manipulate as usual.

Summary

In the previous sections we introduced an object hierarchy for multimedia communication between applications on multimedia workstations. We extend existing visible entities by methods of presenting and controlling video- and audio-streams. The framework proposed provides the application developer with re-usable code for service access, with a consistent graphical interface and an exchangeable 'black box' for link set-up.

References

[1] A. Kündig, "Multimediakommunikation" Eidgenössiche Technische Hochschule Zürich, Jahresbericht 90

[2] U. Röthlisberger, D. Ingold "An Architecture for an Advanced Multimedia- Workstation" Proceedings Multimedia '90

[3] Thomas D. C. Little, Arif Ghafoor "Network Considerations for Distributed Multimedia Object Composition and Communication" IEEE Network Magazine, Nov 90, Vol. 4, No. 6

[4] W. H. Leung et al. "A Set of Operating System Mechanisms to Support Multi-Media Applications"
1988 International Zurich Seminar on Digital Communication

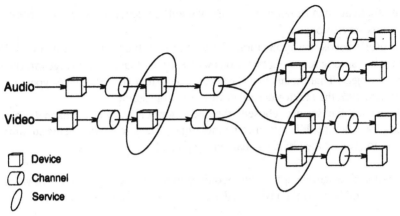

Figure 1: Multicast

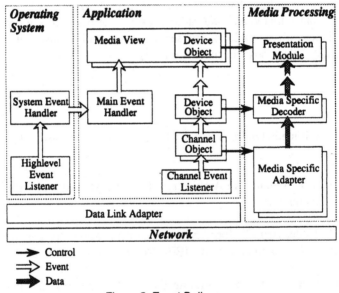

Figure 2: Event Delivery

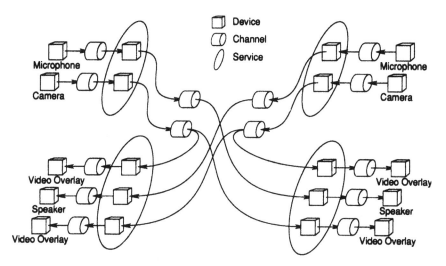

Figure 3: Picture Phone Application

Session VIII: Multimedia Abstractions II

Chair: Venkat Rangan, University of California at San Diego

The eighth session was devoted to new abstractions for programming multimedia applications, and tools for specifying their execution and presentation. Programming abstractions for multimedia have not received much attention before, and are becoming increasingly important as multimedia applications grow in both complexity and number.

In the first paper of this session, Simon Gibbs of the Universite de Geneve together with Christian Breiteneder, Laurant Dami, Vicki de Mey, and Dennis Tsichritzis presented "A Programming Environment for Multimedia Applications." The environment consists of two levels: A systems level environment for handling synchronization between media, and a user level environment for specifying media presentation.

The systems level environment encapsulates hardware dependencies, and enforces synchronization between media objects which can be sources, sinks, or filters, depending on whether they produce or consume media values, or both. The user level environment provides an object-oriented framework in which, application designers can specify in a scripting language the composition, presentation, and communication among media objects. Sequences of presentations constitute activities, and multiple activities can be executed sequentially, concurrently, or periodically.

The second paper was by Gerold Blakowski, Jens Hübel, and Ulrike Langrehr of the University of Karlsruhe ("Tools for Specifying and Executing Synchronized Multimedia Presentations"). Gerold presented tools for managing synchronized presentation of multimedia in distributed heterogeneous environments.

Users can define objects of type MediaInformation, Presentation, or Transport. A graphical synchronization editor helps users to specify both intra-object and inter-object synchronization using a context-free grammar based description language. From these synchronization specifications, the synchronizer constructs a presentation thread of all the media objects, using which it enforces the synchronization relationships. The synchronizer also takes actions such as pause/wait or acceleration/skip in order to speed up a lagging media presentation or slow down a leading media presentation.

The entire media synchronization management system including the graphical editor is being designed as part of the MODE project in the NESTOR authoring-learning environment being developed at the Universities of Karlsruhe and Kaiserslautern.

The last paper by Ralf Steinmetz and J. Christian Fritzsche of IBM European Networking Center and Johann-Wolfgang-Goethe University of Frankfurt ("Abstractions for Continuous-Media Programming"), presents an interesting new concept in which continuous media are treated as data types within programming languages.

Christian elaborated that earlier approaches commonly use multimedia libraries and toolkits for removing hardware and implementation dependencies of multimedia applications. However, neither libraries nor toolkits permit full integration of multimedia into programming environments; the encapsulation they provide frequently comes at the expense of performance, and they are not efficiently supported by the operating system. Simply adding abstract data types to programming languages does not suffice to adequately express synchronization, communication, and parallelism in multimedia applications.

The authors propose rather to use media as fundamental data types within high-level languages, thereby greatly enhancing the expressive power within multimedia applications. Using an object-oriented language, starting with media devices and their data units as fundamental objects, higher-level media abstractions are built using class hierarchies. Media objects can also have a lifetime, in which case they become active objects.

In summary, this session touched upon two significant aspects of multimedia systems: synchronization and programming abstraction. Interesting questions were raised as to whether synchronization mechanisms can assume the existence of globally synchronized clocks or not. Another interesting issue was whether media such as video and audio must be regarded as composed of frames and samples, or higher-level objects such as images and phonemes. These topics will continue to assume increasing importance as the use of computer systems to support digital continuous media becomes more and more pervasive.

A Programming Environment for Multimedia Applications

Simon Gibbs, Christian Breiteneder,
Laurent Dami, Vicki de Mey, Dennis Tsichritzis

Université de Genève[1]

Abstract

A programming environment for the development of multimedia applications is described. The environment is based on a two-level architecture: a systems-oriented *framework-level* concerned with hardware control and synchronization and a user-oriented *scripting-level* concerned with presentation specification. The two levels are outlined.

1. Introduction

It is perhaps self-evident that programming environments providing specific support for multimedia would facilitate the development of multimedia applications. Looking more closely, however, one can discern many reasons why such environments are needed:

First, multimedia involves concepts from audio recording, video production, animation, and music – concepts that are novel to many programmers. Additionally, multimedia operations often involve special hardware, leading to lack of portability and longer development times. Information about media properties and hardware dependencies should be incorporated within the programming environment.

Second, the equipment used by the traditional "composers" of multimedia (video professionals, music editors, etc.) is relying more and more on digital technology and consequently becoming more and more programmable. Until fairly recently, multimedia equipment could be viewed as interconnectable hardware "boxes" (recorders, mixers, monitors, etc.). Now, however, with software controllable components playing a greater role in both producing and transforming multimedia data, such a model breaks down. We must recognize that software-related concepts, such as process and operation, are an integral part of multimedia and a general environment for multimedia must allow for the incorporation of software-based components with the traditional hardware-based components.

A third reason for considering a multimedia programming environment is that complex user interfaces, such as virtual realities can be viewed as multimedia applications. Yet such applications, because of novel interface devices (e.g., stereoscopic displays, head-position and orientation trackers) and the need for real-time performance, are difficult to construct with current programming tools.

Finally, we hope that multimedia programming environments, and their associated concepts, may lead to a general model for describing and developing a wide range of multimedia appli-

---

1. *Authors' address:* Centre Universitaire d'Informatique, 12 rue du Lac, CH-1207 Geneva, Switzerland.
Email: {simon, chris, dami, vicki, dt}@cui.unige.ch
Tel: +41 (22) 787.65.80
Fax: +41 (22) 735.39.05

cations. At present there seems to be a tendency to develop multimedia applications in an *ad hoc* one-of-a-kind fashion – so, we believe, a unifying conceptual model is needed.

This paper describes a programming environment for developing multimedia applications. The requirements for the environment are that it:

- be based on a simple conceptual model of multimedia functionality, yet one which is general enough to capture the variety of multimedia, including sound, video, music, and animated sequences,
- be easy to use and not require expertise in multimedia technology, yet be open and extensible so that more experienced programmers are not constrained,
- encapsulate hardware dependencies,
- allow complex multimedia effects, for example the synchronization of an audio and video signal, or the juxtaposition of two video signals.

One approach for such an environment is a two-level architecture consisting of a system-oriented layer and a user-oriented layer. The first, called the "framework level," is concerned with hardware control and synchronization, the second, called the "scripting level," is concerned with presentation specification.

2. The Framework Level – Multimedia Objects

The framework level has been described elsewhere [5][6], here we provide just a summary[1]. The starting point is the use of data types to characterize media information:

Definition: A *media value*, v, of data type D, is a (finite) sequence d_i, where the encoding and interpretation of the d_i are governed by D. In particular D determines how the *presentation* of v (the physical realization of v, within some medium, over some time interval) can be obtained from the d_i. Presentation of v takes place at a rate r_D, the *data rate* of D. This rate indicates the number of sequence values presented per second.

Media values are related to *media objects*, these are defined as follows:

Definition: A *media object* is an active object which produces and/or consumes media values (of specified types) at their associated data rates.

Active objects, like ordinary or *passive* objects, have state (instance variables) and behavior (methods). In addition, each active object is associated with a process which may be running even if no messages have been sent to the object.

Each media object can be viewed as a collection of *ports*. A port has a (media) data type and is used either for input or output. Media objects are divided into three categories: *sources*, *sinks*, and *filters*. A source produces media values, a sink consumes values, and a filter both produces and consumes.

Informally, *multimedia values* are aggregates of media values, while *multimedia objects* are aggregates of media objects. How these aggregates are formed is discussed in section 2.2.

Media objects and media values make use of two inheritance hierarchies. For example, the class LaserDiscPlayer would be (ultimately) a subclass of MediaObject, similarly LaserDiscVideo would be (again ultimately) a subclass of MediaValue.

---

1. It should perhaps be pointed out that there has been a change in terminology from references [5] and [6]. In particular, what were previously called Multimedia, MultimediaObject and CompositeMultimediaObject are now MediaValue, MediaObject and MultimediaObject respectively.

2.1 Operations on Media Objects

All classes of media objects inherit methods from the class ActiveObject and the class MediaObject. A schematic and partial C++ specification of these classes is:

```
class ActiveObject {
public:
    bool                Start();
    bool                Stop();
    bool                Pause();
    bool                Resume();
};

class MediaObject {
public:
    //
    // temporal coordinates
    //
    objectTime          CurrentObjectTime();
    objectTime          WorldToObject(worldTime);
    worldTime           CurrentWorldTime();
    worldTime           ObjectToWorld(objectTime);
    objectInterval      WorldToObjectI(worldInterval);
    worldInterval       ObjectToWorldI(objectInterval);
    //
    // composition
    //
    void                Translate(worldTime);
    void                Scale(float);
    void                Invert();
    MultimediaObject    Parent();
    //
    // synchronization
    //
    void                Sync(worldTime);
    worldInterval       SyncInterval();
    worldInterval       SyncTolerance();
    SyncMode            SyncMode();
    void                Cue(worldTime);
    void                Jump(worldTime);
};
```

At any time an ActiveObject is in one of three states: IDLE, RUNNING or SUSPENDED. The methods of ActiveObject are used to change state.

The class MediaObject makes use of two temporal coordinate systems: *world time* and *object time*. The origin and units of world time are set by the application. The origin would normally be set to coincide with the beginning of presentation activity. World time would run while the activity is in progress, and be stopped or resumed as the activity is stopped or resumed.

Object time is relative to a media object. In particular, each object can specify the origin of object time with respect to world time and the units used for measuring object time. (Normally these units relate to the data rates of the object's ports.) Furthermore, each object can specify the *orientation* of object time, i.e., whether it flows forward (increases as world time increases) or backwards (decreases as world time increases).

2.2 Composition and Synchronization of Multimedia Objects

The framework level provides a technique for aggregating media objects. This technique, called *temporal composition*, is used to form *multimedia objects*.

The motivation for temporal composition comes from the need to model situations where a number of media components are simultaneously presented. Television and films are two obvious examples, each containing both audible and visual components.

Definition: A *multimedia object* is a media object containing a collection of *component* media objects and a specification of their temporal and configurational relationships.

The two groups of relationships specified by a multimedia object are used for different purposes. In particular:

- *temporal relationships* – indicate the synchronization and temporal sequencing of components.
- *configurational relationships* – indicate the connections between the input and output ports of components.

A composite, c, maintains synchronization by attempting to assure

$$\text{Abs(c.CurrentWorldTime() - } c_i.\text{CurrentWorldTime()}) < c_i.\text{SyncTolerance()}$$

for each activated component c_i. However, because of the varying nature of components, multimedia objects must be flexible and support a variety of synchronization techniques. In the framework, each component has a *synchronization mode* attribute. Depending on the value of this attribute, which can be queried by the method SyncMode, the multimedia object adopts different approaches to synchronization. Presently there are four synchronization modes: NO_SYNC, DEMAND_SYNC, TEST_SYNC, and INTERRUPT_SYNC (see [6] for further details).

3. The Script Level – Scripts and Activities

In the previous section we discussed techniques to combine media objects in order to obtain more elaborate behavior. Such composition techniques are very powerful but their proper application depends on two important constraints. First, the objects to be composed have to be well understood both individually and in partnership with other relevant objects. Second, composition requires programming, i.e, it is both tedious and error prone. In this section we discuss a higher-level way to specify composites called "scripting" [7]. Scripting is based on a *scripting model* which defines the allowed ways that objects can be composed. In this manner many of the details of the composition do not have to be explicitly stated. In addition, scripting smooths over certain incompatibilities between objects and allows the composition of objects which have not been *a priori* designed to work together.

We now show how the notion of scripting can be applied to multimedia. We first provide a definition of "script" for the multimedia programming environment:

Definition: A *script* is an instance of a *script class*. Script classes are specializations of the class of MultimediaObject.

Scripts differ from multimedia objects in that there are constraints on the types of components allowable within a script and, possibly, constraints on their configuration. These constraints are part of the specification of script classes.

A scripting language can be used to specify scripts. Interpretation of such a language relies on the scripting model. For multimedia, the scripting model contains:

1. multimedia hierarchies

   The scripting model knows about MediaValue and MediaObject classes and their respective subclass hierarchies.

2. connection types

   We have not discussed component connections in detail, but the framework level supports a number of different types of connections. Examples are connections corresponding to communication by message passing, buffering, or physical cable.

3. ports

   For each media object class the scripting model contains port descriptions. A port description identifies whether the port is for input or output, the media data type of the port, the connection types which can be attached to the port, and whether the port accepts multiple connections.

4. object interfaces

   Part of the interface of a media object is only used within the framework level whereas other parts are available for scripting (these correspond to the *FII* or *framework internal interface* and the *FEI* or *framework external interface* described in [3]). The scripting model identifies the FII and FEI for each media object class. For instance, "temporal transformations" (e.g., the Scale and Translate methods) belong to the FEI for all media object classes, while synchronization methods belong to the FII.

5. script membership constraints

   The scripting model contains the constraints on component types for the various script classes.

6. script configuration constraints

   Script classes may specify constraints on the configuration of components and connections within their instances. This information is part of the scripting model.

The scripting language contains two main constructs: scripts themselves and *activities*. A script is specified by combining activities; an activity, in turn, is either a script or a media object. (Consequently a script reduces, at the framework level, to a multimedia object.) There are three operators used to combine activities:

- $a_1 >> a_2$: sequential execution. Activity a_2 will be scheduled after the completion of a_1.

- $a_1 \ \& \ a_2$: parallel execution. Activities a_1 and a_2 start together.

- n^*a : repeated execution. Activity a is repeated n times.

Examples of scripting facilities are described in [1][2][4].

4. Conclusion

The above has outlined an approach for constructing an environment for programming multimedia applications. The environment contains two layers: a system-oriented layer consisting of an object-oriented class framework, and a user-oriented layer based on a scripting language. At the moment we are refining the environment's design by developing a demanding "driver" application [8].

References

[1] Dami, L, Fiume, E., Nierstrasz, O. and Tsichritzis, D. Temporal Scripts for Objects. In *Active Object Environments*, (Ed. D. Tsichritzis) Centre Universitaire d'Informatique, Université de Genève, 1988.

[2] Dami, L. Musical Scripts. In *Active Object Environments*, (Ed. D. Tsichritzis) Centre Universitaire d'Informatique, Université de Genève, 1988.

[3] Deutsch, L.P. Design Reuse and Frameworks in the Smalltalk-80 System. In *Software Reusability, Vol. II*, (Eds. T.J. Biggerstaff and A.J. Perlis) ACM Press, 57-71, 1989.

[4] Fiume, E., Tsichritzis, D., and Dami, L. A Temporal Scripting Language for Object-Oriented Animation. *Proc. Eurographics'87*, North-Holland, 1987.

[5] Gibbs, S. Composite Multimedia and Active Objects. *Proc. OOPSLA'91*, 97-112.

[6] Gibbs, S., Dami, L., and Tsichritzis, D. An Object-Oriented Framework for Multimedia Composition and Synchronisation, *Eurographics Multimedia Workshop*, Stockholm, 1991.

[7] Nierstrasz, O., Dami, L., de Mey, V., Stadelmann, M., Tsichritzis, D., and Vitek, J. Visual Scripting: Towards Interactive Construction of Object-Oriented Applications. In *Object Management*, (Ed. D. Tsichritzis) Centre Universitaire d'Informatique, Université de Genève, 1990.

[8] Tsichritzis, D. and Gibbs S. Virtual Museums and Virtual Realities. *Proc. of the International Conference on Hypermedia and Interactivity in Museums*, 17-25, 1991.

Tools for Specifying and Executing Synchronized Multimedia Presentations

Gerold Blakowski, Jens Hübel, Ulrike Langrehr

University of Karlsruhe, Institute for Telematics
Zirkel 2, D-W7500 Karlsruhe, Germany
Phone: (++49) (721) 608-3414, Email: blakowski@ira.uka.de

Abstract

Multimedia applications require the handling of synchronization between media streams. We present tools for creating, editing and presenting synchronized multimedia objects. Specifying synchronization is supported by the graphical Synchronization Editor and execution of multimedia presentations is performed by the Synchronizer.

Major topics cover requirements resulting from underlying distributed heterogeneous environments and the consideration of user interactions. The tools we have developed are not restricted to a fixed set of media, but support the inclusion of arbitrary user-defined media.

The Synchronization Editor and the Synchronizer are part of the MODE project, that is used for the handling of distributed multimedia objects in NESTOR, an authoring-learning environment in development at the Universities of Karlsruhe and Kaiserslautern in cooperation with the CEC Karlsruhe, Digital Equipment.

1 Introduction

The integration of time dependent media in applications requires the synchronization of media streams. In this paper we describe object-oriented tools that support the graphical specification and execution of synchronized multimedia presentations.

The developed components are the user-friendly graphical Synchronization Editor and the Synchronizer covering the requirements resulting from distribution of media objects in a heterogeneous environment. These requirements include the handling of error conditions as well as the support of alternative presentation forms and different presentation qualities for the heterogeneous platforms. Other important requirements met by the tools are the integration of any kind of user-defined media and the support of user interaction.

The system is part of the MODE project (Multimedia Objects in a Distributed Environment) [Bla90, Bla91] that is used in NESTOR (Networked Systems for TutORing) [Mue91], an authoring-learning environment that is in development at the

Universities of Karlsruhe and Kaiserslautern in cooperation with the CEC Karlsruhe, Digital Equipment.

The overall synchronization system in MODE consists of four components:

- the *Synchronization Editor* that is used to create synchronization and layout specifications for multimedia presentations;

- the *Global Synchronization Coordinator* that coordinates the creation of presentation units (*presentation objects*) from the basic information units (*information objects*) and the transport of the objects in the distributed system;

- the *Synchronizer* that receives the presentation objects from the Global Synchronization Coordinator and initiates their local presentation according to the synchronization specification;

- the *Optimizer* that chooses presentation qualities and presentation forms depending on user demands, network and workstation capabilities and presentation performance.

In the following, after an overview about the related work, our synchronization model, the Synchronizer and the Synchronization Editor are described in more detail.

2   Related Work

Synchronized multimedia presentations, called *multimedia objects*, are composed of *single-media objects*, multimedia objects and synchronization information. This allows to create *synchronization hierarchies*. Examples for single-media objects are a video sequence, a piece of text or a picture.

Several ways of describing multimedia synchronization have been published:

Hierarchical synchronization: Multimedia objects are regarded as a tree consisting of nodes that denote serial or parallel presentation of the outgoing subtrees.

Synchronization on a time axis: Single-media objects are attached to a time axis that represents an abstraction of time.

Synchronization at reference points: Single-media presentations are composed of subunits presented at periodic times. A position of a subunit in an object is called a reference point. Synchronization between objects is defined by connections between subunits of different objects that has to be presented at the same time.

Synchronization described by means of hierarchical structures [AFN89, SS89] is based on the two main synchronization operations:

- Serial synchronization of actions

- Parallel synchronization of actions

An action is either an atomic action or a compound action. An atomic action handles a presentation of a single-media object, a user input or a delay. Compound actions are a combination of synchronization operators and atomic actions.

The introduction of a delay as possible action [LG90] allows to model further synchronization behavior like delays in serial presentations and delayed presentation of a single object in a parallel synchronization.

Hierarchical structures are easy to handle and are widely used. Restrictions from the hierarchical structure arise by the fact that each action can only be synchronized at its beginning or end. For example presenting a subtitle at a certain scene of a video stream requires dividing the video stream into serial components. In the same way a synchronized multimedia object used as a component in another synchronization can not longer be regarded as abstract unit, if it has to be synchronized at a point between beginning and end of its presentation. Therefore abstractions from the internal structure of multimedia objects cannot be achieved in general. Additionally, synchronization conditions exist that cannot be represented in hierarchical structures. An example are three objects that are presented in parallel and each object is synchronized once with each other object but independent of the third object.

Besides hierarchical structures synchronization can also be defined by means of a time axis. In the Athena Muse project [HSA89] a synchronization is described by attaching all objects independent of each other to a time axis. Removing one object does not effect the synchronization of the other objects.

With modifications this kind of specification is used in the model of active media [TGD91]. A world time is maintained which is accessible to all objects. Each object can map this world time to its local time and moves along its local time axis. If it detects a discrepancy between world time and local time exceeding a limit it has to synchronize to world time again.

A time axis based mechanism is used in QuickTime [LM91], too.

Synchronizing objects by means of a time axis offers the opportunity of abstractions from the internal structure of single-media objects and multimedia objects used in further synchronizations. Defining the beginning of a presentation of a subtitle relative to a scene in a video stream requires no knowledge about the related video frames. Problems arise if objects have no deterministic time of presentation (like presentations of objects depending on user interactions) because synchronization can only be defined using fixed points of time and if synchronization based on one common world time is not sufficient to express the synchronization conditions between presentation streams. Depending on the coherence of the presentation streams a synchronization by using a common time axis might be to strong or to weak.

A third way of defining synchronization is a description of timing relations between objects without explicitly referencing time. Steinmetz [Ste90] proposes such a model. Dynamic objects such as video or audio are regarded as composed of periodic presentations of subunits for examples video frames or audio samples. In this case synchronization is specified by denoting subunits in objects that shall be presented at the same moment.

Like synchronizing by means of a time axis this description allows synchronization not only at the beginning or end of presentations but also during a presentation. Synchronizing at reference points requires in contrast to hierarchical structures mechanisms for detecting inconsistencies. It is impossible to create delays during a multimedia presentation using only reference points of objects. To solve this problem Steinmetz proposes time specifications. A time specification introduces a delay at a certain point of synchronization.

We are using a synchronization model based on synchronization at reference points extended by means of handling fixed intervals of time, objects of unpredictable duration and conditions resulting from the underlying distributed heterogeneous environment.

3 Synchronization Model

MODE allows to give classes one or more of the attributes *information class, presentation class* and *transport class*. Objects of information classes (*information objects*) are the basic information units. If information objects should be presented, they generate one or more objects (*presentation objects*) of suitable presentation classes. Presentation objects contain all data and methods to display themselves and can exist independently from the information object. Especially they give the choice to transport the basic information unit or to transport its presentation in the case of a remote presentation. If an information or a presentation object has to be transported, it is converted into an object of a transport class (*transport object*) for the duration of the transport and on the other node it is reconverted. Transport objects consist of all methods and data necessary to transport and compress the object.

The generation, display and conversion/reconversion operations are initiated by calling corresponding methods on the objects. Additionally the information objects offer methods that give information about available presentation forms or timing behavior of presentation objects. Using these method interfaces of the information and presentation objects, the Synchronization Editor and the Synchronizer can handle any kind of user-implemented media.

A synchronization specification created with the Synchronization Editor and used by the Synchronizer is stored in text form following a syntax defined in a context free grammar (Synchronization Description Language). This allows usage of the synchronization specification by MODE components independent of their implementation language and environment.

An overview about the architecture is shown in figure 1.

Information objects may support various presentation forms and qualities. Presentation forms supported by a text information class may be a presentation as text on the screen or as audio sequence generated by a speech synthesizer. Presentation qualities of a video sequence are color depth, resolution and frames per seconds. If an information object is used in a multimedia object its properties have to be determined more precisely. In addition an information object should not be influenced by its utilization in a multimedia object. Therefore *basic objects* are introduced. A basic object consists of a reference to an information object and a collection of presentation attributes.

We distinct between *dynamic basic objects* and *static basic objects*. A presentation of a dynamic basic object is composed of a sequence of presentation objects (for example single frames of a video sequence) presented at periodic times. The synchronization of such a single-media sequence is called *intra-object synchronization*. The index of each presentation object is called a *reference point*. A presentation of a static basic object such as displaying a piece of text or a picture has only two reference points, the beginning and the end of the presentation.

The synchronization between presentations of basic objects (*inter-object synchronization*) is defined using reference points. A *synchronization element* is a combination of a reference point and the corresponding basic object (BasicObject.ReferencePoint). It denotes a position in an information object presentation. Two ore more synchronization elements are defining a *synchronization point* that is the base of the inter-object synchronization. The list of all synchronization points specifies the whole inter-object synchronization.

A reference point may be a natural number, referencing for example the number of a video frame, or a fraction between 0 and 1 that we call a *relative reference point*. Before presentation, relative reference points are mapped to their nearest non-relative reference point. Since multimedia objects have no obvious external reference points they may only be synchronized relatively.

Inter-object synchronization at reference points has the following advantages:

- It is possible to choose between loosely and tightly coupled synchronization of presentations of dynamic objects. This is much more flexible than simply using one absolute or virtual timer as common synchronization base.

- The synchronization of objects with non-predictable presentation duration can be handled easily.

- The synchronization points can be maintained if presentation objects are delayed.

- Useful manipulations such as fast-forwarding, rewinding or slow-motion can simply be realized by changing the periodicity of the presentations of the dynamic objects. This does not effect the synchronization points a multimedia presentation consists of.

Two special kinds of objects are supported additionally, timers and interactive objects.

Timers are objects with the ability to synchronize at reference points that occur in 1 ms steps. Timers are used to model absolute delays like in a slide-show where every 10 s the slide changes. Also the synchronization based on one common time axis can be modeled in our synchronization schema by using one timer.

Interactive objects are objects synchronized only at beginning and end that have the additional ability to communicate with the user. They are examples for objects of non-predictable presentation duration. Depending on user interactions interactive objects can for example use methods offered by the Synchronizer. These methods cause a starting or stopping of a presentation of a basic object, simulate fast-forwarding or rewinding of multimedia objects, or move the presentation to a certain point in a multimedia object.

For every basic or multimedia object that is part of a multimedia object one or more *alternative presentations* can be specified that are used if the preferred presentation can not be used due to limitations of the workstation or network resources. If a workstation has for example no audio device an alternative presentation form might be subtitles. Each alternative may add additional synchronization elements that replace those of the preferred object.

For each basic object a presentation quality can be specified. The presentation quality is described by a set of attributes consisting of an attribute name, a preferred value, a value domain that describes all possible values for this attribute and a priority. Before starting a presentation the Optimizer is called that chooses a value from the domain according to user demands and the capacity of network and workstation resources. If possible, the preferred attribute value is chosen. The priority of an attribute describes the importance of its attribute value relative to other attributes.

Three types of actions are used to define a behavior in exception conditions occurring during a synchronized presentation.

Waiting action: A presentation of a dynamic basic object has reached a synchronization point and waits longer than a specified time at this synchronization point. Possible actions are displaying the last presentation object, pausing or ignoring the synchronization point.

Acceleration action: A presentation of a dynamic basic object has reached a synchronization point. Acceleration actions are necessary for other presentations of dynamic basic objects, if they are involved in this synchronization point and will need longer than a specified time to reach it. Possible actions are to temporarily increment the presentation speed or to skip all objects in the presentation up to the synchronization point and to continue there immediately.

Skipping action: A presentation object to present next has not arrived in time. A possible action might be to skip the object and to present the next one.

A basic object may have a priority that indicates its sensitivity to delays of its presentation. For example, audio objects will in general be assigned higher priorities than video objects because a user recognizes delays in an audio-stream earlier than those in a video-stream. Presentations with higher priorities are preferred in presentation and synchronization to objects with lower priorities.

Depending on the timing behavior during a presentation, a graceful degradation or enhancement of presentation quality is possible.

Using synchronization at reference points causes problems if parts of an information object that are used in multimedia objects are changed. Synchronization is lost, for example, if a scene in a video object is inserted or removed. Therefore a kind of version controlling is necessary that allows mapping synchronization from one version to another.

4 The Synchronizer

The task of the Synchronizer [Hue91] is to perform the synchronized presentation (figure 2) according to the introduced synchronization model. This comprises the intra-object and the inter-object synchronization.

For each intra-object synchronization a *presentation thread* is created managing the presentation of a dynamic basic object. The priorities of these threads are used to implement priorities of basic objects. All presentations of static basic objects are managed by a single thread.

Synchronization is performed by a signaling mechanism. Each presentation thread reaching a synchronization point sends a corresponding signal to all other presentation threads involved in the synchronization point. After having received such a signal other presentation threads may perform acceleration actions, if necessary. After having sent all signals the presentation thread waits until it receives all signals from the other participating threads of the synchronization point and may perform a waiting action.

The presentation threads are also responsible for executing skipping actions.

The Synchronizer also supports joining of different basic objects. Joining is useful for example to reduce transport costs, if bitmaps for video frames and subtitles are created on the same node. In such a case they may be joined at the node of creation to a single bitmap containing video and text information.

5 The Synchronization Editor

A specification of a synchronization in a textual form is difficult. In text-oriented languages it is not easy to imagine the timing and layout relations between the objects. The graphical Synchronization Editor [Lan91] eases the specification of synchronization and reduces the time for creating multimedia objects.

Describing multimedia objects the user has to regard three dimensions. Beside the two spatial dimensions that are used in most applications only dealing with static media, the time has to be regarded as third dimension.

Editing a three-dimensional view [OHK90] on the screen raises some problems. The position of three-dimensional elements on two-dimensional input and output devices is not unique. Since multimedia objects consist of a collection of coherent objects, many objects will be hidden by others. To avoid these problems we use three types of two-dimensional views of a multimedia object.

In the first kind of view, the *Presentation View* (figure 3), the user can run single objects independent of each other. Similar to a tape recorder he can stop, pause, restart, fast-forward and rewind them. This presentation is supported by the Synchronizer. By viewing the objects of the synchronization it is easy to find suitable synchronization points. A synchronization point between all currently presented objects is created by clicking on a button. In the same way the user can switch to other synchronization points. To get a quick impression of the edited multimedia object the user can regard it in the same way as the single objects.

For visualizing and directly editing the timing relations between the objects the second view - the *Time View* (figures 4 and 5) - is used. It shows all objects ordered at a horizontal time axis. Objects are represented by rectangles. Their width indicates their expected duration. A synchronization point is denoted by a vertical line between the objects. Editing facilities comprise adding, moving, deleting of synchronization points and basic objects and selection of points of time in the synchronization.

Since not all objects are presented during the whole representation, the layout of the multimedia object changes in time. That is why the layout cannot be regarded independently of the synchronization. A third view - the *Layout View* (figure 6) - is implemented in order to edit the layout at a certain point of time in the synchronization. It uses a layout model [Emm91] that allows to specify absolute and relative distances that are needed if the presentation layout is resized at runtime.

More instances of each type of view can be opened in parallel, for example several open Layout Views may show snapshots at different points of time.

6   Summary

Tools for specifying and executing multimedia presentations have been presented. The Synchronization Editor allows the user to graphically specify the synchronization of multimedia objects. The actual local presentation is done by the Synchronizer that initiates and coordinates the presentations.

The Synchronizer cooperates with the Optimizer and the Global Synchronization Coordinator that is responsible for the transport of the presentation units.

A standardized interface to the media objects enables the tools to integrate any kind of user-defined media.

The object-oriented synchronization model that is used by the tools is an extension of the synchronization at reference points and allows the integration of objects of unpredictable duration, user interactions and timers.

Designing the synchronization system, we have taken care of the difficulties in presenting multimedia objects raising from the underlying distributed heterogeneous environment. These difficulties are different capacities of the available resources and possible errors like network delays and loss of presentation units.

Therefore, on specifying the synchronization the user can determine alternative presentations to objects, that will be used if the first choice is impossible because of resource conditions. The user can also specify waiting, acceleration and skipping actions to handle delayed or lost presentation units.

The synchronizer uses the presentation form and quality, that is selected by the Optimizer regarding the existing resources and user demands. If an exception condition occurs at runtime the synchronizer performs the specified actions.

Our practical experiences have shown the usefulness of the presented schema. Further work is on the full integration with the Global Synchronization Coordinator and the Optimizer as well as an implementation based on realtime scheduling.

References

[AFN89] AFNOR Expert group, *Multimedia synchronization: definitions and model, Input contribution on time variant aspects and synchronization in ODA-extensions, ISO IE JTC 1/SC 18/WG3*, February 1989.

[Bla90] Gerold Blakowski, "Supporting Multimedia Information Presentation in a Distributed, Heterogeneous Environment", in *Proceedings 2nd IEEE Workshop on Future Trends of Distributed Computing Systems, Cairo, Egypt, September 30 - October 2, 1990*, pages 29–35, Los Alamitos: IEEE Computer Society Press 1990.

[Bla91] Gerold Blakowski, "The MODE-FLOW-GRAPH: A Processing Model for Objects of Distributed Multimedia Applications", to be published in *Proceedings International Symposium on Communication*, Tainan, Taiwan, December 1991.

[Emm91] Richard Emmerich, "Design and Implementation of a Geometry Manager in X-Windows". Diploma thesis (in German), University of Karlsruhe, Institute for Telematics, Germany, April 1991.

[HSA89] Matthew E. Hodges, Russell M. Sasnett, and Mark S. Ackerman, "Athena Muse: A Construction Set for Multi-media Applications", *IEEE Software*, pages 37–43, January 1989.

[Hue91] Jens Hübel, "Design and Realization of a Runtime System for the Synchronized Presentation of Multimedia Data Streams". Diploma thesis (in German), University of Karlsruhe, Institute for Telematics, Germany, September 1991.

[Lan91] Ulrike Langrehr, "Design and Realization of a Graphical User Interface for the Definition of the Synchronized Presentation of Multimedia Data Streams". Diploma thesis (in German), University of Karlsruhe, Institute for Telematics, Germany, September 1991.

[LG90] Thomas D. C. Little and Arif Ghafoor, "Synchronization and Storage Models for Multimedia Objects", *IEEE Journal on Selected Areas in Communications*, 8(3):413–426, April 1990.

[LM91] D. Littman and T. Moran, "Quicktime: It's about Time", *Mac World*, pages 80–81, August 1991.

[Mue91] Max Mühlhäuser, "Computer Based Learning with Distributed Multimedia Systems", in Hans Jürg Bullinger, editor, *Human Aspects in Computing, Proceedings of the IVth International Conference on Human Computer Interaction, Stuttgart, Germany, September 1-6, 1991, Volume 2*, pages 953–958, Amsterdam: Elsevier 1991.

[OHK90] Ryuichi Ogawa, Hiroaki Harada, and Asao Kaneko, "Scenario-based Hypermedia: A Model and a System", in *Hypertext: Concepts, Systems and Applications, Proceedings of the European Conference on Hypertext INRIA, France, November 1990*, pages 39–51, 1990.

[SS89] Michael Salmony and Doug Shepherd, "Extending OSI to Support Synchronization Required by Multimedia Applications", Technical Report 43.8904, IBM European Networking Center, Tiergartenstr. 8, 6900 Heidelberg, P.O.Box 103068, Germany, April 1989.

[Ste90] Ralf Steinmetz, "Synchronization Properties in Multimedia Systems", *IEEE Journal on Selected Areas in Communications*, 8(3):401–412, April 1990.

[TGD91] Dennis Tsichritzis, Simon Gibbs, and Laurant Dami, "Active Media", in Dennis Tsichritzis, editor, *Object Composition*, pages 115–132, Genève: Université de Genève, Centre Universitaire d'Informatique 1991.

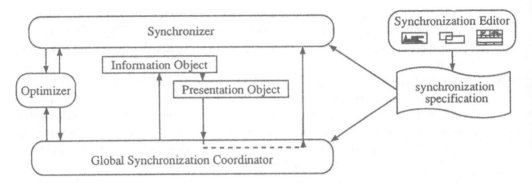

Figure 1: The architecture of the synchronization components of MODE.

Figure 2: The output of the Synchronizer for a subtitled video presentation created with the Synchronization Editor.

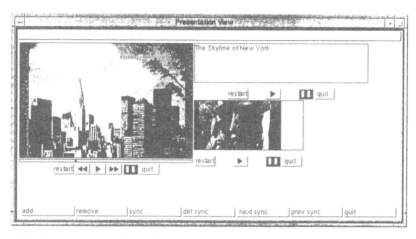

Figure 3: The Presentation View allows to display single objects to find suitable synchronization points.

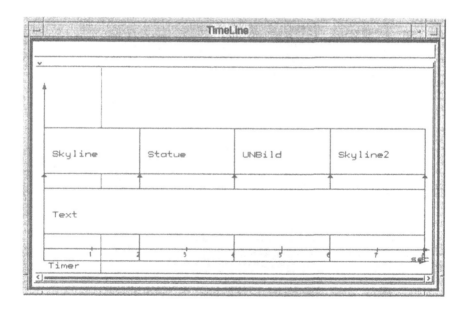

Figure 4: This Time View shows how a slide show is synchronized using a timer object.

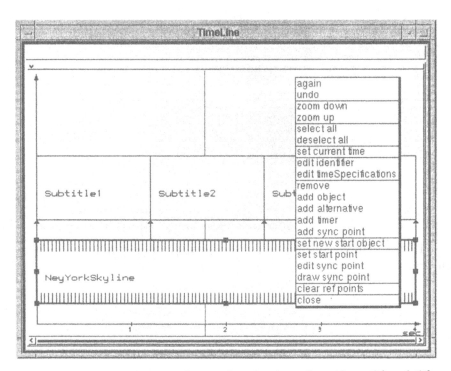

Figure 5: This Time View shows the synchronization of a video with subtitles and a popup menu with some editing options.

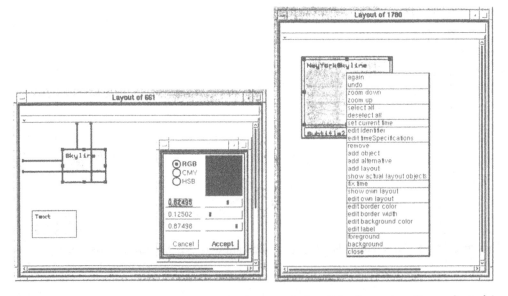

Figure 6: The left Layout View shows the layout of the slide show at a time selected in the Time View and how for example a border color may be specified. The right Layout View shows a video and a subtitle that are grouped to be presented in a single window.

Abstractions for Continuous-Media Programming

Ralf Steinmetz
IBM European Networking Center
Tiergartenstr. 8, 6900 Heidelberg, Germany
Tel: +49 6221 404280, Fax: +49 6221 404450
E-mail: steinmet at dhdibm1.bitnet

J. Christian Fritzsche
Johann Wolgang Goethe-University of Frankfurt
Computer Science Department
Robert-Mayer-Strasse 11-15, 6000 Frankfurt, Germany
Tel: +49 69 7988373, Fax: +49 69 7988353
E-mail: fritzsch at rbiffm.informatik.uni-frankfurt.de

Abstract

This paper surveys different techniques for programming multimedia applications. To the notion of the authors, so far no work on multimedia programming as integral part of high level languages has been performed (as it may be, to treat media as types). Some new ideas and concepts in this direction are presented in this document.

1. Introduction

Work on multimedia computing and communications has been focused on the provision of suitable workstation and network components together with appropriate software technology. The HeiTS ("Heidelberg High-Speed Transport System") prototype under development at IBM ENC in Heidelberg is one of these systems [7]. Today's multimedia applications are usually programmed in conventional languages (such as C), augmented with hardware-specific multimedia libraries. Replacing any underlying continuous-media device, even with a functionally-equivalent component from another vendor, often requires reimplementing a substantial part of the application programs. Some applications may have been produced with tools either generating or providing the code to interface with the multimedia devices. In such a case, any replacement of the multimedia equipment requires major changes in the tools; new interfacing methods; and, at the least, regenerating the applications' executable code.

Thus, the following questions arise:

1. Why are multimedia application so hardware dependent?

2. How can this problem be overcome?

An analogy can be made to techniques for programming with floating-point numbers. The diverse hardware engines for performing floating point processing are also different in terms of architecture, instructions and interfaces. Sometimes RISC architectures or parallel processing are used. Nevertheless, only a few standard representation formats such as the IEEE format are used. Programmers use built-in functions of high-level languages (HLL) for their programming with real numbers. Any change in the hardware would rarely affect, e.g., a Fortran application program.

Compared to our multimedia environment we find relatively well-defined abstractions, usually HLL data types, within the programming languages. It is thus possible to hide the actual hardware from the application without any major decrease in performance.

The research community frequently approaches multimedia programming within object-oriented environments (see, e.g., [2; 3; 6; 16] at one workshop on multimedia [8]or [20] as further examples). We encompass a similar development in the presentation of communication functions to the applications. "Multimedia objects" allow for a fast integration of all kinds of very different capabilities and functions with the environment itself. Unfortunately the class hierarchies encountered are very dissimilar: Today, there is no consensus on a common or "the best" class hierarchy. A multimedia product developed with an object-oriented language is still the exception.

In this paper we describe various possible abstractions for continuous media, and as a new concept we propose to treat "multimedia" as an integral part of a HLLs.

2. Libraries as Programming Abstractions

Let us first consider the most common approach currently encountered: All continuous media processing is based on a set of functions packaged as a library.

In a computing system, each device is accompanied by a device driver and a library to control all the available functions. In DiME, we experimented with a wide assortment of audio and video devices attached to workstations. We found that the libraries are very different with respect to their degree of abstraction. Some can be regarded as an extension to the window system, others simply control collections of bytes to be passed as control block to the respective device.

As an example let us take some functions which support IBM's "Audio Visual Connection" (AVC):

```
acb.channel = AAPI_CHNA
acb.mode    = AAPI_PLAY
...
aud_init(&acb)          /* acb is the audio control block */
...
audrc = fab_open (AudioFullFileName, AAFB_OPEN, AAFB_EXNO,
                  0, &fab, 0, 0, 0, 0);

fork (START IN PARALLEL)
  aud_strt(&acb)
  displayPosition(RelativeStarttime, Duration)
...
acb.masvol = (unsigned char)Volume
audrc      = aud_crtl(&acb)
...
```

Libraries are very convenient at the operating system level, but, there is no consensus (and we assume there will never be one) about which are the most convenient functions for the various devices to be supported. As long as there is no proper operating system support for multimedia and no integration into programming environments, a variegated multitude of functional call interfaces will remain.

A more structured approach interfaces audio and video through "toolkits" (see e.g. [1; 2]). These toolkits are used

- to abstract from the physical layer

- to introduce client-server paradigms, i.e. to hide communication

- as interface for quality of service parameters

It is also known that toolkits facilitate the hiding of implementation aspects such as the process structure. Out of our experience this "encapsulation" of the basic implementation architecture is only possible at the expense of performance and within one type of system (e.g. only one operating system).

3. Multimedia-Specific Abstractions at System Level

Some dedicated abstractions, such as "time capsules" [9], are seen by a multimedia system as extensions to files. These extended files are used for the storage, exchange, and accéss of continuous media. Individual data items in a time capsule have a "life span" or "duration", in addition to being associated with an indication as to the type of data and the actual data.

This concept is easier to understand and more useful for video than for audio. In the case of full-motion video (25 frames per second), each frame has the duration of 40 msec. In a normal presentation, the read access is performed at this rate. For fast-forward, slow-forward, fast-backwards, etc., the presentation rate is changed. This can be accomplished in one of two ways:

1. The presentation duration of the data items (i.e., the video frames) can be altered. In the case of slow-motion video individual consecutive frames become valid for a longer duration.

2. The duration is not affected, but instead the selection of the segments to be delivered by the time capsule is influenced. At the fast forward mode some data items will be skipped, but the duration for each frame remains. In slow motion mode, frames may be delivered twice.

Herrtwich's work could be extended by taking into account the granularity of the data items (pixel, video frame, sequence) for time capsules. The change of rate should not be performed on a per-sample basis, but instead should be applicable to sequences of samples. With respect to the video presentation hardware, each video frame must still have the same duration; a change in the rate is not applicable. The same is true for audio. For intermediate processing this may not apply. A similar approach combining the data with the rate is presented in [6].

4. Abstract Data Types as Programming Abstractions

As we will use the answering machine example throughout this paper, let us specify this application with Abstract Data Types (ADT) which is one example of a formal specification. This formal definition - also called called abstract type definition - is an interface specification without any description of the internal algorithms.

Using the ADT definition, we encountered some severe restrictions: Communication and synchronization can not be expressed directly, and synchronization of multimedia data is essential for most multimedia applications. These applications also include a certain degree of parallelism which can only be expressed with an ADT by using some tricks, as shown in the following example. (The order of the functions' execution is not defined since this does not make sense in an ADT specification.)

The following example describes the answering machine which we used as a preliminar example to study and compare the various programming interfacing techniques:

```
ADT Answering Machine
operations:
        create        :                        -> CALL_LIST;
        toggle_state  :                        -> {.collect_calls,report_calls}
        answer        : CALL x CALL_LIST -> BOOLEAN x CALL_LIST;
        play          : CALL_LIST        -> BOOLEAN x CALL_LIST;
        volume_control: INTEGER

semantics:
        create  = create_list();
        answer(call,call_list)
                  = IF collect_calls
                    THEN((accept_call AND play_infomation_message AND
                            record AND disconnect),
                          enqueue(call,call_list))
                      ELSE (FALSE,call_list);

        play(call_list)
                  = IF report_calls
                    THEN((play_voice_mail(head(call_list) AND
                            display_relative-position),
                          dequeue(call_list))
                      ELSE (FALSE,call_list);
```

Although the ADT definition of the answering machine gives a good preliminary idea of the actual application, it excludes the description of some essential features and thus, it is not a real substitute for the required multimedia programming abstractions. This formalism looks like a program, but, it does not define the communication required between the functions. State selection and volume control are possible at any time during operation; these operate in parallel. At this level of abstraction we cannot describe details such as how to play voice mail or how to determine the average volume level. Even the required synchronization between play_voice_mail and display_relative_position cannot be specified.

5. Object-Oriented Approaches as Programming Abstractions

In object-oriented environments −according to Wegner's definition [21; 14]−, multi-media programming is approached by the implementation and expansion of class hierarchies. However, very different types of such class hierarchies can be built:

An **application related class hierarchy** introduces abstractions conceived specificly for one application or a well defined set of applications and it's environment. This is the most commonly used approach and leads to the actual variety of class hierarchies.

Let us now focus on the view of objects in a more physical sense: Let us associate **devices with objects** as done, e.g., in the DiME project [17]. Common methods should be usable in a device-independent fashion. Synchronization methods are applicable to many devices and may be mapped onto, e.g., start/stop/on/off operations. Some devices may comprise various media: e.g., a computer-controlled VCR or Laser Disc Player (LDP) are storage devices combining video and audio. Within multimedia systems an abstract type definition of devices such as cameras and monitors can be provided. However, nothing is said about the actual implementation. It turned out to be a rather difficult task to define a common interface across several similar audio or video, input or output devices as shown in the following example:

```
class media_device
{char* name;
 public:
    void on(), off();
};/* end media_device*/

class media_in_device :
public media_device
 {private:
        DATA data;
  public:
        refDATA get_data();
 };/* end media_in_device */

class media_out_device :
public:
        void put_data(refDATA dat);
 };/* end media_out_device */

    class answering_machine :
public media_device
 {private:
        list  my_list; // class for ADT  list
        media_in_device  recorder;
        media_out_device message_for_caller,
                        message_from_caller;
        refDATA information; // text a caller hears
        void display_position ();
```

```
public:
    void answer()
            {message_for_caller.on();
             message_for_caller.put_data(information);
             message_for_caller.off();
             recorder.on();
             my_list.enqueue(recorder.get_data());
             recorder.off();
             }
    void play()
    {

    message_from_caller.on();
            message_from_caller.put_data(my_list.head());
            display_position();
            message_from_caller.off();
            my_list.dequeue();
            }
};/* end answering_machine */
main(){};
```

The concept of devices as a class hierarchy offers the possibility of parallelism by a simple parallel execution of the methods. Synchronization is not defined in this hierarchy and must be provided from elsewhere. Multiple inheritance was often required in implementing the answering machine.

Initial concepts of DiME were based on a **data flow principle** with sources, sinks and intermediate processing components. Similar approaches are recently discussed in [2](see the comet's node types) [6](sinks, sources and filters) [18] (module with variable number of input and output data channels). This "Lego" model allows to assemble the data flow path by chaining the object or connecting input ports to the respectice output ports of other objects.

The **media class hierarchy** is a special data type structuring method which defines classes correspondent to the different attributes of the individual kinds of media. The following class hierarchy is extracted from the code (see Appendix A) and denotes only a part of the whole hierarchy:

```
media
  audio
    music
      opus
        note
          sample
    speech
      ...
```

```
visual
  video
    image
  animation
    ...
  text
    ...
raw data
  ...
```

We defined methods 'get' and 'put' for these classes. Related to the discussion about granularity of media, we introduce a second kind of relationship, apart from the 'is-a' hierarchy of classes. The new relationship is the 'is-sequence-of' relationship to model the granularity. This offers the chance to define synchronization in terms of granularity.

A unique property of multimedia objects is their lifetime as discussed in [17] and denoted by [6] as **active objects**: The processing is performed during as long as the connection exists or data is transferred even if no method (apart from a "new" and/or "init") is invoked. Typical methods are "play" and "stop". Gibbs' multimedia programming environment is extended towards the user-oriented interface by a "scripting" language with constructs for parallel, sequential and, e.g., repetitive processing ("a > > b", "a&b", "n*a").

Communication-oriented approaches incorporate **objects in distributed environments** by explicitly defining classes and objects for communication. In [3] information, presentation and transport classes are distinguished. Information encapsulated by the information objects can generate presentation objects to be played or displayed. Information objects can also be converted into transport objects for the purpose of communication, and transformed into presentation objects afterwards (see [3] for the complete state transition graph). We could imagine this model to be extended by a storage class as information is processed/coded differently for communication, presentation as well as storage purposes. And, storage formats are essential as they rely on, e.g., database, CD- and compression specific coding (plain CD-ROM ISO 9660, CD-ROM XA, DVI, CD-I, ... formats).

Another approach we studied is known as **application models**: Applications are derived from a generic application **class hierarchy**. We can either derive models from the basic functions of the applications [20] or understand media as perception, storage, transmission, and presentation media (see MHEG). There are three fundamental combinations of these media: The first we call "live presentation", which means that a live scene is perceived, the data are transmitted and then presented, e.g. a live TV broadcast that you watch on your TV at home. The second is a "recording medium": a combination of perception and storage media. The third combination is a (re-)play medium that presents stored data. From these three classes more specialized classes can be derived, e.g. video observation, video recorder, audio player, and projectors for film or slides.

Related to the inheritance within all above mentioned concepts are the advantages of *polymorphism*, i.e. the same function call or method can be applied to different objects. We may use "play" with e.g. audio and video data, there will be different implementations to perform this command, the data may be a file in the local file system or some audio/video sequence on a remote file server. Within the object-oriented framework "play" may be defined in various classes, according to the object to perform this operation the respective method is selected. Apart from code reuse, this concept is very useful considering the ease of system's use reducing the complexity of the various underlying systems and approaching uniqueness.

Our examples and the actual implementation were done in C++, but the results are not at all dependent on this specific language. For the future we see a coexistence of various class hierarchies with complex interrelationships. Since this complexity in not easily manageable, let us shift our focus to conventional HLLs.

7. Programming Abstractions within High-Level Languages

In procedural HLLs, multimedia functions may be issued by a set of uniform, i.e. relatively device-independent function calls. This leads to a certain desirable abstraction and assures better programming style and productivity. However, programs must be able to manipulate multimedia data very efficiently. Thus, in a procedural HLL the program will typically directly access multimedia data structures or control the attached processing engines via device-dependent system calls. In Appendix B we included some of the C-Code (it is C with C++ notation) of the answering machine. In the following example, we just show typical programming statements making use of a similar notation to OCCAM 2 [11; 19] (derived from CSP −Communication Sequential Processes− [10] as the language for programming transputers [22]) due to it's simplicity and inherent parallelism.

```
a,b REAL;
ldu.left1, ldu.left2, ldu.left_mixed AUDIO_LDU;
...
WHILE ...
  COBEGIN
    PROCESS_1
      input(micro1, ldu.left1)
    PROCESS_2
      input(micro2, ldu.left2)
    ldu.left_mixed := a * ldu.left1 + b * ldu.left2;
END_WHILE
...
```

For HLLs, an alternative to libraries is to consider **the media as data types**, e.g. a data type for video. In the case of text, a character would be the "atomic element" (bits and bytes aside). A program would manipulate characters: They can be copied, compared with others, deleted, generated, retrieved from a file, stored somewhere, be part of a data structure, etc. Why not permit the same operations on continuous media LDUs (to the extent that it makes sense)? Viewing media as a data type, we can distinguish different approaches based on the granularity of the media to be addressed by the functions (e.g. pixel, whole picture or sequence of pictures; audio sample or audio block). So far we experienced:

- If these sequences are too small, e.g. individual audio samples, real-time processing becomes difficult (DSP algorithms must be generated, changed, enhanced [15]). If the granularity is too coarse, individual items are not longer accessible.

As a practical solution, the programming capabilities should be restricted (i.e. pixel manipulation for DCT or FFT are not the domain of HLLs).

- The meaning of the operators $"+"$, $"-"$, etc. is not only media dependent but also application specific: The addition of two video pictures may be a superposition (with some transparent color) or just and addition of the luminance values. An agreement on the common interpretation is required.

- The compiler generated heap size is restricted (for efficiency purposes chaining of audio/video LDUs is not practical). Careful allocation and manipulation of buffer space can be reached through a system wide homogeneous buffer management and/or applying dedicated optimizing steps for the code generation.

```
file_h1 = open (MICROPHONE_1, ...)
file_h2 = open (MICROPHONE_2, ...)
file.h3 = open (SPEAKER, ...)
...
read (file_h1)
read (file_h2)
mix (file.h3  file_h1, file_h2)
activate (file_h1, file_h2, file.h3
...
```

Instead of extending the notion of data types we could try to follow the approach of looking at **continuous media streams as files**. By opening files we associate the physical files with file names and, the program uses file handles. In our case we will associate a device generating or consuming continuous media with a file name. Read and write functions describe what will happen if data items are available. By a seek function we could position at individual items, but in the following we will typically consider sequences of such items. Often such continuous media may also be derived from a source like a microphone or camera, in such a case a seek function will not be applicable. This is similar to discrete data derived from a keyboard. This approach is very convenient in UNIX environments because there devices are often treated as files at the application programming interface. We could then extend the notion of a device to Leung's "active devices" [12]: All file related functions are applicable and in addition a device could be activated and deactivated. The activation means that the actual data transfer is initiated and it is stopped by issuing the deactivation command. Less operations (than in the case of "media as data type") are applicable as natural extension to the file system.

```
    PROCESS cont_process_a:
      ...
      On_message_do
        set_volume ...
        set_loudness ...
        ...
      ...
    [ main ]
    pid = create (cont_process_a)
    send (pid, set_volume, 3)
    send (pid, set_loudness)
    ...
```

Protocols for continuous media involve time-dependent processing. If we consider the lifetime of a process to be equivalent to the lifetime of the respective connection between the source and the sinks, then another way of incorporating media processing in the HLL is to look at **continuous media as processes**. Creation of the process identifies and reserves the respective physical devices. The interface to continuous media is through IPC. For example, the transfer of continuous media-data can be controlled by issuing signals or messages. The continuous-media process itself determines what actions should be carried out. So far we experienced that there is no one, single approach to be the solution. Note, this report discusses on-going (and not completed) research.

8. Conclusion

Our first, very preliminary, impression was that there exists very little or none work in this area of HLL abstractions for continuous media. And, too many different object-oriented approaches exist. This impression was correct.

Let us come back to the two questions posed within the introduction:

1. *Why are multimedia applications so hardware dependent?*
    The novelty and the diversity of device-specific multimedia functions are the main reasons for the insufficiency of actual programming abstractions for continuous media. Receiving data from a multimedia device and controlling the device (which may be a a camera) is different than simply receiving characters from a keyboard. There is an increase in complexity. Also, different devices (a camera and a microphone or, just two different CD-based storage devices) differ in their functions.

2. *How can this problem be overcome?*
    With proper multimedia specific operating system extension AND programming abstractions, this insufficiency can be tackled. In this paper we discussed and proposed some approaches for continuous media abstractions. However, we do not believe that any one approach is universally applicable? Nevertheless, we doubt that there is one single solution. HLLs, e.g., require enhancements and we are presently experimenting with some of these abstractions.

Summarizing our view of programming multimedia systems: Applications may directly invoke libraries or even call device drivers, a more convinient approach is to use toolkits or even HLLs or object-oriented environments. We see a coexistence of this various approach with a lack of research and development in some of them (e.g. HLLs). We are currently investigating in more detail the integration into HLLs.

Martin Zimmermann supported our practical work with many valuable advices concerning object-orientation and C + +, Manny Farber devoted considerable time for commenting and improving the quality of the whole work. Thank you.

References

[1] *David Anderson, Ramesh Govindan, George Homsy.* **Abstractions for Continuous Media in a Network Window System;** Technical Report UCB/CSD 90/596, UC Berkeley, Sep. 1990.

[2] *David Anderson, Pamela Chan;* **Toolkit Support for Multiuser Audio/Video Applications;** 2nd International Workshop on Network and Operating System Support for Digital Audio and Video, Heidelberg, Nov. 18-19, 1991.

[3] *Gerold Blakowski;* **Concept of a Language for the Description of Transport and (Re-)presentation Properties of Multimedia Objects;** (in German) Informatik Fachberichte no.293, Spriger-Verlag, 1991, pp.465-474.

[4] *Magdalena Feldhoffer;* **Communication Support for Distributed Applications;** International IFIP Workshop on Open Distributed Processing, Berlin, Oct. 1991.

[5] *E. Fiume, D. Tsichritzis;* **Multimedia Objects;** in: Active Object Environments, D. Tsichritzis (Ed.), University of Geneva, June 1988, 121-128.

[6] *Simon Gibbs, Christian Breiteneder, Laurent Dami, Vicki de May, Dennis Tschichritzis 91;* **A Programming environment for Multimedia Applications;** 2nd International Workshop on Network and Operating System Support for Digital Audio and Video, Heidelberg, Nov. 18-19, 1991.

[7] *D. Hehmann, R.G. Herrtwich, R. Steinmetz:* **Creating HeiTS: Objectives of the Heidelberg High-Speed Transport System.** GI-Jahrestagung, Darmstadt, Oct. 1991.

[8] **2nd International Workshop on Network and Operating System Support for Digital Audio and Video;** Heidelberg, Nov. 18-19, 1991, Proceeding to appear as Lecture Note on Computer Science, Springer Verlag, 1991.

[9] *Ralf Guido Herrtwich;* **Time Capsules: An Abstraction for Access to Continuous-Media Data;** IEEE Real-Time Systems Symposium, Orlando, December 5-7, 1990, pp.11-20.

[10] *C.A.R. Hoare;* **Communication Sequential Processes;** Prentice-Hall International, 1985.

[11] *Inmos Limited;* **Occam Programming Manual;** Prentice-Hall International, 1988.

[12] *W. H. Leung, G. W. R. Luderer, M. J. Morgan, P. R. Roberts, S.-C. Tu;* **A Set of Operating System Mechanisms to Support Multi-Media Applications;** Proc. Intern. Seminar on Digital Comm., Zurich, Mar. 1988, pp. 71-76.

[13] *T.D.C. Little, A. Ghafoor;* **Synchronization and Storage Models for Multimedia Objects;** IEEE Journal on Selected Areas in Communication, vol.8, no.3, Apr. 1990, pp. 413-427.

[14] *O.M. Nierstra.* **A Survey of Object Oriented Concepts;** SIGMOD record, vol. 18, no. 1, March 1989.

[15] *L.R. Rabiner, L.W. Schafer;* **Digital Processing of Speech Signals;** Prentice Hall International Inc., 1978.

[16] *Lillian Ruston, Gordaon Blair, Geoff Coulson, Nigel Davies;* **A Tale of Two Architectures;** 2nd International Workshop on Network and Operating System Support for Digital Audio and Video, Heidelberg, Nov. 18-19, 1991.

[17] *Ralf Steinmetz, Reinhard Heite, Johannes Rückert, Bernd Schöner;* **Compound Multimedia Objects - Integration into Network and Operating Systems;** International Workshop on Network and Operating System Support for Digital Audio and Video, International Computer Science Institute, Berkeley, Nov. 8-9, 1990

[18] *Daniel Steinmberg, Josh Sirota, David Berry;* **A Multimedia Application Programming Interface Paradigm;** 2nd International Workshop on Network and Operating System Support for Digital Audio and Video, Heidelberg, Nov. 18-19, 1991.

[19] *Ralf Steinmetz;* **Occam 2: The Programming Language for Parallel Processing;** (German), Hüthig Verlag, Heidelberg, 1988.

[20] *Ralf Steinmetz, Thomas Meyer;* **Modelling Distributed Multimedia Applications;** to appear at IEEE Int. WS on Advanced Communications and Applications for High-Speed Networks, München, March 1992.

[21] *Peter Wegner;* **Dimensions of Object-Based Language Design;** Proceedings of OOPSLA'87, October 4-8, 1987.

[22] *Colin Whitby-Strevens;* **Transputers - Past, Present, and Future;** IEEE Micro, vol.10, no.6, December 1990, pp.16-19 & 78-82.

Appendix A: Part of the answering machine's C + +-Code with Media as Class Hierarchy

```
class   media
        {private:
                // Attribute Set
         public:
                refDATA get();
                void    put(refDATA dat);
                refAttributes information();
        };/* end media */

class   audio :
        public media
        {private:
        //>>>> local Attributes
         public:
        //>>>> public methods
        };

class   music :
        public audio
        {private:
        //>>>> local Attributes
         public:
        //>>>> public methods
        };

class   opus :
        public music
        {private:
          sequence_of note;
          //>>>> local Attributes
         public:
          //>>>> public methods
        };

class   note
        ...
class   sample
        ...
```

```
class   speech :
        public media
        {private:
          //>>>> local Attributes
         public:
          //>>>> public methods
        };

class   discourse :
        public speech
        {private:
          sequence_of sentence;
          //>>>> local Attributes
        //public:  ....
        };

class   sentence :
        public discourse
        {private:
           sequence_of word;
         //public: ....
        };

class   word
        ...
class   phonem
        ...

class   visual :
        public media
        {private:
           //>>>> local Attributes
        //public: ....
        };
        ...
```

Appendix B: Extract of our C-Code (in C + + notation) of the Anwering Machine

```
//Answering_Machine
#include <string.h>
#include <stream.h>
enum state {collect_calls,report_calls};
```

```
//functions:
state select_state ()
      {/* collect_calls or report_calls*/
       char selectionffl14";
       int i = 0;
       cout<<"select operation state: collect_calls or report_calls\n";
       cin>> selection;
       while (selectionffli++");
       cout<<"your choise:"<<selection<<"."<<i<<"\n";
       return (selectionffl0" == 'r'/*eport_calls "*/);};
int volume_control;
state sele;

//operating:
main()
{
      sele = select_state();
      switch(sele)
      {
      case collect_calls:
              {/*accept_call;
               play_infomation_message;
               record..determines average volume level..;
               enqueue_call;
               disconnect*/
               cout<< "collecting\n";
               break;
              }

      case report_calls:
              {//play_voice_mail..adjusts volume.. and
               //display relative position in parallel
               cout<<"reporting\n";
               break;
              }
      default: cout<< "none\n";
      }
}
```

Session IX: Projects II

Chair: Hideyuki Tokuda, Carnegie-Mellon University

Session IX continued the presentation and discussion of multimedia research projects from Session VI. It showed a very diverse set of approaches, but integration of new media in the existing workstation environment was a common theme to all of them. Only some fraction of the session can be captured in this summary because three of the four speakers made use of video presentations (Andy Hopper did not even have a paper) or even plugged their PC into the projector (Klaus Meissner).

Jim Hanko from SUN Microsystems started the session with a talk about "Integrated Multimedia at SUN Microsystems," joint work with David Berry, Thomas Jacobs, and Daniel Steinberg. SUN has a wide range of multimedia product, advanced development, and research efforts underway. These efforts are focused on developing basic technologies to be used as a fundamental platform for multimedia applications. This platform shall enhance work-group collaboration. Jim talked briefly about SUN's Audio Platform, VideoPix, and the Advanced Multimedia Platforms.

The Audio Platform provides tools to edit audio annotations to mail and other documents. Audio data is portrayed graphically as a set of bars and lines, representing segments of sounds and silence. Regions of audio may be selected and manipulated using an interface paradigm similar to text editing. Another subproject of the Audio Platform is Radio Free Ethernet where audio is distributed across the Ethernet. A source node collects audio information and then uses IP multicast to transmit it across the network.

VideoPix is a frame grabber for NTSC or PAL video, realized as an S-Bus card. It can also be used for low-quality motion video and was applied by SUN for first experiments in workstation conferencing.

In the Advanced Multimedia Platforms program SUN has two projects. DIME (the Digital Integrated Multimedia Environment, not to be confused with DiME, the Distributed Multimedia Environment developed at the IBM European Networking Center) investigates the integration of multimedia in all layers of system hardware and software. As timing constraints often make it impossible for an application to directly handle multimedia data, the project concentrates on investigating ways in which the application can control such manipulation, but all data handling takes place at the lowest possible level in the operating system. The COCO (Conferencing and Collaboration) project is concerned with support for collaborative work. It uses the DIME platform.

For most of his talk, Jim concentrated on his own project, the High-Resolution Video Workstation. This DARPA-funded research project is developing a testbed for the investigation of new hardware and software architectures that will enable the use of HDTV-quality video for workstations. The group has developed their own hardware for a SUN 4/470 consisting of a video input processor, algorithm accelerator processors, bulk frame memory, and a video output processor, all connected by a high-speed bus. SUN sees this project as a technology driver. As Jim put it: "There is so much data, you can't play tricks."

During the discussion, the use of HDTV resolution in a workstation environment was questioned. The point was made that too many applications will not require high resolution, for the remaining others it will be too costly, and for some applications (e.g., in medicine) HDTV resolution may not be sufficient.

Klaus Meissner from Philips-Kommunikations-Industrie in Siegen, Germany, gave the second talk of the session titled "Architectural Aspects of Multimedia CD-I Integration in UNIX/X-Windows Workstations." His talk was different from most of the other presentations in that he focused on a system built around a storage utility. The findings presented come from the ESPRIT project MultiWorks which is concerned with developing a multimedia workstation based on high-end PC technology and CD-I. On the PC side, the X Window System with a Motif interface was used on top of the UNIX operating system.

Key to the effort is to use an off-the-shelf CD-I system and to integrate it into a UNIX workstation. The advantage of such an approach is that inexpensive CD-I equipment can be used (marketed at $799 since October 1991 in the U.S.) and that regular CD-I titles can be played by the system. The system connects CD-I system and the workstation with a SCSI interface. Digital overlay functions for video also had to be provided: The CD-I system plays back analog video which is clipped onto the workstation display. At the highest level, CD-I system and workstation communicate through regular SUN RPC.

Bernd Lamparter of the University of Mannheim reported on "X-MOVIE: Transmission and Presentation of Digital Movies under X," a topic he has work on together with Wolfgang Effelsberg. The general idea of the system is to store, transmit, and present digital films in a computer network. The hardware used in the system is standard hardware as found in typical workstations today. The movies are displayed in a window of the X Window System. This allows full integration with the classical components of computer applications such as text, color graphics, menus, and icons.

One major problem in displaying movies under X is to properly deal with X's color lookup tables (CLUTs). If two subsequent frames are coded with independent CLUTs, the CLUT of the second image has to be loaded into the video adapter card just before the second image is displayed. This causes the first image to be displayed briefly with the colors of the second image. This is visually disturbing even if the images have similar colors because similar colors do not necessarily use the same CLUT entries. The problem is solved if all images of the movie use the same CLUT. This restricts the colors of the entire movie to 256, which is insufficient.

Bernd has developed an algorithm which overcomes this problem by smoothly adapting CLUTs to the changing colors of the movie. In any CLUT, 32 entries are left free. These can be loaded with new colors before a new image is displayed. The second image can use the new colors and all old colors except 32 which are left for loading the next CLUT update. This algorithm is implemented and results in good video quality.

In X-MOVIE, a movie consists of a header, an initial CLUT, and a sequence of images with their CLUT updates. The Movie Server operates on such movie data structures. On request from the Movie Client (= X Server) it can store or retrieve information about a movie, play, stop, move faster or slower, etc. If Movie Server and Movie Client run on different machines they communicate with MTP, the Movie Transmission Protocol, which provides jitter control at the transport interface. Bernd also briefly discussed the transmission protocols used for X-MOVIE.

Last speaker of the session was Andy Hopper of Olivetti Research Center and the University of Cambridge. Andy showed a new video demonstrating his multimedia system, Pandora, in use. Pandora has previously been described in the literature and is one of the most prominent multimedia system today. It is used for audiovisual conferencing and multimedia mail. Pandora relies on special hardware that is added to each multimedia workstation in the form of a "Pandora's Box" which also connects to the network (the Cambridge Fast Ring in ATM mode).

Andy presented some findings from the use of Pandora. He noticed that most people using the system initially complain that the image they see of themselves (in a local loop-back) is not a mirror and that a slight delay occurs which makes the system appear to not lip-sync properly. Users, however, quickly get used to these things.

Pandora users mostly use the system for video mail. Videophone conversations are less popular, the majority of conversations involves two people only. Permanent video links between two offices are rarely used; a browsing feature that allows to look into other people's offices (without listening) is more popular. Live television and radio are not popular, but people frequently access a recording of the latest news which the system automatically provides. This changes somewhat during major news events (earthquakes, wars, etc.). The most interesting finding was that hardly anyone uses video in combination with text − mail remains mono-media.

Note: Andy Hopper's paper is not contained in these proceedings. For information about his work contact ah@cl.cam.ac.uk.

Integrated Multimedia at Sun Microsystems

James Hanko, David Berry, Thomas Jacobs, and Daniel Steinberg

Sun Microsystems has a wide range of multimedia product, advanced development, and research efforts underway. These efforts are focused on developing fundamental platform-level technologies for enabling applications that use multimedia capabilities to enhance work-group collaboration. This paper describes some of Sun's multimedia work: VideoPix, Sun Audio, the Advanced Multimedia Platforms project, and the High Resolution Video Workstation project. It also discusses some lessons derived from them that can be applied to future hardware and software architectures.

Introduction

The current generation of personal computer-based multimedia products focuses primarily on playback technology. This technology is useful in the areas of information archival (e.g. video encyclopedias) and in many instructional applications. However, the playback model is but a subset of what is needed in a workstation environment.

Workstations are frequently used in a work-group environment, where information is shared among a physically dispersed set of collaborating users. For multimedia to achieve its full potential in such an environment, it must support the interactions between a set of users. Therefore, we are developing enabling technologies for interactive real-time multimedia. Furthermore, we assert that in order to be successful, multimedia technologies must provide capabilities beyond the mere duplication of services provided by existing, discrete products (e.g. televisions, telephones, video conferencing devices, or videotape recorders).

In order to properly integrate multimedia capabilities into the workstation, it is essential that media data, such as audio and video streams, be available for real-time manipulation by application programs. Although such manipulation may seem infeasible at the current time, advances in processor and I/O architectures will make it possible in the future. It is important that a foundation for these capabilities be designed into multimedia architectures, so that applications can make use of them as soon as these operations become feasible. Otherwise, the computational element will be forever relegated to function merely as a controller, unable to add real value. Furthermore, it must not be necessary to sacrifice essential workstation qualities, such as interactive multitasking, in order to provide the user with multimedia capabilities.

Sun Microsystems has an interdependent set of multimedia product, advanced development, and research efforts underway. Current products, such as Sun Audio and VideoPix were developed to meet specific user needs, and later integrated to provide limited multimedia capabilities on standard desktop platforms. Present advanced development work demonstrates support for the integration of many aspects of the multimedia problem (such as those used to support video conferencing), by means of add-on cards. Ongoing research efforts are investigating those features that are required to truly integrate multimedia into the workstation environment. A key aspect of the approach taken by Sun's research groups is the introduction of a new data type — Time Critical Media; it differs from the normal data found in workstations (e.g., text or graphics) in that the data needs to be manipulated and presented correctly with regard to time. For example, in order to preserve the fidelity of a movie, it is necessary to present the visual and audio data back to the viewer in sequence, and with regard to the temporal relationship of one value to the next.

Sun Audio Platform

Sun workstations are currently equipped with voice-quality audio I/O capabilities. Voice messages can be recorded and used for electronic mail messages, document annotation, and sound-tracks for presentations. Audio digitization is performed by an AMD79C30 Coder/Decoder (CODEC). The CODEC samples incoming analog audio data at 8000Hz, with approximately 13-bit precision, and encodes the digital audio data using the CCITT μ-law standard. Simultaneous input and output of μ-law data is supported. All SPARCstation desktop units include a built-in speaker and a break-out cable for plugging in a microphone and headphones.

Audio can be integrated with desktop applications in a number of ways. Authoring applications allow direct manipulation of audio data by providing record, playback, and editing features. Presentation applications, including document viewers and hypermedia tools, often use visual objects (e.g., glyphs) to indicate the presence of an audio attachment. These objects can be activated, by a mouse click for instance, in order to hear the underlying audio data. Other applications may use audio for the delivery of alerts or background status information. Audio can also be used to enhance traditional user interfaces, providing audible feedback that is synchronized with mouse events, for example.

Audio Interfaces

The many ways in which audio data is integrated into applications has some interesting implications in the area of human interface. Consider, for example, the problem of output volume control. A common instinct is to provide a volume slider as part of the graphical interface to programs that use audio. This fits in with the model of the transistor radio: if the volume is too soft, turn it up. This approach, however, cannot support applications that produce audio output without direct user actions (like an appointment announcer or print queue monitor).

One model is that of the component audio system: all audio data is routed through a single control amplifier and therefore share a single volume control, speaker switch, headphone interface, etc. This model can be realized on the desktop by providing an Audio Control Panel application that acts as a configuration controller for desktop audio capabilities. Users then learn to go to one specific place to adjust the output volume, rather than require that every application present its own volume control. One exception to this approach might be found in the control of audio recording levels. Because audio recording generally requires more active user participation, it is usually more convenient to locate the volume control for record level adjustment in the audio recorder application itself.

Another important human interface consideration is that of uniform recording levels. The audio production and broadcast industries go to great lengths to ensure that the volume levels of all audio data conform to a set of conventions. If this were not the case, you would probably have to adjust the volume of your radio for every new song, commercial, or live announcement. In the present anarchy of workstation audio, it is difficult to enforce a standard recording level for all voice messages and annotations. It is important, therefore, that audio recording utilities make some effort to help users obtain recordings at suitable signal levels.

AudioTool

AudioTool is a desktop utility for audio recording, playback, and simple editing. It is integrated with Multimedia MailTool to provide voicemail capabilities. Audio data is portrayed graphically as a set of bars and lines, representing segments of sound and silence. Regions of audio data may be selected and manipulated using an interface paradigm similar to text editing. A selected region may be inserted, deleted or dragged out to another application. The AudioTool controls include buttons for Play and Record, as well as Fast Forward and Rewind. An animated marker indicates the current position during playback operations and a level meter displays instantaneous volume information.

One sub-panel of AudioTool acts as the desktop audio controller, and includes an output volume slider and a speaker/headphone switch. Another sub-panel contains controls for record level configuration, including an Auto-Adjust feature that samples the audio input to adjust the record gain to achieve a normalized signal level.

Radio Free Ethernet

Radio Free Ethernet (RFE) provides a demonstration of one way in which the network can be used to distribute audio information. Based on unpublished work by Steven Uhler and Peter Langston at Bellcore, this facility enables packets of audio data to be broadcast around a local area network, using Internet Protocol (IP) Multicast technology. A workstation can act as a radio transmitter by collecting audio data from its input port and broadcasting RFE packets of data onto the network. Any workstation on the network may then tune in to a particular station by extracting the RFE packets from that station and queueing the audio data to its output device. The network load amounts to roughly 8100 bytes/sec. per active station, though this figure can be reduced by using simple compression and by squelching the broadcast of silent data.

Radio Free Ethernet has been a useful prototype for testing multimedia networking implementations. In addition, it has been used sporadically for the broadcast of radio programs and other music sources, and extensively during significant events such as the Gulf War, U.S. Senate hearings, and the Super Bowl.

VideoPix

VideoPix is a single slot SBus frame grabber for NTSC or PAL video [Sun91a]. The software provided with VideoPix allows users to view the digitized images on their workstation screen locally, or across a network via a video server program. The principal goal for the VideoPix project was to produce a low cost still-frame video digitizer (rather than continuous video). Therefore, the workstation CPU performs most of the processing necessary to produce displayable images.

Hardware and Basic Software

The VideoPix card is an SBus slave device. The hardware digitizes the incoming composite video and places it into two serial access memory buffers, one for each field of the video data. (This operation can be performed at 30 frames per second.) Once the image data has been placed in the VideoPix memory, the host reads it and performs the transformations necessary to generate an image displayable on a frame buffer. To date, the highest rate achieved for this step is approximately 15 fields per second.

A loadable Unix device driver is provided, along with a software library to aid in programming the VideoPix hardware and two applications: a viewer program, *vfctool*, and a (network) video server program, *nvserver* (similar in concept to [Arons 89]).

The viewer program has two modes: in the first it controls the VideoPix hardware directly through the VideoPix library; in the second it acts as a client viewer for *nvserver*, (taking the image data over the network). The tool can display images in true color, index color (pseudo-color), grayscale, or monochrome (bitmap) depending on the host frame buffer's capabilities. The tool also provides mechanisms to save digitized images as Sun Raster, TIFF or Postscript files.

The Video Server

The video server allows multiple users to independently view images from the VideoPix hardware at different resolutions and image types. For example, user A can decide to view images in grayscale at a resolution of 360×240 pixels, while user B chooses to view the images in pseudo-color at a resolution of 180×120 pixels. In this case, the video server converts the data into both formats and sends the data to the clients. The video server also allows clients to remotely control the VideoPix hardware (e.g., a client may select a different input port or adjust the hue of an image).

The video server is organized in three layers. The top layer handles client connections, using separate control and data channels (sockets) for each client. Asynchronous I/O facilities are used to optimize the data movement. The middle layer transforms the image data from a generic format (4:1:1 YUV) to the form requested by each client. Caching is used, so that an identical transformation is not repeated. The bottom layer is responsible for obtaining the generic image data and managing the hardware. If only one image format is needed by clients, the VideoPix library provides a means for accelerating the computation by combining the capture and conversion process.

Although VideoPix was intended for still-frame capture only, it is capable of reasonable performance on limited-motion video over Ethernet connections. The results in Table 1 show the maximum rates currently achievable on uncompressed grayscale frames using the VideoPix hardware. (The amount of data per frame varies with the size and type of image requested.) The two rows show the performance (frames per second) measured for various frame sizes when two SPARCstation 2 systems were on the same network and when they separated by one gateway machine.

The VideoPix hardware and software has provided the opportunity to experiment inexpensively with networked digital video. Experiments that have been performed include the use of the technology for limited motion video conferences. For example one VideoPix customer held a video conference between sites in Vancouver and Toronto, Canada over a T1 link. When used in conjunction with the Radio Free Ethernet program described in the Sun Audio section, VideoPix provides a means for broadcasting digital video and audio across Ethernet. The combination of the two has been used to broadcast live television news within the Sun campus.

Advanced Multimedia Platforms

The Advanced Multimedia Platforms (AMP) program was formed to investigate the issues involved in integrating multimedia technologies (OS, networking, application support, conferencing, collaboration, etc.) into UNIX workstations. The two projects currently underway in AMP are DIME and COCO.

DIME

Currently, most multimedia applications must deal with data and devices directly, and must be written using intricate knowledge of the specific media. A bridge is needed between the application and the operating system (OS), because each has different views and requirements. The application is interested in dealing with abstract objects and is generally not concerned with how multimedia data is handled, while the operating system has the job of moving the data and managing the underlying hardware and network usage.

The Digital Integrated Multimedia Environment (DIME) project has been investigating paradigms to support the integration of video and audio Time Critical Media within a UNIX environment. The DIME project is approaching the problem as a whole: from the hardware up through device-independent distributed services, application toolkits, and new interfaces for the end-user. Therefore, the support necessary for multimedia applications has been considered from both the OS and application points of view. This broad approach has led to an understanding of how each layer interacts with the others. Furthermore, the DIME project has developed initial implementations at each level of abstraction to validate its architectural model (see Figure 1).

The time constraints imposed by Time Critical Media, and the amount of data associated with it, often make it difficult for the application to deal directly with the data (especially with current OS and I/O architectures). Therefore, the project has been investigating ways in which the application can control the manipulation and movement of data (that takes place within the OS) as if the media data were manipulated by the application program itself. An application can still directly access and process the data, if necessary (incurring the additional overhead).

To gain experience with the problems involved in writing applications which access multimedia, a video editor has been developed. The goal was to produce an editor that was both simple to implement and easy to use. To this end, a number of features have been assembled to allow the user to both view and edit stored video.

The DIME project has produced an experimental SBus card which contains the Intel i750 video processor chip set [Hartney 91]. The i750 provides the ability to capture, compress, and decompress NTSC video in real-time. The compressed video bandwidth (200KBytes/sec.) is manageable within current workstations and networks. This card is used as the hardware testbed for the development of the project's other components.

DIME Support for the Multimedia Application

A major goal of this project is to provide application writers with an environment that supports the following concepts: distribution and location transparency, device independence, resource sharing, and adaptability.

Distribution and Location Transparency — In trying to locate a device (such as a camera or microphone), a user should be able to specify information such as the workstation to which is connected, the user to whom it belongs, or the room in which it is located. Once an application has found a device, its location should be transparent; a remote device should be as easy to use as a local one.

Device Transparency — Similarly, it should not be necessary for the application to understand the peculiarities of each device. For example, there should be a standardized interface for a "camera" to which the handler for each type of physical camera will conform. If the application wants to use a feature of a particular camera which is not found in the standard interface, then it would have to specify the use of an enhanced interface when the device is requested.

Resource Sharing — Even when restricted to a local machine, the user will frequently be running multiple applications which will compete for access to the limited resources on the machine. For example, it is usually necessary to multiplex a speaker device over time in order to service competing applications (i.e. serial exclusive access). However, some workstations might be equipped with hardware which allows audio sources to be mixed. In that case, it may not be necessary to suspend one application to service another, since the output of the each application could be mixed together.

Adaptability — The underlying system should assist in gracefully handling overload periods when it is not able to cope with all of the Time Critical Media resource requirements. The goal is to allow the system to adapt to the changing availability of resources on both the local workstation and across networks, so that acceptable service to all applications can be provided, based upon application defined policies. An alternative system model of peak resource reservation and service denial (for the late-coming applications) was rejected because many applications have variable service requirements (e.g. video can be displayed at less than 30 video frames/sec. and still be functional). With today's powerful UNIX workstations, it is desirable to allow as many applications as possible to run simultaneously with adequate service, rather than restrict users to a few applications with "perfect" service.

To provide this functionality, **resource** and **appliance** abstractions have been developed. A resource represents any hardware device or software module that needs to be shared in any way. An appliance is an instance of a resource; it presents a standardized interface to applications. The Multimedia Data and Services Manager (MDSM) provides the support for these abstractions. It acts as a bridge between the high level interface that the application uses (via the Multimedia Application Kit) and the actual hardware and software components corresponding to the appliances.

The Multimedia Application Kit

The multimedia appliance abstraction is used by the Multimedia Application Kit (MAK). Applications can allocate, control, and destroy appliances, as well as connect them together to form networks over which Time Critical Media will flow. The application does not have to deal with the media itself. Instead, the OS provides the means for transporting the media data between appliances.

The MAK also enables the application to access control panels that it can present to users. For example, within an application controlling a digitizer board, the user may select video characteristics via a visual interface. The application might present sliders and buttons on a panel which correspond to the options for that board. However, the panel has to be independent of the application because options will vary from board to board. The application can ask the MAK for the proper panel once it has located the corresponding appliance, and if one is available it can be presented to the user.

Multimedia Operating System

The goal of this effort is directed toward providing system support for Time Critical Media. The general problem can be expressed as that of enabling Time Critical Media to be transferred, processed, and delivered to the consumer (either within a single system or across a network) within the media's time constraints. The DIME and High Resolution Video Workstation groups have together developed programming abstractions, called **conduits and transducers**, to address these issues. These abstractions are described in the High Resolution Video Workstation section of this paper.

COCO

The Conferencing and Collaboration (COCO) project has been exploring the opportunities for multimedia-equipped workstations to support and augment collaborative work. The project is investigating the needs of the people engaged in collaboration, and developing prototype technology to meet those needs. To this end, it has built a development platform and applications that demonstrate the value of multimedia technology in supporting collaborative work.

The COCO project has chosen to investigate distributed, real-time, computer-integrated, small-group (two to ten people) collaborations using extensible platforms. That is, the COCO project is focusing on augmenting the interactive aspects of the collaborations that occur when a few people come together to tackle a problem, rather than the organizational coordination that often dominates large groups.

The COCO group aims to support "immediate" interaction among collaborators, where actions taken by each participant are presented to all participants as they occur in real time, rather than the asynchronous "off-line" interaction of electronic mail. Video conferencing is a crucial component of the project as it provides some of the immediacy of face-to-face meetings. In contrast to the traditional approach of using computers to "control" dedicated conferencing systems, the COCO project makes use DIME technology and networking capabilities to permit the integration of video, audio, and textual data within a workstation, and enable real-time conferencing between widely distributed sites. This level of integration permits new functionality including: novel multimedia, multiuser interfaces, synchronizing events in the computer application with the audio/video conference; and journaling audio, video and computer conference text for annotation, storage, and retrieval. The result of this work is an extensible platform and basic applications that allow new or different tools to be quickly prototyped and integrated into the system.

A pilot desktop video conferencing application has been produced using the DIME platform and a "shared whiteboard" application. The environment is also being extended to support generic "collaboration aware" applications. The intent is that applications will be able to support collaborative use without much additional programming effort. Finally, surveys of users (at Sun) of commercially available video teleconferencing systems have been conducted to better understand how well these systems address the users' needs and how future computer based conferencing can be improved.

High Resolution Video Workstation

The High Resolution Video (HRV) Workstation project (funded in part by the U. S. Defense Advanced Research Projects Agency) is developing a research testbed for the investigation of new hardware and software architectures that will enable the manipulation of high-bandwidth, time-critical data by standard workstations. The HRV workstation project's goal is to integrate HDTV-quality video into the workstation environment in such a way that the essence of the workstation is not compromised.

HRV Hardware

The HRV team's goal in designing its hardware platform was to provide a testbed that is sufficiently flexible so that many concepts and trade-offs of integrating high-resolution video into workstations could be explored. Therefore, the hardware is capable of supporting a wide array of video formats (NTSC, PAL, SMPTE 240M) on several types of displays (HDTV monitors, flat panel displays), and can do so at full frame rates on uncompressed high-resolution video data (e.g., 2 million pixels at 72 Hz).

The HRV workstation consists of a Sun 4/470 augmented with several special-purpose units that are connected by a high speed bus, as shown in Figure 2. These units are:

Video Input Processor — A functional unit capable of capturing, digitizing, and transporting high-resolution SMPTE 240M video (as well as NTSC and PAL) at full resolution and frame rate.

Algorithm Accelerator Processors — A computing resource consisting of four Intel i860 processors, running at 40MHz, each with 4 MB of local memory [Sun 91b].

Bulk Frame Memory — A large (256 MB per board), fast memory that can be accessed via the system's high speed bus. This memory is intended to provide, among other things, storage for just over one second (32 frames) of uncompressed HDTV-quality (2K by 1K pixels at 32-bits/pixel) video.

Video Output Processor — A double-buffered, multiple plane-group frame buffer, fully integrated into a standard window system, designed to facilitate the simultaneous display of multiple full-frame video streams. The Video Output Processor (VOP) contains two 24-bit true color image plane groups and two 8-bit index color plane groups. A window ID plane group combined with a window ID lookup table provides pixel-by-pixel selection of the image source and facilitates rapid switching between double-buffered image frames. The VOP is capable of producing non-interlaced high-resolution (2K by 1K pixels) output at up to 72 frames per second. The VOP also contains a 40MHz Intel i860 processor which can be used to control video transfers and accelerate window system and low-level graphics operations.

The high speed bus consists of a front-plane containing a 128-bit wide data path that is clocked at 20MHz (peak throughput is 320 MB/s). The high-speed bus contains no address lines; transfers are set up at the source and destination in a manner analogous to DMA transfers. The bus controls are routed through a set of FIFO's so that the setup for subsequent transfers can be overlapped with data transfers in progress. A separate control bus is provided to streamline this setup process. These facilities are necessary to provide actual throughput that approaches the theoretical peak.

Note that the high-speed bus is (barely) sufficient to update each pixel of a high definition display 30 times per second (at 24 bits/pixel true color), which requires 240MB/s. Although this interconnect is sufficient to perform meaningful experiments in the near-term, it is inadequate for long-term developments. For example, if real-time image processing were to be performed, two bus transactions per frame would be required, for a total bandwidth requirement of 480MB/s. Clearly, the ultimate answer is a switch-based interconnect which can allow many simultaneous high-bandwidth connections [Tennenhouse 89, Hayter 91].

HRV Workstation Software

The main thrust of the HRV software effort is to develop system software that supports multiple, continuous, high-bandwidth, time-critical data streams (e.g., digital video and audio Time Critical Media) in the workstation environment. In order to meet the needs of workstation-based media applications, the project's software efforts have concentrated on the area of system support for time-critical information. This includes the development of new programming abstractions for time-critical applications, and operating system, window system, and network software extensions to support them.

Programming Abstractions for Time-Critical Applications

To provide a natural means for applications to specify time-critical data manipulation requirements, the HRV workstation and DIME groups have developed new abstractions that elevate such data streams to the level of first-class system data types. Just as files provide access to byte streams and sockets encapsulate networks links, the new programming abstractions of **conduits** and **transducers** [HRV 90b, HRV 91d] provide a means for manipulating time-critical data sources, sinks, and their connections.

A transducer represents a logical source or sink of continuous time-critical data. It can be a real device (e.g., a microphone or speaker), a virtualized device (e.g., a window within a workstation display), or an endpoint of some process (e.g., a digital filter).

A conduit represents a time-critical connection between transducers, and is responsible for maintaining the temporal coordination within the connection (and between connections). A conduit can be a simple connection of a source to a sink, or it can represent a composition of several simpler conduits. In that case, the (outer) conduit is responsible for temporal coordination between the underlying connections, even when the connections are distributed over many workstations. For example, it can ensure that a video stream remains coordinated with a separate audio stream. Conduits achieve this distributed synchronization via a software analog of a Phase-Locked Loop, which uses a stable local time base and error feedback to provide synchronization with an external signal.

Operating System Software

In order to arbitrate among the competing demands of individual applications using time-critical data streams, system resources are managed via an approach we call Time-Driven Resource Management (TDRM) [HRV 90a]. That is, all resources are allocated according to application-defined time constraints. Further, when resource demands cannot be met, the system chooses which activities to perform and which to defer based on application-defined policies [Northcutt 87]. When a time constraint cannot be met, the application is notified by means of an exception mechanism, and it can choose to accept a degradation in service, to modify its requests, or to abort its activity.

Information about the timeliness requirements of time-critical information streams can be extracted from media conduits and transducers, in most cases obviating the need for explicit control by an application developer. For example, an application may specify its requirements in a form that is natural within its domain (e.g. 30 frames/sec. NTSC video), and this can be translated at the conduit level into a form that is suitable for controlling system resource management decisions.

As a vehicle for experimentation in the HRV workstation, a mini-kernel (called the HRV Exec [HRV 91c]) has been developed. In order to facilitate the migration of successful results into mainstream workstation products, the Exec is a stripped-down version of a future Solaris (SunOS) operating system release that includes POSIX real-time extensions. In addition, a lightweight message-based IPC facility has been developed to support time-critical applications [HRV 91c]. The IPC facility introduces minimal overhead message passing, and is designed so it can be implemented in shared memory systems in a way that avoids all data copying operations (without requiring virtual memory system intervention). It also supports the notion of time-constraints as explicit attributes of messages, in order to support TDRM in distributed applications.

The HRV workstation project has an ongoing research effort into the practical applications of TDRM. Among the areas of current research are processor and system bus scheduling. Some of the pieces of information necessary to properly schedule activities with respect to time are: *timeliness* — when the activity must be completed; *importance* — how important completion of the activity is to the application or user; and *resource requirements* — how long the activity will take [Jensen 75]. This is different than traditional real-time approaches that attempt to combine into a single value (a priority) the orthogonal attributes of importance and timeliness.

Extensive simulations have been performed on processor scheduling algorithms. The preliminary results show that there are efficient TDRM-based heuristics which, when resources are available, can complete all activities within their allotted times, and when an overload occurs, can ensure that the important activities are completed (by deferring or aborting activities of lesser importance). However, the results also show that the effectiveness of these methods in inherently non-deterministic environments (e.g., a workstation) depends on having distinct importance and timeliness attributes directly available to the scheduling algorithm.

Window System Software

Another focus of the project's software effort is to fully integrate video into the standard window manager. Unlike text and graphics windows, windows which display video information may not be paused arbitrarily while a window server performs screen management functions. The HRV group, in conjunction with other groups at Sun, is defining a uniform way to identify video window attributes which can be exploited by window systems to handle video windows more appropriately.

Because the Video Output Processor contains a general-purpose CPU, the HRV workstation team has also taken the opportunity to experiment with the optimal functional split between window servers and accelerated frame buffers. An interesting result of this effort is that the general-purpose i860 processor, running straightforward C-language code, has achieved better 24-bit text and 2D graphics performance than 24-bit frame buffers using specialized hardware and hand-crafted microcode.

Network Software

Because current computer communications networks have not been designed to support media transport, resource management decisions, such as bandwidth or buffer allocation, are made without regard to the needs of time-critical applications. The HRV workstation project is investigating the application of Time-Driven Resource Management techniques to network resources. This should enable networks to meet the applications time constraints when possible, and deal with overload conditions in a manner consistent with the requirements of the applications. However, the lack of a global time base complicates the direct application of techniques developed for local environments. To address this issue, the HRV team has begun a collaboration with research teams at BBN [Escobar 91] that have been looking into similar problems. Another collaborative effort is underway with a group at USC-ISI to extend their work in audio and video network protocols [Topocic 91] with concepts from Time-Driven Resource Management.

Conclusion

Sun Microsystems has a wide-ranging program of research and development in multimedia issues. Many compromises were made in current products' hardware and software architectures due to cost and time-to-market pressures, and these ultimately affected their performance. Within the advanced development and research projects, other trade-offs have been made due to development cost or component availability, or in order to provide sufficient flexibility so that various approaches may be fully explored. Nonetheless, these projects have produced valuable insights into the proper approaches to multimedia system architectures.

One interesting result is that specialized hardware to accelerate certain operations can rapidly become irrelevant due to algorithmic advances or improved general-purpose processor speeds. Another is that, in order to achieve the number of concurrent operations necessary to support advanced multimedia applications, workstations must evolve from bus-based architectures to switch-based architectures.

References

[Arons 89]   B. Arons, C. Binding, K. Lantz, and C. Schmandt
               The VOX Audio Server
               Multimedia '89: 2nd IEEE COMSOC International Multimedia
               Communications Workshop, April 1989.

[Escobar 91]   J. Escobar, D. Deutsch, and C. Partridge
               A Multi-Service Flow Synchronization Protocol
               Technical Report, Bolt Beranek and Newman, March 1991.

[Hartney 91]   K. Hartney, *et al*
               The i750 Video Processor: A Total Multimedia Solution
               Communications of the ACM, April 1991.

[Hayter 91]   M. Hayter and D. McAuley
               The Desk Area Network
               Technical Report No. 228, University of Cambridge Computer
               Laboratory, May 1991.

[HRV 90a]   High Resolution Video Workstation Project
               *High Resolution Video Workstation: Requirements and Architectural
               Summary*
               HRV Project Technical Report #90051, May 1990.

[HRV 90b]   High Resolution Video (HRV) Workstation Project
               *System Support for Time-Critical Media Applications: Functional
               Requirements*
               HRV Project Technical Report #90101, November 1990.

[HRV 91c]   High Resolution Video (HRV) Workstation Project
               High Resolution Video Workstation: Executive Software Specification
               HRV Project Technical Report — in preparation, August 1991.

[HRV 91d]   High Resolution Video (HRV) Workstation Project
                 High Resolution Video Workstation: Video Programming Abstractions
                 HRV Project Technical Report — in preparation, September 1991.

[Jensen 75]   E. D. Jensen
                 Time-Value Functions for BMD Radar Scheduling
                 Technical Report, Honeywell System and Research Center, June 1975.

[Northcutt 87]   J. D. Northcutt
                 Mechanisms for Reliable Distributed Real-Time Operating Systems: The
                 Alpha Kernel
                 Academic Press, Boston, 1987.

[Sun 91a]   Sun Microsystems, Inc.
                 VideoPix Technical Information Document
                 Part No: FE297-0/15K, January 1991.

[Sun 91b]   Sun Microsystems, Inc.
                 VX/MVX Visualization Accelerator
                 Sun Technical Manual #800-6432-10, August 1991.

[Tennenhouse 89] D. Tennenhouse and I. Leslie
                 A Testbed for Wide Area ATM Research
                 In Proceedings ACM SIGCOMM, September 1989.

Tables and Figures

Bytes Per Frame	153600	86400	21600
SS2 to SS2: local net	2.4 fps	5 fps	10 fps
SS2 to SS2: via 1 Gateway	1.5 fps	3 fps	9.5 fps

Table 1 Video Server Frame Rates

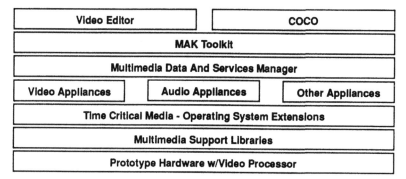

Figure 1 DIME Architecture

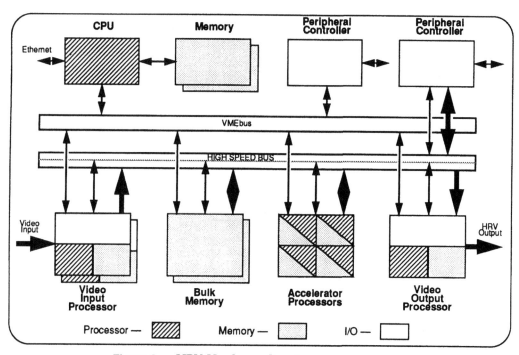

Figure 2 HRV Hardware Structure

Architectural Aspects of Multimedia CD-I Integration in UNIX / X-Windows Workstations

Dr. Klaus Meißner

Philips Kommunikations Industrie
PoINT Software & Systems
D-5900 Siegen
Germany
Phone : (+49) 271 / 380 - 2574
E-Mail: meissner@pki-si.philips.de

Abstract:

This paper describes the integration of CD-I systems in UNIX and X-Windows/Motif workstations. It is based on results of the ESPRIT project MULTIWORKS. First a short overview of the main CD-I features is given. A description of the basic requirements for the integration of CD-I in professional workstations follows. One of the major requirements is the support of off-the-shelf CD-I titles running in a Motif window concurrent with other UNIX applications and the support of new UNIX/Motif multimedia applications, which use the CD-I player as a digital video, sound and CD-ROM sub-system. The client / server architecture and the separation and distribution of time critical multimedia API functions over both systems (UNIX WS and CD-I system) are presented in more detail with special emphasis on the critical problems concerning real-time handling. Finally some experiences with the system are described.

Keywords:

Multimedia, Optical Storages, CD technology, CD-I, DVO, Multimedia API, UNIX, X-Windows, Motif

1. Introduction

New market research about trends in professional multimedia solutions [1, 2] shows that CD based multimedia business and kiosk (point of information / sales) applications and systems will be the fastest growing segments in this market. The precondition for this commercial success is the availability of at least de-facto standards on media, data format, system and application programming interface levels and also the availability of sophisticated (authoring) tools supporting these standards. For CDs several world wide (de-facto) standards are in place [2], including those for the physical level (IEC 908, ISO 10149), the logical level (ISO 9660, CD-ROM XA = Yellow Book, CD-RDx), for data formats (CD-ROM XA, MPEG, JPEG), for the API level (X/Open XCDR, CD-I API) and for system level (Green Book = CD-I Specification). With the Multimedia PC (MPC) initiative and the release of Microsoft's Multimedia Windows another de-facto standard on system level for PCs will be established soon. This standard will support CD-ROM XA technology. All these platforms including the Macintosh and NeXt workstation have different multimedia characteristics and currently different multimedia application programming interfaces (MM API) and tool environments. This is a problem for the publishing industry because the investment for a multimedia title is very high.

The driving force behind the CD technology was and will continue to be the consumer market. End-user prices of below $400 for a CD-ROM player and starting with $800 for a CD-I system with high quality DA sound and optional full motion video support can only be reached if the quantities are comparable with today's CD-DA players, i.e. much higher than PC quantities. The key factor for a customer to buy such equipment besides the price is the availability of several interesting applications with real added value. Publishers are willing to invest in new technologies if the business is large enough (consumer market) and the infrastructure (standards, authoring tools, knowledgeable developers) is in place. Large business perspective justifies the investment in sophisticated authoring tools, which is the precondition for involving creative title developers. Both the manufacturer of consumer products and the publishing industry has an essential interest in developing sophisticated development tools for CD titles.

Several end-users are interested in professional and consumer multimedia titles, especially for training and education. They are willing to train at home but they would like to have equivalent help support if they are doing the business. E.g. recently an insurance company decided to develop a training application based on CD-I, to pay the price difference between a CD-I and a CD-DA player for the employees if they would use the system at home and they decided to integrate comparable on-line help into future releases of their business applications.

This paper describes the integration of CD-I systems into professional UNIX and X-Windows/Motif based workstations. It is based on results of a feasibility study we performed in the framework of the ESPRIT project MULTIWORKS (MULTimedia Integrated WORKStation, project no. 2105 and 2713). The Multiworks high-end workstation is a 486 PC with an EISA bus, a Super VGA display system, ISDN WAN and Ethernet (also FDDI) LAN connections. A basic requirement for the integration of multimedia CD-technology in Multiworks was to support off-the-shelf CD-I titles running in a Motif window concurrent with other UNIX applications and to support new UNIX/Motif multimedia applications, which use the CD-I player as a digital video, sound and CD-ROM sub-system.

2. CD-I characteristics

CD-I is a family of hardware and software components [3] that covers not only the delivery of audio and visual information, but also a CD volume and file format standard. The heart of the CD-I system is the Multimedia Controller (see Fig. 1), based on an MC68070 microprocessor on which the real-time operating system CD-RTOS is running. The current family of stand-alone CD-I devices (CD18x series) can easily be equipped with various communication paths (RS232, SCSI, Ethernet) to interface the system to peripherals or other computers. All time critical operations in CD-I are supported by special hardware (e.g. the video system that is configurable for PAL or NTSC or the audio processing and output controller), thus enabling many visual effects, moving image display, advanced graphics functions and audio delivery in different quality levels. Interactivity is supported by homogeneously integrated input devices like mice, keyboard and optional extensions like graphics tablet, joy-stick and others. CD-I offers four video planes that can be mixed in various ways: a cursor plane, a foreground and a background digital video plane and a background plane that can either show a constant colored background or an external video signal. Video information (i.e. still images or image sequences) can be encoded in five different schemes:

- differential encoding for luminance (Y) and two chrominance components (U, V) for high image quality, called DYUV.

- Color Lookup Table encoding (CLUT8, CLUT7, CLUT4) for computer graphics.

- run-length encoding for animation, which can compress CLUT coded images further.

- natural representation in red, green and blue color components with five bits per pixel (RGB555).

- (optional) full motion video decompression based on a subset of the MPEG ISO standard [4, 5].

CD-I offers four different audio quality levels: CD-DA (which in stereo mode uses 100% of the bandwidth of about 150 kbits/sec.), ADPCM level A (HiFi quality stereo, data rate 50%), level B (FM quality stereo, data rate 25%) and level C (AM quality stereo, data rate 12,5%). Depending on the encoding level up to 16 mono or 8 stereo audio channels can exist simultaneously in the same CD-I track.

The CD-RTOS operating system is an extended version of OS-9 specifically tailored to CD-I. Because of its modular design the basic system is stored as a ROM module, additional software modules (e.g. I/O handler) can be loaded at runtime. The inherent real-time capabilities guarantee proper handling of all time critical events and operations, while the operating system kernel and system services provide a programming interface very closely resembling to UNIX.

A standard CD-I application is a program (e.g. written in C) developed for the CD-RTOS operating system that can be played on any stand-alone CD-I system that fulfills the CD-I base case requirements [6]. Such a program is normally stored on a CD-I disk.

The multimedia capabilities of the CD-I hardware are used and controlled by using the native CD-I application function calls from the CD-I Presentation Support Library (Native CD-I API = NAPI) [3, P. 230ff]. These functions are an integral part of the CD-I system; the functions, their parameters and the data structures they interact with are described in the GREEN BOOK [6]. Since these functions are elementary, a high level Application Programming Interface (CD-I API) and development environment, called BALBOA [7], is available on top of the NAPI. BALBOA is organized into a group of manager (library functions), each handles a CD-I element, such as video display, animation, sound, cursor controls, timer or combined multimedia objects.

3. Requirements on the Integration of CD-I in UNIX / X-Windows

As described above, it should be possible to start and control standard CD-I titles fully executed on the CD-I subsystem by using the keyboard and mouse of the UNIX workstation exclusively. But it should also be possible to develop new multimedia UNIX applications that use the CD-I subsystem as a high performance video, sound and CD-ROM player that offers functions like:

☐ special visual effects, e.g. continuous dissolves, wipes, curtain and mosaic effects.

☐ output of audio signals in different quality levels, ranging from CD digital audio quality to mono AM broadcast quality.

☐ playing short audio segments (soundmaps) from memory.

☐ instantaneous switching between audio channels during a running application.

☐ full true color representation (e.g. effectively 3 x 8 bits = 16 million colors simultaneously when using the DYUV image coding method, or 256 out of 16 million colors for the CLUT8 coding method).

☐ arbitrarily shaped image regions for matte functions.

☐ instantaneous switching between two video planes.

☐ full screen full motion video decompression.

This requires on the host side, beside the communication aspect,

☐ the support of digital video overlay functions (controller) and their integration in the window system,

☐ mouse, keyboard and screen redirection. If the mouse cursor is moved to the CD-I window, the coordinates have to be mapped to the full CD-I screen size and transmitted with possible keyboard input to the CD-I subsystem.

☐ a high level multimedia application programming interface (MM API) based on a high level multimedia peripheral interface (HPI). All real-time critical functions must be executed on the CD-I subsystem by using the HPI functions. The host HPI handlers have to manage the execution of these functions or a sequence of these functions.

☐ CD-ROM file service support. It should be possible to access database or text objects stored on a CD, mounted on the CD-I system, by standard UNIX volume and file operations. This requires the mapping of ISO 9660 structures, read from the CD-I system as physical blocks, to UNIX objects on the host system.

☐ down loading of CD-I system software.

To respond to multimedia events on the CD-I subsystem quickly enough and to read large data objects in reasonable time from a CD, a high speed physical connection between the host and the CD-I system is required, which is normally SCSI.

4. Architecture

The architecture of the distributed approach (Fig. 2) is based on the client / server paradigm, with the host being the client and the CD-I subsystem being the server. Fig. 3 shows the components and their interrelationships relevant for multimedia host applications. A major design goal was to support in a later step an integrated CD-I solution based on a (PC) plug-in controller with all CD-I specific hardware / software components and connected to a standard CD-ROM drive. Therefore it was necessary to layer the system in such a way that this integrated approach can as much as possible be based on components of the distributed approach without change of at least the upper software layer. This results in the usage of standard computer communication interfaces and protocols. Three groups of new software components can be identified in Fig. 3:

☐ modules on the client and server side implementing the communication service and providing the HPI at the client side. These are the "Remote Procedure Call" layer (RPC), Transport layer (TL), Data Link layer driver (streams modules, file manager and SCSI drivers) and the Server Process (SP) on the CD-I side. Most of these components have to be changed or can be dropped in the integrated approach.

☐ components implementing the High Peripheral Interface (HPI) at the server side. This consists of the CD-I API Library, which is normally the BALBOA library with extensions, and a support module (CD-I API Extensions) that permits the installation of time critical application specific routines (callbacks) on the CD-I side at runtime (dynamic linking, etc.).

☐ the Multimedia Application Interface (MM API) Layer on client side. This set of high level library functions and widgets available in the programmer's development kit is realized by (a sequence of) HPI function calls.

The envisaged method of interaction between host and CD-I subsystem is as follows:

a. The application program on the host invokes a multimedia function (HPI function) like "create image" or "play real time record" by calling the appropriate entry of the RPC client library at the client side. This includes the delivery of several parameters like image coding method, file name on the CD etc.

b. The RPC component converts this request into a message which is then sent to the CD-I system by the communication software.

c. The communication parts on the CD-I side receive the message and hand it over to the RPC server component, which runs under the control of the server process.

d. The RPC routines decode the message and do the appropriate call to the HPI, i.e. to the CD-I API or to the CD-I API Extensions, respectively.

e. The HPI functions are mapped to the low level operations of the native CD-I API and many time critical actions are done in the context of the CD-I API components and CD-RTOS (by virtue of the BALBOA software). Application-specific time critical routines are loaded on top of the HPI at the CD-I side (application specific callbacks).

f. Return values are sent back to the application stepping backward through the same components through which the call was propagated.

Since the host waits for the completion of the operation it has requested from the CD-I device, this interaction mode is called a "synchronous service request". In case of "asynchronous service requests" an acknowledgement value is returned to the application, but after that, the function continues to operate at the CD-I side. Examples are "play real time record" or "start animation sequence". There may be a need to send a signal back to the application, when the started operation terminates or on similar events. This can be done by the server calling a callback routine that the client has specified before.

For understanding of the characteristics of communication between the UNIX host and the CD-I subsystem, it is useful to list the different cases where information has to be passed between both systems:

□ exchange of control information and small amounts of data (remote procedure call).

□ parallel virtual input channels for the CD-I device mouse and keyboard.

□ exchange of massive amount of data (especially read operations from the CD disk).

□ audio stream from CD-I device.

□ video stream from CD-I device.

These logical communication channels have to be mapped onto physical communication media with regard to the expected transferred data volume and the tolerable transmission delays. Apart from the implementation effort, it would be desirable to have a single digital communication path (e.g. a parallel bidirectional data bus) that carries all the types of data. However, since the video and audio streams available at the output connectors of the current (table top) CD-I systems are both analog, they have to be kept separate from the rest. Therefore an analog video channel is connected to the analog video input connector of the digital video overlay (DVO) board and a twin analog channel for stereo audio is connected with a separate stereo amplifier or headset. In a future integrated approach the video (and audio) interface to the host system must be digital.

The purpose of the Transport Library is to provide a software interface for the transparent sending and receiving of messages (byte blocks) while hiding the details of the used physical medium (SCSI, Ethernet) as well as the related protocol layers below. Such lower protocol layers are responsible for synchronization, error detection and handling, packaging etc. Since the connection between host and server is a point-to-point link, the software needed will be simple (e.g. no network layer). However, multiple virtual channels must be supported at least for the transmission of remote procedure calls in parallel with keyboard input and mouse events regarding the CD-I window.

The term "Remote Procedure Call" denotes a concept that allows one process (the client process) to have another process (the server process) execute a procedure call. Client process and server process are separate and may even reside on different machines. From the viewpoint of the client process, the remote procedure call behaves exactly like a local procedure call, i.e. as if the procedure was executed in the client process's own address space. To implement a remote procedure call, several types of data have to be exchanged between both processes (i.e. have to be transferred across the underlying communications medium or network):

□ control information (identification of client and server processes, identification of the procedure to be called etc.),

□ the input arguments passed to the called procedure and

□ the result of the procedure, passed back to the client process.

Since the representation of data types (byte order, word length) may be different on different machines, the arguments and results to be transferred have to be transformed into a universal representation. Basis for the RPC implementation was the public domain RPC package developed by Sun Microsystems [8]. This package uses a general scheme for encoding RPC arguments, the XDR external data representation scheme, which is also able to transfer complex data structures including e.g. data referenced by pointers. The RPC package consists of two main components:

□ an RPC stub routine generator (rpcgen) that processes a formal C-like specification of the data types and routines to be implemented by RPC. The rpcgen pre-compiler generates the source code of the client and server stub routines and associated header files also XDR routines for the transformation of those data types defined in the specification file.

□ a set of library functions that are called by the stub routines on client and server side (for simplification not drawn in Fig. 3).

The client stub routines library resides between the HPI and the RPC library at the client side. It contains a stub routine for each HPI function and an XDR routine for each user-defined parameter type, while the proper RPC library is the generic RPC/XDR part. Similarly, the RPC library at the server side forms the generic part, while the RPC/XDR stub routines reside between the server process and the HPI at the CD-I side.

The intention of defining an HPI is to simplify applications programming and to remove real-time interactions from the interface between client and server. This requires a set of generic high level functions handling often used multimedia effects like cut, fade in / fade out, dissolve (cross fade) wipes and blends. Most of these effects are fully supported by primitives of the native API of CD-I. A prerequisite for these functions is to describe the properties and relations of audio-visual "objects" in a way that permits the server program to perform autonomously much detail work that otherwise would have to be done by the application programmer. The abstraction from machine details, a major goal of the HPI, means that some very specialized effects that are possible for a standalone CD-I device cannot be realized in the host-server configuration. An overview of HPI functions and a detailed discussion about their realization in a client/server environment is given in [7 and 9]. Not only the intention to provide an abstract machine, but also some limitations in the real-time behavior, caused by the communication necessary between both machines prevent the programmer of a client/server application from doing everything that is possible on a standalone CD-I system.

An important aspect of the definition of the HPI is the way real-time events are handled. The standard way of achieving fast communication between different application parts and processes on a CD-I device is the usage of signals. This mechanism is used for instance for informing the application process that the play-back of an audio track (which is played asynchronously, i.e. only triggered by the application) has finished. Other examples are timer events, generation of signals on video blanking intervals and others. In the client/server environment the only way of handling time-critical events is to do the processing of signals entirely on the CD-I server side, i.e. the client has to instruct the server to link specific action routines to the occurrence of signals. This means the server has to be configurable and extensible at application runtime by user-provided routines. This may happen in two ways:

a. the host downloads a module of type "Program" that can be executed (chained or forked) by the server as a subprocess. This module can also be loaded directly from the CD-I disk.

b. the host downloads a subroutine module that is dynamically linked to the server and is executed in the context of the server process.

The server process on CD-I side represents in principal the host application on the server side. Under the assumption that the host runs a multiuser/multitasking operating system, the user can have more than one application started simultaneously. He can switch between them at his will. Similarly, it may be desirable to start several multimedia applications. Of course, only the one which currently has the input focus, will be active (running) producing audio/video output. If concurrent multimedia applications should have access to the same CD-I player, a listener - server model has to be implemented on the CD-I system. The listener process listens for incoming connection requests from host applications. When a request arrives, the listener spawns a server process that services the application. The listener continues awaiting further connection requests. Such a configuration requires that the complete hardware and software context be automatically changed when the user switches between applications, i.e. when the corresponding server process is changed. Whether the CD-I software below the native CDI API (Green Book) does support this, requires further investigations.

In the model described so far, the host system triggers the CD-I system to perform specific operations with the multimedia material stored on the CD-I disk. However, many possible applications (e.g. travel information systems, sales support systems) also need direct access to the data stored on the CD-I disk (airline time tables, order numbers, prices and many more). Therefore it is necessary to have access to the files on the disk from the host. This requires the integration of the CD file access into the file system structure of the host in a homogeneous way. For UNIX this means that the file and directory structure of the CD should become a part of the normal UNIX file system tree.

It has already been mentioned that the current implementation covers the integration of the CD-I video output into the X windowing environment using an analog video path. The analog CVBS signal is fed into digital video overlay board, also a Multiworks result. The display parameters like image size (upscaling and downscaling is possible), position on the screen, brightness, contrast and color saturation can be controlled by the X application using the call interface that has been provided by the VEX (video extensions for X) working group. If, additionally, live video input from a camera, television receiver, VCR or videodisc is needed, these signals can be fed into the external video input jack of the CD-I system. The external video signal is then displayed in the CD-I backdrop plane, i.e. behind the CD-I generated screen contents (the video planes A and B of the CD-I system can be set transparent for the CD-I screen also for arbitrarily shaped regions).

Fig. 4 shows the global architecture and the integration of the above described components in the UNIX WS.

The architecture described so far reflects the situation that the multimedia application is running on the UNIX workstation and the CD-I subsystem is used as multimedia peripheral. A major requirement is also to support standard CD-I titles executed on the CD-I system but controlled by the host input devices and displayed in a window of the host monitor. If the host mouse is moved or a mouse key is pressed, a corresponding mouse event is buffered in an event queue of the X window system. The output of the CD-I system will be shown in a window on the host screen. As long as the cursor location is inside this window, mouse events are transmitted to the CD-I system, where they cause a CD-I signal to be generated or simulated. The mouse coordinates transmitted are relative to the CD-I window's origin and in units of the CD-I coordinate system. The continuous transfer of mouse events to the CD-I system is needed for two classes of operations:

- the mouse movements control real-time video or graphics operations (drawing programs, or video games where visual objects are to be moved across the screen).

- some instantaneous actions have to be performed depending on the current mouse position ('hot spots').

Considering the handling of (host) keyboard input, the situation is very similar. When the CD-I window has the input focus, the keyboard events are redirected to the CD-I system. Fig. 5 shows the global architecture of such a configuration.

5. Conclusions

The above described concepts were evaluated in a prototype implementation. The results are positive especially regarding performance and the usability. The efficiency of the HPI functionality was tested by the development of a UNIX demonstration application. It became obvious that multimedia applications which are based only on the functions of the host HPI and so do not use any application specific call back routines on the CD-I subsystem, are easy to develop. Moreover it was proved that the architecture supports general multimedia applications as well those in which time critical procedures are executed on the CD-I system. But in this case the realization also showed that the development process of multimedia applications grows much more complex, as the programmer has to think and work in two very different development environments. So the question arises whether the commercial benefit of playing consumer titles in professional workstations and so offering binary compatibility to the GREEN BOOK can compensate for these disadvantages. If this requirement is eliminated and the CD-I hardware concept (chip set) is realized e.g. on a PC board and if the CD-I API is fully integrated into the host operating system or window system, the authoring process will be essentially simplified. The integration of the CD-ROM XA in UNIX / Motif is based on this concept and realized in the product XGRIP 3.0 which has been proven successful. The shortly specified concept of the "Bridge Disc" by Philips and Sony takes this problem into account in so far as the same CD can be used on a CD-I and a CD-ROM XA system if the specific application programs are developed for both platforms.

6. Acknowledgements

Significant ideas realized in the prototype of CD-I integration in UNIX / Motif were developed by A. Wepner and Dr. R. Mesters, both authors of the feasibility study report. Particularly the experiences A. Wepner gained during the implementation of major parts of the system software and the test application were essential improvements of the concept. This applies also for many constructive contributions of B. Klee and W. Kramer during the design phase. The author would like to record his thanks for these contributions, for the commitment during the implementation and for the constructive comments on this paper.

7. References

[1]   Inteco Europe Corp., Surrey GU21 1JD, UK;
        Multimedia in Europe;
        Volume 1, Section F; July 1991

[2]   Meißner, K.;
        Aspects of Multimedia Application Support in UNIX;
        Multimedia in Action, Fifth International Protext Conference;
        Luxembourg; November, 20 -22, 1991

[3]   Philips International;
        Compact Disc-Interactive: A Designer's Overview;
        Kluwer Technical Books, Deventer-Antwerpen; 1987

[4]   Dorerty, R.;
        Full-Motion Video Technique for Interactive CD Player;
        Electronic World News; June 17, 1991; P. 12

[5]   Sijstermans, F.; Meer, J. van der;
        CD-I Full-Motion Video Encoding on Parallel Computer;
        Communication of the ACM; Vol. 34, No. 4; April 1991; P. 81 - 91

[6]   N. V. Philips; Sony Corp.;
        Compact Disc Interactive Media Provisional Specification (GREEN
        BOOK);
        May 1986

[7]   Philips International;
        BALBOA Run-Time Environment: Programmer's Guide
        August 1990

[8]   Sun Microsystems;
        Network Programming, Revision A;
        May 9, 1988

[9]   Mester, R.; Wepner, A.;
        Feasibility Study: Integration of CDI Devices into a Multiworks
        Workstation;
        ESPRIT Project 2105, Multiworks; Document Id. PKI9006BP8.3;
        September 30, 1990

8. Trademarks:

BALBOA: trademark of Philips, The Netherlands
CD-I: trademark of Philips, The Netherlands
MC68070:   trademark of Motorola, Inc., USA
OS-9:   trademark of Microware Systems, De Moines, USA
UNIX: trademark of AT&T Bell Laboratories, USA

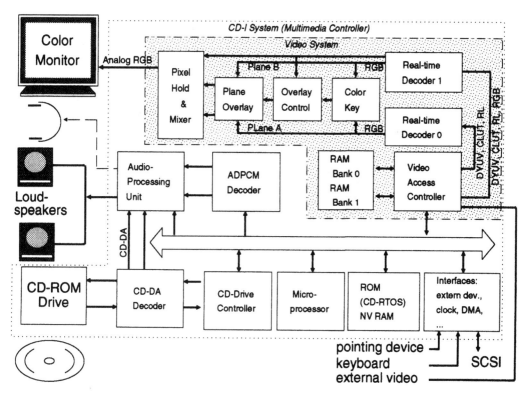

Fig. 1: Hardware components of a CD-I system [3]

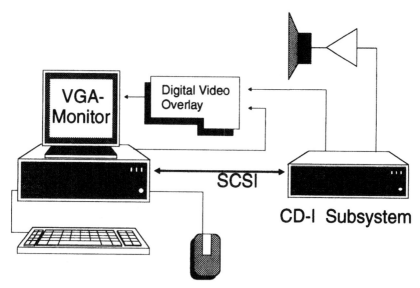

Fig. 2: System Configuration of the Prototype

Client (UINIX WS)   Server (CD-I)

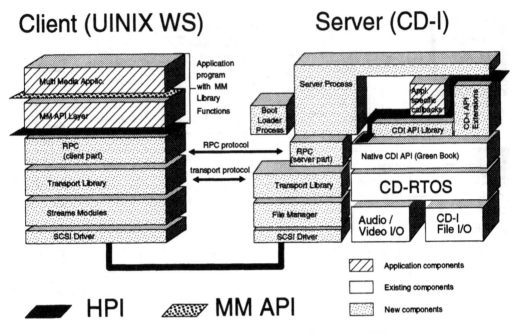

Fig. 3: Client / Server Architecture [9]

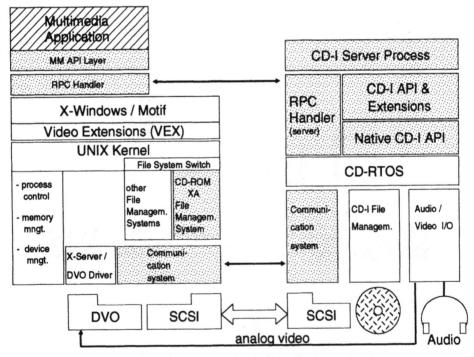

Fig. 4: Architecture with MM Host Applications

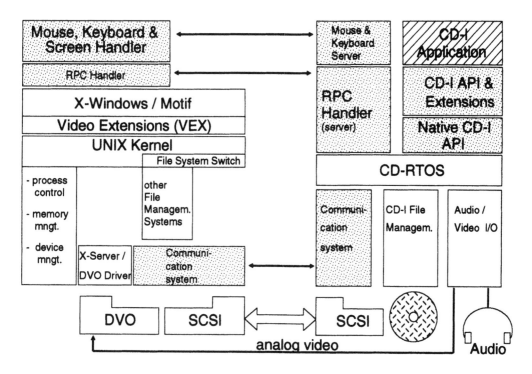

Fig. 5: Architecture with MM CD-I Applications

X-MOVIE:
Transmission and Presentation of Digital Movies under X

Bernd Lamparter   Wolfgang Effelsberg

Praktische Informatik IV

University of Mannheim

lamparter@pi4.informatik.uni-mannheim.dbp.de

Abstract

We describe a system for storing, transmitting and presenting digital films in a computer network. The hardware used in the system is standard hardware, as found in typical workstations today; no special hardware is required. The movies are shown in a window of the X Window System. This allows full integration with the classical components of computer applications such as text, color graphics, menues and icons. The X-MOVIE system is based on color lookup table technology. We present a new algorithm for the gradual adaptation of the color lookup table during the presentation of the film.

1   Introduction

Modern workstations have fast processors, high resolution color graphics displays and high performance network adapters. For all three components a considerable performance improvement can be expected for the next years. Such a powerful hardware scenario enables new computer applications. Perhaps the most important one is the integration of digital films into a classical data processing environment.

For many years, computers have been used for the processing of digital films. The most important directions in research and development can be classified into three categories: computer animation [LMS91], computer simulation and digital video [LD87, MD89]. This paper concentrates on the transmission and presentation of sequences of digital images in a distributed system of standard UNIX workstations and servers. We present an experimental system called X-MOVIE which is operational at the University of Mannheim. Our goal is the presentation of digital movies in a window of the X-Window-System under UNIX [Jon88, SGN88]. Unlike other systems reported in the literature the X-MOVIE system does not require special hardware.

In Chapter 2 of the paper we describe the architecture of the X-MOVIE system. Chapter 3 classifies formats of digital still images and digital movies. In Chapter 4 we concentrate on the network requirements for the transmission of digital movies.

The integration of a movie window into the X-Window-System is described in Chapter 5. We report first experiences with our implementation in Chapter 6. Chapter 7 concludes the paper.

2 Architecture of the X-MOVIE System

The X-MOVIE System is a distributed test bed for integrated transmission and presentation of digital movies. It consists of interconnected UNIX workstations. In our current implementation, the network is a standard Ethernet. An FDDI-ring has been installed recently, and X-MOVIE is currently being ported to FDDI. The architecture of the system is shown in **Figure 1**.

The main components of X-MOVIE are the Movie Server, the Movie Client and the X-Client. These three components can run on one, two or three different UNIX systems depending on the requirements of the application.

The **Movie Server** is able to store and replay sequences of digital images (digital films). It maintains a movie directory. On request of the Movie Client a sequence of images is sent over the network to the Movie Client. The transmission protocol is called MTP (Movie Transmission Protocol). It was developed specifically for this purpose. The **Movie Client** is an extension of the standard X-Server of the X-Window-System. The extensions implement a new set of functions for the purpose of displaying movies in a window on the screen. Examples of new functions are XMOpenMovie, XMPlayMovie, and XMShowSinglePicture. These extensions were integrated into the source code of the X-Window-System. The third component of the system is the **X-Client**. Similar to the X-Server, it is an extention of the standard X-Client. The set of new functions mentioned above was integrated into the Xlib function library of the X-Client so that a programmer can now invoke the new movie functions just like other Xlib functions.

In this paper we concentrate on the transmission and presentation of digital movies; we do not discuss the creation of such movies here. Digitized video films can be used as well as computer generated digital movies.

3 Formats for Digital Images and Movies

3.1 Formats for Digital Images

An image digitized with high quality ("true colors") uses 24 bits of storage per pixel (one byte for each of the color components red, green and blue). About 16 million different colors can occur in one image. Experience shows that the human eye cannot recognize smaller color differences [Lut89]. However, "true color" representation needs too much storage to be practical (720 kbytes for a 600×400 pixel picture), so image compression is needed.

A very widely used form of image compression is the color lookup table technology. The graphics adapters of PC's and workstations can typically display 256 different colors at a time. The color components for each color are stored in a color lookup table (or "color map") with 256 entries. Each pixel of an image is represented by one byte which is an index into the color lookup table (CLUT) (see Figure 2).

For a specific image the colors in the CLUT can be optimized by choosing the most appropriate 256 colors out of a set of 16 million. So only 256 colors can be presented at one time, but out of a large set of colors. With each image the corresponding CLUT must be stored. The color lookup table reduces the amount of data to one third without much loss of quality.

3.2 Compression of Images

Even with color lookup tables images are still very large, and further compression is desirable. We distinguish between lossless and lossy compression. Lossless compression implies that the decompressed image data is an exact copy of the orignal image data. Lossy compression allows small differences, and usually provides higher compression factors.

Compression algorithms can also be classified according to their time behavior. Symmetric algorithms use the same amount of time for compression and decompression whereas asymmetric algorithms use more time for compression than for decompression. Asymmetric algorithms are often useful for applications where compression can be done off-line, and the compressed images are stored for later use. Decompression is often done much more frequently, and can be done in real-time.

Image representation with a CLUT can be considered as a simple, lossy compression. Decompression is done in real-time in the graphics adapter. More sophisticated compression needs more computing power for decompression. This computing power can be provided with the following hardware approaches:

- Use a faster main processor
- Provide special hardware for image decompression in the workstation or in the color graphics adapter
- Use multiple parallel processors, i. e. transputers.

An interesting compression scheme for still images is the Color Cell Compression algorithm (CCC, [CDF+86]; an extension can be found in [Pin90]). CCC compresses to 2 bits/pixel. Because of the blockwise nature of the algorithm, compression and decompression can be parallelized well. The algorithm is lossy and symmetric.

ISO is standardizing schemes for compression of both still images and movies [Gal91]: JPEG (Joint Photographic Expert Group) and MPEG (Motion Picture Expert Group). Both algorithms are lossy and symmetric. The compression factors are very high due to the Discrete Cosine Transformation step, but compression and decompression require considerable computing power. Therefore the current trend is to build special compression chips. JPEG chips are already available in the market, and MPEG chips are expected soon.

3.3 Formats for Digital Movies

A digital movie can be stored as a sequence of images, either with 24 bits per pixel or with 8 bits per pixel (as indices into a CLUT). In order to provide the impression of continuous movement, at least 25 images per second must be shown.

3.3.1 Compression for Movies

The storage requirement is already a problem with still images, and a much greater problem with movies. Fortunately new possibilities for compression arise for movies. Due to the fact that the color difference of a pixel over time is often small, differential pixel coding can be used. Two-dimensional compression algorithms can be extended into three-dimensional ones, with time as the third dimension. Another simple scheme is to compute the difference image between two frames and then compress the difference image.

3.3.2 Movies and Color Lookup Tables

When showing a movie as a sequence of still images, each coded with a CLUT, the following problem arises: If two subsequent frames are coded with independent color lookup tables the CLUT of the second image has to be loaded into the adapter card just before the pixels of the second image. This has the effect that the first image is visible in the colors of the second image for a short time. This is visually very disturbing, even if the images have similar colors, because similar colors do not necessarily use the same CLUT entries. The problem is solved if all images of the whole movie are coded for the same CLUT. However, the whole move will then only have 256 different colors. Although 256 colors are sufficient for a single image, with an optimized CLUT, they are not sufficient for a long sequence of images, perhaps starting with a blue sky and white clouds and ending in a room with green and brown interior. A new solution must be found for showing movies with color lookup table technology.

We propose to use the following algorithm to solve this problem: For each image some of the 256 CLUT entries are not used. For example 32 CLUT entries are left free. Before loading the next image we can now load 32 new colors into the CLUT without changing the colors of the current image. The second image can use all of the old colors and 32 new colors. It will again leave 32 entries free for its successor. This leads to continous updating of the CLUT while the movie is showing (see Figure 3). In fact the amount of free entries can vary from image to image depending on the actual difference in colors between the images. From our experience the loss of a few out of 256 colors has no visible effect at a rate of 25 frames/s. This algorithm was implemented successfully in the X-MOVIE system.

4 Transmission Protocols for Images and Movies

In a UNIX environment the first attempt with network protocols will always be the Internet protocols TCP/IP [Com88].

IP is the protocol of the Network Layer (layer 3 in ISO/OSI); it offers a connectionless packet transfer between hosts. The protocol can loose packets and deliver packets out of sequence. Bit errors are detected, and the faulty packet is discarded.

TCP offers a reliable connection-oriented transport service over IP. It uses standard technology for error detection and recovery (sequence numbers and retransmission with go-back-n), and for flow control (sliding window).

A third protocol in the Internet ist the user datagram protocol (UDP). Basically it offers a programming interface for IP. Most of the applications in the Internet use TCP, some use UDP (see Table 1).

4.1 The Movie Transmission Protocol MTP

In the X-MOVIE environment a movie consists of a header, an initial CLUT and a sequence of images with their CLUT updates. The Movie Server operates on such movie data structures. On request from the Movie Client (= X Window Server) it will provide the following services:

- Store/retrieve information about a movie
- Play
- Stop
- Step forward/step backward
- Show picture $< n >$
- Slower/faster

If Movie Server and Movie Client run on different systems, a Movie Transmission Protocol (MTP) is required to transfer the sequence of images. The MTP is an application layer protocol in OSI terminology. The transmission of MTP Protocol Data Units must be accomplished by mapping them to the services of the transport subsystem. Unfortunately the transport subsystems available today are not appropriate for movie transmission because they do not provide **isochronous data flows** (sometimes called Continuous Media Data Streams, [Her91]).

4.2 Isochronous Data Flow

The term "isochronous" means a data flow between end systems with minimal delay jitter. It is not the delay as such causing the problem, but the variance in delay between subsequent packets. Two major obstacles prevent isochronous data flows: The operating systems of workstations and hosts cannot handle data traffic in real-time, and today's network protocols cause variable delays.

In order to solve the operating system problem the OS scheduler must be modified to take the requirements of real-time processing into account. Early work on this topic can be found in [ADH91, ATW+90].

As far as the network protocols are concerned, the problems are much more difficult to solve because almost all layers introduce variable delays. It starts out with media access protocols in LANs. For example, in a CSMA/CD-based protocol, carrier sensing can lead to variable delays, depending on traffic from other stations on the bus. Also, collisions can delay packet transmission in an unpredictable way. Similarly, a Token Ring LAN will have variable packet delays depending on where the token happens to be when transmission is requested, and how many priority stations are waiting for transmission. The same is true for the token-based media access control protocol of the high-speed LAN FDDI.

Conventional transport protocols are not suitable for isochronous transmission either. Both TCP and ISO TP4 use time-outs and retransmission for error correction. Of course retransmitted packets will have a longer delay than first-shot packets. Fortunately the new high-speed networks now under development, such as DQDB or Broadband ISDN based on ATM, will provide protocols for isochronous data flow [Kil91, Fun91]. In parallel new high-speed transport protocols are under development [HS91].

4.3 MTP Design Issues

For our X-MOVIE test bed we have tried to use existing transport protocols and networks for movie transmission in order to gain early experience. A Movie Transmission Protocol MTP was developed and tested over the Internet protocols. MTP is a hybrid protocol: some transmissions are reliable (i.e. acknowledged), others are unreliable. All data packets from the Movie Client to the Movie Server (such as "stop") are acknowledged. Control packets from the Movie Server to the Movie Client are also acknowledged, but image transfers are not. They require continuous, fast transmission. Data loss for a part of an image is not critical since an update will follow after 1/25 s.

In conventional transport protocols flow control techniques such as "Sliding Window" are used to prevent the receiver from being overrun by a faster sender. In isochronous transmission window-based flow control is inappropriate, and rate-based flow control must be applied. Sender and receiver have to work at the same data rate, they have to be synchronized [Ste90]. Movies arriving at the display from a network have to be synchronized with real-time. Hence a synchronisation mechanism between Movie Client, Movie Server and real-time must be implemented. Synchronization is not yet implemented in MTP.

5 Integration of Movies into a Window System

For the workstation user the movie should appear in a window on the screen, without disturbing other windows. The handling of the movie window should be similar to the handling of other windows. Movie windows are not yet supported in current window systems, such as the X Window System. An integration of movie windows can be done in the following ways:

- Hardware can be added (blue-box or realtime digitizing of analog video sources) [Bru89, Lut89]

- The window software is accelerated so that a sequence of still images can be passed to it, and it will draw them at 25 images per second

- The window system is extended by a movie service.

The goal of the X-MOVIE project is to avoid special hardware; hence we only elaborate on the other two possibilities.

5.1 Fast Transmission of Single Frames

Each window system has a mechanism to draw raster images, so in principle we can show movies by drawing images fast, one after the other [Lof90]. But each image causes considerable overhead because it invokes a large number of subroutines of the window system until it is finally presented. Also the programmer of the client has to control the flow of images explicitly. Therefore we considered this solution to be unacceptable.

5.2 Movie Service as an Extension to the Window System

A more promising approach is the integration of a movie window as a primitive of the window system. In case of X this implies full control of the movie window by the X Server process. The X Server process manages the connection establishment and release to the Movie Server and controls the movie transmission. We see many advantages compared to the solution above:

- The communication overhead between X Client and X Server is lower

- The code for movie presentation (i.e. presentation of sequences of images) does not have to be repeated in each client program

- The X server can be better adapted to the hardware, so the performance and hence the movie quality is better.

- The manipulation of the movie window (open, close, zoom, ...) works as usual (for the user and for the programmer).

The prototype of the X-Movie System is based on the latter solution. The subroutine library of the X Window System (Xlib) was extended to provide the following new functions [Kel91]:

XMListMovies(): Requests the X-Server to send a list of stored movies and images

XMFreeMovieInfo(): Frees the memory allocated with **XMListMovies**

XMOpenMovie(): Opens a movie connection

XMPlayMovie(): Starts a movie in a window

XMShowSinglePicture(): Shows a single frame of a movie

XMStopMovie(): Stops a running movie

XMDestroyMovie(): Frees all resources allocated for the movie (memory, communication channels)

6 Experience

The X-MOVIE system was implemented on IBM PS/2 Model 80 workstations under the AIX Operating System. All three components, the Movie Server, the Movie Client and the extended X Client, are now operational on PS/2s. The Movie Server was also ported to a DECSystem 5400 under ULTRIX. The PS/2s have 8514A color graphics adapters. In the current configuration the network is an Ethernet.

In our first experiments we used the TCP/IP protocols for digital film transmission. Our measurements showed that the throughput was only about 150 kbit/s. This was much lower than expected. The low transmission rate is mainly due to some of the TCP protocol mechanisms. One example is the Sliding Window flow control mechanism which adapts very slowly to the differences in speed of two participating systems. In our next experiment we used the UDP Protocol over IP. Here we could measure a throughput of approx. 2 Mbit/s. UDP has no error recovery, and approximately 10% of the packets were lost in the network. In a third experiment we used UDP and IP over a 16 Mbit/s IBM Token Ring Network. All three X-MOVIE components ran on PS/2 systems. The throughputs measured were approximately the same as with Ethernet.

In a fourth experiment all components were installed on the same workstation. Now no real time parallelism between the different processes was possible. The frequent process switches between the UNIX processes for the three components created a processor bottleneck, and the data rate between Movie Server and Movie Client dropped to below 1 Mbit/s [Sut90, Zim90].

The color graphics adapter 8514A has a color lookup table with 256 entries (8 bits per pixel). Our experiments were done with two digital films: Film 1 had an image size of 128 by 128 pixels, film 2 had 220 by 220 pixels. The films were dithered (converted to color lookup table technology) off-line and stored on the disk of the Movie Server. Film 1 could be shown at 10 frames/s, film 2 at 3 frames/s.

In a more detailed performance analysis we investigated the chain of components in the movie data path: the magnetic disk, the main memory, the bus and the network adapter card of the Movie Server; the network adapter card of the Movie Client (= X Server), the bus, the main memory and the graphics adapter of the Movie Client. Our results were that in all configurations the network adapter cards and network transmission protocols were the bottleneck of the system.

7 Conclusions and Outlook

The X-MOVIE system is a test bed for digital image transmission and presentation in a network of UNIX workstations. It is entirely based on standard hardware and uses standard network technology and standard graphics adapters with color lookup tables.

We have presented an algorithm for the continuous adaptation of a color lookup table for digital movies. It allows the presentation of films with good color quality using standard color lookup table graphics adapters.

Experience has shown that standard Ethernet technology and TCP/IP protocols are not suitable for movie transmission. In particular the Sliding Window flow control and the Go-Back-n error recovery scheme in TCP are inappropriate. The transmission of digital movies will require new protocols which are at the same time isochronous, i.e. have a very low delay jitter, and provide high throughput.

For the presentation of movies on the workstation screen, extentions to the X-Window-System were presented. It was shown that these extensions can be implemented and used without major problems. First experience with X-MOVIE has shown that the presentation of digital movies in a network of standard UNIX workstations is feasible.

Recently we have installed RISC workstations and an FDDI ring, and we are currently porting the X-MOVIE components to this new environment. We expect that films with image sizes of 256 by 256 pixels can then be shown at 25 frames/s. In the long run we intend to integrate JPEG and MPEG compression chips into the system.

Acknowledgment

We wish to thank Prof. A. Schmitt, W. Leister, and A. Stößer from the Institut of Operating and Dialog Systems of the University of Karlsruhe from where we got film 1 (generated with the VERA Raytracing System).

References

[ADH91] D.P. Anderson, L. Delgrossi, and R.G. Herrtwich. Process Structure and Scheduling in Real-Time Protocol Implementations. In *Kommunikation in Verteilten Systemen*, Mannheim. Informatik-Fachberichte, Springer-Verlag, 1991.

[ATW+90] D.P. Anderson, S. Tzou, R. Wahbe, R. Govindan, and M. Andrews. Support for Continous Media in the DASH System. In *10th International Conference on Distributed Computer Systems* Paris, May 1990.

[Bru89] T. Brunhoff. *VEX: Video Extension to X, Version 5.5*. Textronix, Inc., 1989.

[CDF+86] G. Campbell, T.A. DeFanti, J. Frederikson, S.A. Joyce, A.L. Lawrence, J.A. Lindberg, and D.J. Sandin. Two bit/pixel full color encoding. *Computer Graphics*, 1986.

[Com88] D. Comer. *Internetworking with TCP/IP*. Prentice-Hall International Editions, 1988.

[Fun91] Oswald Fundneider. Breitband-ISDN auf Basis ATM: Das zukünftige Netz für jede Bitrate, 1991. In *Kommunikation in Verteilten Systemen*, Mannheim. Informatik-Fachberichte, Springer-Verlag, 1991.

[Gal91] Didier Le Gall. MPEG:A Video Compression Standard for Multimedia Applications. *Communications of the ACM*, 34(4):46–58, 1991.

[Her91] R.G. Herrtwich. Zeitkritische Datenströme in verteilten Multimedia-Systemen. In *Kommunikation in Verteilten Systemen*, Mannheim. Informatik-Fachberichte, Springer-Verlag, 1991.

[HS91]   L. Henckel and H. Stüttgen. Transportdienste in Breitbandnetzen, 1991. In *Kommunikation in Verteilten Systemen*, Mannheim. Informatik-Fachberichte, Springer-Verlag, 1991.

[Jon88] O. Jones. *Introduction to the X-Window-System*. Prentice Hall, 1988.

[Kel91] R. Keller. Erweiterung des X11-Servers zur Bewegtbilddarstellung. Master's thesis, Universität Mannheim, 1991.

[Kil91] U. Killat. B-ISDN und MAN: Konkurrierende Netztechnologien. In *Kommunikation in Verteilten Systemen (Tutorial)*, Mannheim. Deutsche Informatik-Akademie, 1991.

[LD87]   L. F. Ludwig and D. F. Dunn. Laboratory for Emulation and Study of Intergrated and Coordinated Media Communication. *Proc. of ACM SIGCOMM 87*, Stowe, Vermont, pages 283–291, Aug. 1987.

[LMS91] W. Leister, H. Müller, and A. Stößer. *Fotorealistische Computeranimation*. Springer-Verlag, (Heidelberg, Berlin), 1991.

[Lof90]   G. Loff. Konzeption, Entwurf und Implementierung eines netzweiten Videodienstes. Master's thesis, Universität Karlsruhe, 1990.

[Lut89]   A. Luther. *Digital Video in the PC Environment.* McGraw Hill Book Company, New York, 1989.

[MD89]   W. E. Mackay and G. Davenport. Virtual Video Editing in Interactive Multimedia Applications. *Communications of the ACM*, 32(7):802–810, Juli 1989.

[Pin90]   M. Pins. *Analyse und Auswahl von Algorithmen zur Datenkompression unter besonderer Berücksichtigung von Bildern und Bildfolgen.* PhD thesis, Universität Karlsruhe, 1990.

[SGN88]   R. W. Scheiffler, J. Gettys, and R. Newman. *X-Window-System, C-Library and Protocol Reference.* Murray Printing Company, 1988.

[Ste90]   R. Steinmetz. Synchronisation Properties in Multimedia Systems. *IEEE Journal on Selected Areas in Communications*, 8(3):401–412, 1990.

[Sut90]   S. Sutter. *Datenformate bei der Bewegtbildübertragung über ein lokales Netz.* Studienarbeit, Lehrstuhl für Praktische Informatik IV, Universität Mannheim, 1990.

[Zim90]   R. Zimmermann. *Übertragungsprotokolle für die Bewegtbildübertragung über ein lokales Netz.* Studienarbeit, Lehrstuhl für Praktische Informatik IV, Universität Mannheim, 1990.

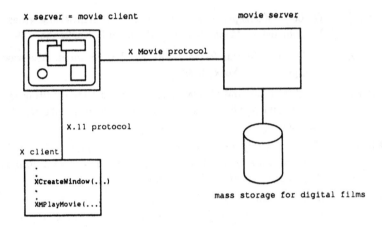

Figure 1: Architecture of the X-MOVIE system

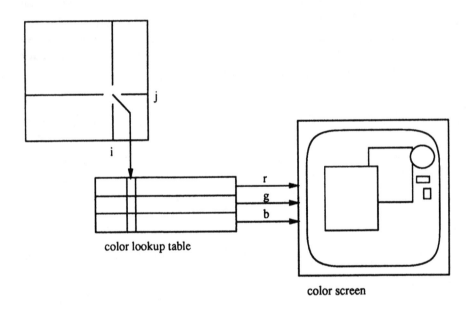

Figure 2: color lookup table technology

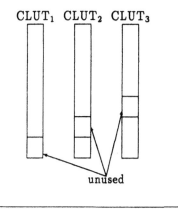

Figure 3: Dynamic adaptation of the CLUT

telnet, ftp, mail, rlo-gin, X-Windows, ...		nfs, X-MOVIE	
TCP		UDP	
IP			
Ethernet	Tokenring	...	

Table 1: Protocol stack in the Internet

Session X: Communication III

Chair: Duane Northcutt, SUN Microsystems Laboratories

The issues related to system support for digital audio and video can only be properly explored within the context of actual systems; it is rarely possible to effectively study a single aspect of the system support problem (e.g., process scheduling, buffer management, network control, etc.) in isolation. This is because of the interactions between the individual resource management activities that must be performed in the end-to-end process of communicating digital audio and video. Therefore, it is necessary to construct complete systems in order to validate proposed solutions and achieve the needed level of understanding of the relationships among various system-level trade-offs. All of the papers in this session describe work that has gained benefits from such empirical systems engineering efforts. These papers address various communications-related aspects of the general problem of transporting digital audio and video across computer communications networks.

The focus of the first paper is on the problems faced in delivering interactive compound documents (which include a rich variety of information types, including continuous media) to end-users via a digital communications network. The paper defines an interesting and technically challenging application area, and describes various difficulties encountered in the authors' attempt at emulating the computing environment needed to support such applications. The described work is intended to serve as a stepping stone for more ambitious experiments with continuous-media distribution and presentation. The second paper describes a CCITT standard protocol designed to meet the needs of videophone teleconferencing. This protocol is aimed at limited-bandwidth transport services and has provisions for interoperability among videophone units with differing capabilities. The author also indicates that his research institute has implemented working prototypes of videophones which use this protocol. In the third paper, key aspects of the design and implementation of a professional-grade digital audio system is described. This all-digital audio system is based on FDDI network technology and makes use of techniques for the synchronization of multiple independent streams of digital audio. The fourth and final paper describes a valuable technique for making effective use of the synchronous mode of transfer in FDDI systems for the transport of digital video information. The proposed network management technique is aimed at minimizing the total amount of buffering required by the system, while ensuring that the time constraints of digital video are met. Analytical techniques are used to prove that the proposed technique provides the desired properties, and work is reportedly underway to validate this work through empirical means.

In his presentation of a paper he wrote together with Gil Cruz and Thomas Judd, "Presenting Multimedia Documents Over a Digital Network," Jonathan Rosenberg of Bellcore described the general problem of distributing interactive compound documents to a large community of users over limited bandwidth connections. He defined the type of information to be distributed to be composed of a wide variety of different information types (e.g., text, graphics, audio, still images, video, synthetic imagery, etc.), that is stored at some remote server site, and accessed interactively by a (potentially large) set of (potentially widely geographically distributed) independent users. The transport service for the delivery of this information to a single user was assumed to be limited in terms of its available bandwidth, in that it would provide much less than is required to transmit a single stream of uncompressed video. Specifically, the assertion was made that the "last mile" problem would persist for the next few decades, and therefore the authors assume a data rate on the order of 1 Mbps.

Jonathan also described the three main areas that are being addressed in this work, and the approach they are taking on each of them. These key areas are:

- Media compression − a direct implication of the limited bandwidth assumption, and a means of effectively expanding the amount of transport bandwidth that is available. Their plan is to make use of existing hardware and software technology for reducing the bandwidth demands of media transport.
- Structural transmission − a means of making more effective use of the limited transport bandwidth by taking advantage of the structure inherent in compound documents and segmenting them into their (independent) constituent components and transmitting them separately (according to their individual requirements). The techniques being explored involve new partitionings of applications between source and destination.
- Time shifting − another technique for the efficient utilization of the limited transport bandwidth, that also uses the structure of the presentation material to predict access patterns and distribute the transport of desired information over a longer period of time (thereby reducing instantaneous bandwidth demands for the transfer of specific information items). In effect, this amounts to the prefetching of information.

Jonathan went on to describe the experimental (PC- and Ethernet-based) system that has been developed as part of project DEMON (Delivery of Electronic Multimedia Over the Network) at Bellcore. He presented a system architecture for an initial prototype that embodied three simplifying assumptions: single source/destination document delivery, circuit-switched network connections, and that the machine at the destination is dedicated to document presentation. Jonathan presented a taped demonstration of some of the initial results of this effort, which took the form of an interactive travel guide to Seattle. This demonstration showed a large (fairly static) image as a background, with a number of text, graphic, and image windows being dynamically created, deleted, and moved on the screen. The overall screen resolution was reported to be 900×675 pixels, with 16 bits/pixel, and a video rate of eight frames/second was shown.

Jonathan concluded by saying that this exercise had proved to his group's satisfaction that a practical system that can support their desired class of applications can be built. He went on to say that this experience has served to reinforce the notion that some form of hardware assistance will be needed to properly support such applications. In particular, he noted that the delivered video frame rate was almost half that predicted based on the theoretical hardware throughput figures.

An interesting discussion followed Jonathan's presentation, which centered on the question of regulatory restrictions on the ability of common telecommunications carriers to provide the type of services described in this presentation. Jonathan noted that while some restrictions have recently been lifted, there are still a large number of regulations and regulatory agencies to be dealt with, so the question is far from being resolved.

In response to questions regarding the assumptions of limited bandwidth to a wide client base, Jonathan noted that field trials are underway with digital service at rates of 1.5 Mbps into homes and 16 Kbps back out. However, this is being done over copper and fiber is a long way off. He noted that New Jersey Bell has estimated that it will take 30 years to completely replace their 56 Million miles of copper with fiber.

The paper "The CCITT Communication Protocol for Videophone Teleconferencing Equipment" was presented by Ralf Hinz of Daimler-Benz AG Ulm, Germany. It describes the recently defined protocol for videophone conferencing developed by CCITT Study Group XV. This protocol supports from one to six channels of synchronized audio, video, and data, all on a single 64 Kbps (ISDN-B) channel. While the protocol is designed to be network-independent, the intention is that this protocol should work with existing, low bandwidth (i.e., from 64 Kbps to 2 Mbps) digital communications paths (and not ATM or gigabit networks).

Ralf provided an overview of the CCITT recommendations that are applicable to this protocol (which included H.261 as a possible means of compression), and described the key features of the protocol. These include: in-band signaling, support for (de)multiplexing of streams, support for both point-to-point and multi-point connection topologies, and allowances for both centralized and distributed network synchronization mechanisms. Furthermore, the protocol encodes eight sub-channels in each transmitted octet; on basic rate ISDN channels, this provides 8 Kbps per sub-channel. In addition, a connection can be composed of up to six ISDN-B channels, so channel numbering has been added to allow for the synchronization of multiple sub-channels from different ISDN channels.

The protocol Ralf described includes a means for endpoint device capability exchange during the setup phase among potential communicating partners. This exchange permits the negotiation of the type of service to be used for the new connection. As part of the negotiation process, each station in the desired connection indicates its capabilities − e.g., number of frames per second of video it can source/sink, audio/video data formats supported, type of compression, type of codec, etc. In addition, the communications mode for a connection can be renegotiated during the course of a session. This capability permits a single communication stream to be used alternatively to send video and audio, FAX, or data, without having to acquire additional capacity or re-establish the connection.

Ralf went on to indicate that, in order to gain experience in the area of videophone conferencing, his research institute has produced working prototypes of videophones that use this protocol. These videophones incorporate the necessary codec's and line interface units, and include mechanisms to balance out differences in delays between different paths (e.g., the audio and the video in one channel of a teleconference).

It was noted that by considering data rates of up to 2 Mbps, it would be possible to use something other than the CCITT standard compression scheme (i.e., H.261) – for example, MPEG.

In addition, questions were raised about the error recovery capabilities of the protocol, especially with respect to the transfer of compressed video. Ralf pointed out that the protocol has a number of features for error recovery, including synchronization markers for video frames that will allow recovery following data loss.

Udo Zölzer presented in "FDDI-Based Digital Audio Interfaces" joint work with N. Kalff at the Technical University of Hamburg-Harburg, Germany. He gave a description of a rather large and sophisticated digital audio system that is based on FDDI technology at the data link layer. This system provided the capability to acquire, process, and present multiple, concurrent streams of digital audio, as well as distribute them among a distributed suite of audio sources, sinks, and processing units. This system manages the distribution of multiple digital audio channels in a variety of different formats – i.e., one of three different sampling frequencies (48 kHz, 44.1 kHz, and 32 kHz), and two different transmission formats (AES/EBU and MADI).

As this is a specialized system, it consists of dedicated-function units that use a TDM bus to connect individual audio channels to the network, and use TAXI parts to interface to the fiber. Each functional unit's fiber connection leads to a distribution switch, and all functional units (as well as the switches) in the suite make use of a common reference clock (i.e., the "house clock"). These data paths contain pure audio data and all of the control functionality exists in separate controllers, connected by traditional LANs. There is no general purpose computing done within the functional units, and no control traffic coexists with the digital audio samples.

Because of the dedicated-function nature of the system and the existence of a common, global clock, inter-sample and intra-sample synchronization is quite straightforward. However, when asynchronous inputs are introduced into the system (i.e., audio streams that are not sampled with the house clock – e.g., from remote studios, vans, etc.), a more difficult form of synchronization is required. Asynchronous inputs must undergo a process of sample rate conversion upon introduction into the system. In addition, when multiple streams of audio are to be synchronized, there are two types of issues to be addressed – one, for the case where the audio streams were acquired at the same nominal sample rate, but have different phasing, and the other for the case when the audio streams were acquired at different sample rates. All of these synchronization issues are addressed in this system, by way of sophisticated DSP techniques for re-sampling of the data streams.

Given the approach to re-sampling used in this system involves a form of interpolation, a question was raised as to what amount of delay is induced in the process. Udo indicated that only a small amount (on the order of one sample time) of delay is introduced.

Another discussion centered on the question of the practicality of house clocks in a more general computer networking environment, and helped draw distinctions between the type of dedicated-function system described here and more general distributed computing environments.

A further discussion was held that attempted to establish the potential viability of implementing traditional DSP functions (such as re-sampling) with high-performance (i.e., high clock-speed, high instructions-per-clock), general-purpose processors, as opposed to DSPs. From the discussion, it became clear that while these algorithms are costly, increases in general-purpose processor performance should make it practical to implement such functions for small numbers of streams, but (as in this system) support for large numbers of streams will require more cost-effective dedicated-function solutions.

The final presentation of this session was given by Bernard Cousin of ENSERB-LaBRI, Bordeaux, France on "Digital Video Transmission and the FDDI Token Ring Protocol." He described a technique for effectively distributing digital video using the synchronous mode of transmission defined in the FDDI specification. The aim of this work is to simultaneously allow the time constraints of the video to be met and arrive at an effective balance between the amount of buffer space needed at sources and destinations and the overhead induced by the passing of the network access control token. In the scheme presented here, this is achieved by choosing the size of transmission units and send/receive buffers in concert with the bandwidth needs of the streams and the network's chosen token rotation time.

By controlling the amount of data that is sent at each opportunity (i.e., token rotation), it is possible to ensure the effective use of a minimal amount of buffering at both the source and destination. In his presentation, Bernard indicated that he has analytical proofs that the proposed transport control scheme can meet the temporal constraints of video streams with a given (minimum) delay bound.

Bernard also discussed the importance of properly choosing the token rotation time for the network – in effect, it should be fast enough to provide good interactivity (i.e., granularity of access) and minimize worst-case buffer size requirements, but not so fast as to induce significant network overhead (and thereby lower the effective bandwidth of the interconnect).

In conclusion, Bernard noted that his group is in the process of constructing a prototype network that uses this control method to transport video over FDDI.

Initial discussions revolved around the question of whether it would be possible to achieve similar delay bounds with FDDI's asynchronous mode. In the course of this discussion, it became clear that the proposed scheme relies quite heavily on the hardware-provided periodicity of token rotation in the synchronous mode of operation.

It was also noted that, while the FDDI specification includes the synchronous mode, most of the currently available implementations only support the asynchronous mode of operation. Bernard replied that this is exactly why they were compelled to construct their own FDDI network interface units.

Another discussion followed that centered on the notion that the proposed scheme involves a three-way trade-off between buffer sizes, token rotation time, and transmission unit sizes. Which of these variables are changed and which are given can differ based on the available degrees of freedom in a given system. For example, given a specific negotiated token rotation time and a given transmission unit size, this scheme can be used to determine the amount of buffering needed to meet the temporal demands of a given video stream.

Presenting Multimedia Documents Over a Digital Network

Jonathan Rosenberg
Gil Cruz
Thomas Judd

Bellcore
445 South Street
Morristown, NJ 07962-1910
USA

October 25, 1991

Abstract
This paper discusses an experimental prototype system for presenting integrated multimedia documents over a digital network. This prototype is the first in a series investigating the requirements placed on the network in support of applications presenting multimedia information. The information consists of multiple media in digital form, including multi-font text, geometric graphics, photographic images, audio and motion video.

The paper describes the motivation for this line of research and the initial focus and goals of our first prototype. We describe the hardware comprising this prototype and the current status of our efforts. This is followed by a discussion of some early results we have obtained in dealing with integrated digital media, including motion video, with off-the-shelf components. Finally, we draw some conclusions about required support for digital media.

1. Introduction
The Information Networks Research group at Bellcore is investigating the network requirements for supporting near-term (five to ten years from now) multimedia information applications. Many researchers believe that multimedia information applications will be important in business, educational and residential environments. Applications might include interactive training, remote classroom lessons and home information services.

We are initially investigating the network requirements of residential electronic multimedia information applications. The home is appealing because the public switched network provides an individually addressable, electronic connection to virtually every residence in the United States. On the other hand, the home is a challenge because the local access loop does not presently support the bandwidth required by multimedia applications, especially those utilizing photographic images or motion video.

Although there are many open research questions, our primary focus is on techniques to enable the presentation of high-quality multimedia documents over networks with *limited bandwidth*. We consider limited bandwidth to be any rate at least an order of magnitude less than required to present uncompressed full-motion video. As we will discuss later, we are initially focusing on a bit rate of 1.5 Mbits/sec.

We are investigating these network presentation techniques in a project known as Delivery of Electronic Multimedia Over the Network (DEMON). As part of DEMON, we will be building prototype applications that present interactive, multimedia documents over a network. The construction of complete end-to-end experimental prototypes will allow us to fully explore the requirements that this class of applications places on the network and display devices.

The documents we are interested in include multi-font text, graphics, photographic images, audio and motion video. We are concerned with the presentation of these media when represented digitally (which we will call *digital media*) over a single network providing a fixed bandwidth connection. We are not considering adjunct analog networks. This is appropriate because the local access loop of the public switched network is expected to evolve to provide digital transport.

Our early work on an experimental prototype application has shown that it is possible to present high-quality multimedia documents, including embedded motion video, over a network at 1.5 Mbits/sec. The relative ease with which we were able to accomplish this is due to two factors. First, we have identified practical techniques for presenting multimedia documents over limited bit-rate connections. Second, by constraining the bandwidth to 1.5 Mbits/sec we have largely avoided pushing against the limits of off-the-shelf hardware and software, including networks, CPUs, operating systems, compression hardware and graphics systems. Despite this, we appear to have hit the wall in dealing with low-rate digital video on current workstation architectures.

This paper is primarily a discussion of the experimental prototype system we are constructing and what we have learned about the network delivery and presentation of digital media. The next section provides a short introduction to our view of multimedia document presentation over networks: why such presentation is important, why we are initially focusing on 1.5 Mbits/sec, why we believe this work will have wide applicability and the techniques we have identified. This is followed by sections briefly describing the DEMON project, the prototype[1] and some early results and conclusions.

2. Document Presentation Over a Network

One might ask why we should worry about presenting multimedia documents over a network. After all, there are standalone platforms available that can present multimedia documents delivered from local storage media, such as CD-ROM, laser disc or even magnetic disc. For some applications, of course, these platforms will be perfectly suitable.

However, network document presentation will be essential when the timeliness of information or the cost of acquiring and storing information locally make the use of a standalone system infeasible. For example, access from homes to a large collection of multimedia encyclopedias may only be economically practical if the documents are stored remotely and information presented over the network on demand. Other applications include daily in-house multimedia

---

[1] It is the policy of Bellcore to avoid any statements of comparative analysis or evaluation of products or vendors. Any mention of products or vendors in this paper is done where necessary for the sake of scientific accuracy and precision, or for background information to a point of technology analysis, or to provide an example of a technology for illustrative purposes, and should not be construed as either positive or negative commentary on that product or that vendor. Neither the inclusion of a product or a vendor in this paper, nor the omission of a product or a vendor, should be interpreted as indicating a position or opinion of that product or vendor on the part of the authors or of Bellcore.

newsletters for geographically distributed businesses and personalized information packages delivered to homes. In these examples, the cost and time required for traditional publication could be prohibitive. Network presentation may be an attractive alternative.

A major roadblock in the way of network presentation of multimedia documents is the bandwidth required by some media. The presentation of uncompressed, NTSC-quality digital video, for example, requires in excess of 80 Mbits/sec.[2] There is much work in the development of high-speed transmission for the public switched network that promises to alleviate this bandwidth bottleneck. Unfortunately, networking capabilities with sufficient bandwidth to easily support presentation of multimedia documents are unlikely to be available to any significant number of residences before the year 2010 [9].

There have, however, been recent advances into attaining increased bit rates over the existing copper loop facilities, including Asymmetric Digital Subscriber Line (ADSL) technology [6]. This technology allows a 1.5 Mbits/sec channel to a customer with a simultaneous low bit-rate return channel (suitable for, at least, a voice connection), over most of the existing copper-loop plant.

The 1.5 Mbits/sec rate is also interesting because of recent standards activity in full-motion video compression. Two emerging schemes, MPEG [4] and H.261 [5], can compress full-motion video for transmission at 1.5 Mbits/sec. Of the two standards, we are considering only MPEG because it is designed to support the the manipulation of digtal video. Recent studies have indicated that MPEG-encoded entertainment video at 1.5 Mbits/sec may be nearly equivalent in quality to VHS video [2].

Despite our current focus on residential applications, we believe that our techniques will also be applicable in non-residential applications. In particular, some techniques for supporting multimedia applications over packet-switching networks operate by reserving resources to guarantee a minimum bit rate connection to an application [1]. In these systems, an application needs to reserve the peak bandwidth required by the document, thus wasting much of the bandwidth (and resources). Our techniques will allow an application to reserve a lower bandwidth, of which little will be wasted.

In addition, even if a workstation is attached to a high-speed network, only a small fraction of the total bandwidth will, in general, be available. For example, on an FDDI network providing total throughput of 100 Mbits/sec, an application might obtain a channel guaranteeing 14 Mbits/sec. This is still nowhere near the bandwidth required for uncompressed full-motion video. Furthermore, although network speeds will increase, we can expect users to never have enough bandwidth, much in the same way that users run out of disk space and RAM, no matter how much you give them.

We have argued that our techniques will be useful not just in residential document presentation, but also in presenting documents over packet-switching networks providing support for multimedia applications. In the remainder of this section, we will describe three techniques we will use in presenting interactive multimedia documents over networks with fixed moderate bit rates: *media compression, structural transmission* and *time shifting*.

---

[2]This figure assumes an uncompressed 480×640 pixel image with 8 bits of color per pixel [3].

Media compression makes use of medium-specific and human cognitive properties to compress bandwidth-intensive media like photographic images and full-motion video. JPEG [8], MPEG and H.261 are examples of media compression techniques. For our purposes, we view media compression as available technology.

Structural transmission makes use of the structure of a document to save transmission bandwidth. Consider a presentation that has a fixed pixmap background upon which parts of the presentation will appear and disappear. Using structural transmission, the background is sent only once. The display machine understands the semantics of a background and will repaint parts of the pixmap as necessary, without the need for network communication. Another kind of structural transmission makes use of media semantics. For example, rather than using several thousand bytes to transmit a blue rectangle as a pixmap, it is possible to send several bytes representing a request to draw that rectangle and color it blue.

Time shifting makes use of periods of low channel utilization to deliver some media ahead of their presentation times. An example of this is the playing of CD-quality audio during a video presentation over a 1.5 Mbits/sec network channel. Using time shifting, the audio would be delivered before it was required during a part of the presentation that left spare channel capacity.

The use of structural transmission breaks a document into its constituent media and logical components. This tends to create peaks and valleys in the bandwidth requirement of the document as its presentation progresses. Media compression is used to squash down some of the peaks. Time shifting chops off peaks that are still too high and places them in earlier valleys. The result of these techniques is to reduce and smooth the bandwidth requirements of a document to make better use of a fixed bandwidth allocation.

3. The DEMON Project

The primary goal of the DEMON project is to study the network requirements of applications for presenting interactive multimedia information. Our initial focus is on the presentation of personalized, interactive information packages to residences. The packages will consist of articles chosen according to a subscriber's interests. The articles will contain little text, consisting primarily of photographic images, full-motion video, audio and geometric graphics. Like television, the articles in DEMON will be highly dependent on temporal media, such as audio and motion video.

Because we work in the context of end-to-end applications, our research will also involve areas besides network document presentation. We are also investigating interfaces for residential users and authoring tools to ease the task of creating multimedia articles. More details on DEMON — including motivation for its user model, the system architecture and examples of articles — can be found elsewhere [7].

4. The First DEMON Prototype

We are planning on building a series of experimental prototype systems exploring and demonstrating issues related to network presentation of multimedia documents. The first prototype, described in this section, is constrained to simplify many aspects of the problem. Future prototypes will explore increasingly complex models of network document presentation, including those required by non-residential applications.

For the first prototype, we are restricting our attention to the presentation of documents from an Information Provider[3] machine to a single residence machine over a circuit-switched connection with ADSL characteristics: 1.5 Mbits/sec available to the customer and a low-bandwidth return channel. This architecture is illustrated in Figure 1.

This architecture makes several simplifications to the general problem of network document presentation:

- *Documents are delivered from a single source machine to a single destination machine.* An economically realistic scenario might require that machines support multiple connections.

- *The network connection is circuit switched.* In many environments, a packet-switching network is the only network available.

- *The destination machine is dedicated to the task of document presentation.* In many environments (for example, the office), a workstation must perform many tasks simultaneously.

Note that these restrictions simplify the presentation of temporal multimedia documents by eliminating much of the resource contention typical in computer systems. We have effectively eliminated contention for network, operating system and some workstation resources, such as screen space. This was done in the interests of making some progress on this problem. We expected network document presentation even under these conditions to be rather difficult and we were determined to make our first step manageable.

In addition, because we are exploring new territory, the first prototype is being built using off-the-shelf hardware. The hardware comprising this prototype is illustrated in Figure 2. Although we are most comfortable with Unix[4] workstations, we have chosen a 486, 33 MHz PC running MS-DOS[5] as the receiving machine. We made this decision for several reasons:

- MS-DOS is a uniprocess operating system, and we could carefully control the scheduling of time-critical activities.

- There are a large variety of graphics systems available for PC platforms (in contrast to the meager choices available for Unix workstations). Our prototype is currently outfitted with a Matrox Image Series 1280[6], which provides powerful graphics and image processing capabilities and a flexibly configurable 8 MB of screen-mapped on-board RAM.

- Hardware support for compression invariably appears for PC platforms well before it is available on Unix workstations. The prototype incorporates hardware JPEG support in the form of a Squeeze-AT board[7] from Rapid Technology.

---

[3]*Information Provider* is our generic term for the entity responsible for the end-to-end delivery of documents.

[4]Unix is a registered trademark of AT&T.

[5]MS-DOS is a trademark of Microsoft.

[6]Matrox Image Series 1280 is a trademark of Matrox.

[7]Squeeze-AT is a trademark of Rapid Technology.

The PC, augmented with compression and graphics hardware, matches our functional view of the TV or information/entertainment appliance of the future. We expect this device to be like a TV, but with RAM and significant media formatting and networking capabilities.

We are running standard operating systems on both machines: SunOS 4.1 on the Sun workstation and MS-DOS 5.0 on the PC. At the moment, we are using TCP/IP as our transport protocol. The software for delivering documents and for displaying them was written by us.

The prototype is simulating a circuit-switched connection by making use of a dedicated Ethernet network. We have used Ethernet technology because it permits us to experiment with a range of bandwidths, expanding our research in both directions from 1.5 Mbits/sec. We will do this by installing a *throttle* (see Figure 2) in the network driver code on the Sun SPARCstation 2.[8] Applications will specify a maximum bit rate, which will be enforced by the throttle.

As discussed in the Introduction, one of our goals for DEMON is to investigate the integration of digital media. To us, this means treating all media, including temporal media, equivalently. This implies that all media (as far as feasible) utilize the same storage devices, network facilities and processing paths. We are, therefore, avoiding such typical tactics as utilizing an adjunct analog network for video transmission or installing hardware to allow video to bypass the operating system and be displayed directly on the monitor.

The first DEMON prototype is in an early stage of construction. We are, however, able to present some documents consisting of multi-font text, geometric graphics, photographic images and motion video (all in 16-bit color) over the Ethernet at approximately 1.5 Mbits/sec.[9]

Currently, the documents presented by the prototype have to be preprocessed by hand for effective network presentation. The initial software was written to demonstrate the general validity of our ideas and to gain some experience into our hardware infrastructure. The next implementation phase will generalize and automate the document presentation process.

Although we are early in the development of the first experimental prototype, we have learned some things about integrating digital media using conventional hardware and software. The next section provides some details about the media characteristics we are dealing with and discusses our experience so far.

5. Early Results from the DEMON Prototype

The first thing we considered for multimedia document presentation at 1.5 Mbits/sec was whether it would be satisfactory to treat an entire presentation as video and use MPEG to compress it. We did this to a short segment of one of our documents.[10] The MPEG process

---

[8]SPARCstation 2 is a trademark of Sun Microsystems.

[9]The throttle is not yet implemented, so we are measuring the attained bit rate, which is currently averaging near 1.5 Mbits/sec.

[10]We took the display output as video, digitized it and used a software system at Bellcore that simulates MPEG encoding and decoding.

produced what we characterize as a"fuzzy" version of the document segment. The motion video looked fine, but much of the text was blurry and difficult to read and the sharp lines comprising the geometric graphics were lost. We consider this quality unacceptable as it severely limits the amount and kinds of normation that can be usefully included in a document.

The documents presented in the prototype were preprocessed by hand to use media compression and structural transmission. Our use of media compression was limited to JPEG (discussed further below), but we found that these techniques were easy to use and produced the desired effect: we were able to present multimedia documents over a network at 1.5 Mbits/sec. The documents were equivalent in visual quality to those we can produce on a standalone workstation.

As mentioned previously, the documents we have been presenting contain multi-font text, geometric graphics, photographic images and motion video. As might be expected, we experienced several levels of difficulty in working with these media. We discuss our experiences with each class of media in the remainder of this section.

The simplest class of media to work with included multi-font text and geometric graphics. This comes as no great surprise since these media typically have low bandwidth requirements. Furthermore, our use of structural compression (sending a command to draw and color an object instead of sending the pixels, for example) lowered the requirements even further. In addition, the powerful capabilities of our graphics system meant that little processing was required by the PC upon receipt of the information.

Photographic images required more effort, although they were still relatively simple to transmit effectively. We have a JPEG board that provided us with flexible input and output. The board allows us to specify arbitrary routines to direct both (compressed) input and (uncompressed) output, so that we are able to feed the board from anywhere and send the output to anywhere. This flexibility is important because it allowed us complete freedom in the graphics systems and peripherals we can use. There were faster boards available but they attain this speed at the cost of being tied (via hardware) to a particular graphics system, and none of these graphics systems provided the power and resolution we needed. Of course, the flexibility we obtained was bought at a cost, as we shall discuss below.

Our strategy for photographic images is to compress each image only as much as necessary to enable the image to be transmitted, decompressed and displayed according to the document presentation schedule. This strategy caused us a problem since it required us to be able to attain a specified compression ratio. Unfortunately, the JPEG board only allows the specification of a *Q factor* during the compression of a image. The Q factor, which ranges from 1 to 255, is a rough idea of how much compression should be applied.

In practice, the compression ratio obtained for a specific image at a given Q factor is highly dependent on the image itself. This meant that we had to try several Q factors for each image until we obtained a compression ratio near the one desired. This is an inherent property of the JPEG compression algorithm. It is interesting to note that we typically were able to use compression ratios of 25:1 without producing any objectionable degradation of an image.

It will be no surprise to hear that transmitting and displaying motion video gave us the most trouble. We are counting on the availability of hardware support for MPEG to allow us to achieve good-quality video at 1.5 Mbits/sec. Unfortunately, no MPEG-decoder boards are currently available for purchase.

We are, therefore, currently making do with JPEG. We decided to do this, rather than waiting for hardware MPEG support, in order to gain some experience in working with digital video. The idea is to compress individual frames off-line, then transmit, uncompress and display them as rapidly as possible. Given the uncertain availability of MPEG, this technique is beginning to appeal to many researchers, developers and manufacturers. Using this technique, we have been able to achieve reasonable-quality motion video with a displayed size of 900×675 pixels and a rate of approximately 6.5 frames/sec over Ethernet at 1.5 Mbits/sec. This has required some tricks, however, as we discuss below.[11]

First, some figures on the performance of the JPEG board and our PC. We can decompress a 300×225,[12] 16-bit image (compressed at approximately 17:1 to about 8 Kbytes) from RAM and deliver it to the graphics system in about 115 ms. This allows us to display about 8 frames/sec from RAM.

Experimentation has shown that we can decompress those same images in only 85 ms if we throw the resultant bits away instead of shipping them to the graphics system. The raw speed of the board should, therefore, allow us to display about 11 frames/sec, but we have not been able to attain that rate. Our unsubstantiated belief is that we are suffering the effects of bus contention. Consider that, for each frame we are transferring bits from memory to the JPEG board, from the board back to memory and then from memory to the graphics system.

If we analyze this process, we see that for each frame we are transferring 8 Kbytes (the size of a compressed frame) from memory to the JPEG board, 135 Kbytes (the size of an uncompressed frame) from the board to memory and 135 Kbytes from memory to the graphics system. This is close to 300 Kbytes for each frame. For a rate of 11 frames/sec, this would be about 3 Mbytes/sec. The graphics system will accept data at the rate of 4 Mbytes/sec, although 11 frames/sec would only require about 1.5 Mbytes/sec. We believe, therefore, that this bit rate is greater than the bus can handle due to contention. It is for exactly this reason that some JPEG boards bypass the system bus to achieve higher effective rates.

Although 8 frames/sec is fast enough to achieve reasonable motion, the frame size of 300×225 pixels is disturbingly small in our application.[13] We would really like something two to three times that size in each dimension. Unfortunately, if we use source images of these sizes, the frame rate slows to the point of unacceptability.

It also turns out that compressing the images at greater than 17:1, thus reducing their size in RAM, does not allow a higher frame rate with larger images. The greater compression clearly

---

[11]There are a series of tricks we use, but do not discuss here. These involve the way input video frames are converted to digital frames. For example, for a video source converted from film, we take every other field and double each field to form a frame. This avoids the blurry frames caused by the 3:2 pulldown during conversion from film.

[12]The 300×225 size is somewhat arbitrary, being a convenient size for the capture system we are using. In addition, it is in the 3:4 aspect ratio used for movies.

[13]We are displaying the documents on a large screen display (60" diagonal) with 1280×1000 pixel resolution.

must reduce the time required to transfer the images to the JPEG board. Unfortunately, the time required to decompress an image is essentially independent of the compression ratio, depending only on the resultant size of the image. Therefore, we reduce somewhat the time to transfer the compressed image to the board, but the decompression time and the time to transfer from the board to the graphics system is effectively unchanged.

The trick we have come up with for increasing the effective size of video is to make use of the hardware zoom capability of the graphics system. This allows us to transmit the same number of bits to the graphics system and let the explosion in bits happen on the board. Of course, the video suffers some from blockiness caused by the expansion (via pixel replication) of the image. We believe that for motion video, a larger frame size is more important than smooth spatial resolution.

We have experimented with sending these images over the Ethernet, then decompressing them and displaying them. With this setup we were able to achieve about 6.5 frames/sec. This still gives reasonable-quality motion video, but calculations tell us that we should be achieving a greater frame rate. Since we are obtaining at least 1.5 Mbits/sec over the Ethernet, we should be able to transmit about 20 frames/sec (each frame is about 8 Kbytes in compressed form). We know that currently we can not do better than 8 frames/sec due to limitations on the PC, but the rate we obtain is consistently less than this. The compressed frames are stored in Unix files and we are not being clever in opening and reading these files. We suspect, therefore, that the time is being eaten up in file access overhead.

We are continuing to push our current configuration as far as possible to support motion video. We will continuing to investigate network delivery and the behavior of the system bus. In addition, we will be looking for methods of reducing the bus load.

An ongoing task for us is the attempt to draw general conclusions from the specific work we are performing. We expect our continued work to yield insights about networks, operating systems and machine architectures for digital media.

6. Conclusions

Even the small amount of work we have done is enough to validate our premise that high-quality multimedia documents can be presented over a 1.5 Mbits/sec network. These documents can contain multi-font text, geometric graphics and photographic images. Although we have not yet worked with digital audio, we do not anticipate any major problems in using it in documents.[14] While we would not claim to have achieved success with motion video, we are quite optimistic given our early experience.

It appears possible to press JPEG into supporting motion video with a minimum of effort, although full-screen, full-motion video is probably infeasible with any reasonable image quality. This is important since it appears that MPEG boards will not be available for some time.

---

[14]We do, however, expect that the inclusion of audio will force us to face the issues involved in close media synchronization.

The ease with which we have obtained our first results is partially due to the fact that we are working with a data stream traveling at only 1.5 Mbits/sec. This low data rate means that we do not push much against the limits of the network, storage media, operating systems and CPU. Presumably, this would not be true were we dealing with a higher bandwidth network. If our data were moving at 10 Mbits/sec we would undoubtedly be experiencing bottlenecks at more places, including the operating system and processor.

The problem we have been studying is a constrained version of the general problem of dealing with networked, integrated digital multimedia. Even with limited bit rates and the elimination of most resource contention, we have begun to hit some of the limits of conventional workstation architectures and operating systems.

The most obvious example of this is the system bus, which appears likely to become the bottleneck for our system (and others) as we process digital audio and video. This confirms our belief that current workstation architectures are inappropriate for supporting multimedia applications that deal with digital media. Workstations are designed to be computation intensive, which is fine for traditional data processing applications. Multimedia applications require architectures that are communication intensive and allow the fast establishment of high bandwidth connections among arbitrary subsystems.

Many researchers and manufacturers attempt to work around this problem by building hardware that bypasses the standard data paths. For example, video is usually delivered in analog form. The video may be displayed on the screen by digitizing it in real time and feeding the bits directly to the frame buffer. Another technique keys the video to the analog signal driving the display. Both techniques avoid the system bus and the operating system.

Unfortunately, by bypassing the standard system data paths, these solutions tend to have limited functionality and it becomes difficult to integrate them with other hardware and software components. This can lead to unpleasant side effects like the disappearance of the screen cursor when it enters a video window or the inability to overlay video with graphics.

Finally, we believe that the time is ripe for the investigation we have begun into network document presentation. This is due largely to the recent advances in compression technology and processor speed and the plummeting costs of memory, storage and graphics engines.

We make use of these advances by trading local processing power for network bandwidth using a variety of techniques. In addition, recent research on improving the bandwidth of cooper loop facilities and in providing multimedia support on packet-switching networks has made it desirable to pursue the network presentation of multimedia documents at moderate bit rates.

Acknowledgements

Abel Weinrib gave me useful comments on an earlier version of this paper and Rong Chang and Bob Kraut provided criticisms of this paper. Paul Crumley was helpful in discussing the PC architecture.

References

[1]   Domenico Ferrari.
        Real-Time Communication in Packet-Switching Wide-Area Networks.
        Technical Report TR-89-022, University of California and ICSI, 1989.

[2]   Robert S. Fish and Thomas H. Judd.
        A Subjective Visual Quality Comparison of NTSC, VHS, & Compressed DS1-
            Compatible Video.
        Proceedings of the Society for Information Display , May, 1991.

[3]   Howard P. Katseff, Robert D. Gaglianello, Thomas B. London, Bethany S. Robinson,
        Donald B. Swicker.
        Experiences with the Liaison Network Multimedia Workstation.
        In *Proceedings of Distributed & Multiprocessor Systems (SEDMS II).* 1990.

[4]   Didier Le Gall.
        MPEG: A Video Compression Standard for Multimedia Applications.
        Communications of the ACM 34(4), April, 1991.

[5]   Ming Liou.
        Overview of the px64 kbit/s Video Coding Standard.
        Communications of the ACM 34(4), April, 1991.

[6]   Earl E. Manchester.
        New uses for residential copper.
        Telephony , June, 1991.

[7]   Jonathan Rosenberg, Robert Kraut, Louis Gomez and C. Alan Buzzard.
        Multimedia Communications: a User-centered Viewpoint.
        IEEE Communications (Special Issue on Multimedia Communications) , May, 1992.

[8]   Gregory K. Wallace.
        The JPEG Still Picture Compression Standard.
        Communications of the ACM 34(4), April, 1991.

[9]   R. S. Wolff.
        What's Ahead for Copper?
        Telephone Engineer and Management , October, 1988.

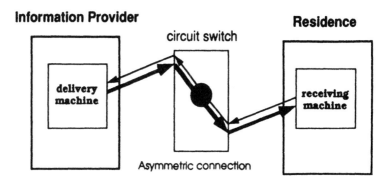

Figure 1: Architecture of First DEMON Prototype

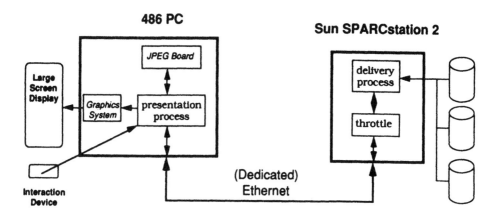

Figure 2: Configuration of First DEMON Prototype

The CCITT Communication Protocol for Videophone Teleconferencing Equipment

Ralf Hinz
Daimler-Benz AG
Institut für Informationstechnik
Tel. 0731 / 505 – 21 32
Fax. 0731 / 505 – 41 04
Wilhelm-Runge-Str. 11
7900 Ulm

Abstract

During the last years, the CCITT has defined the communication protocol for videophone teleconferencing equipment. This paper gives a summary and a survey of the concepts and properties of this protocol.

1   Summary

In the last years, some effort has been made to define a protocol stack which can be used by different terminal equipment for audio, video, and data transmission. This paper gives a summary and a survey of the recommendations of the CCITT communication protocol for videophone teleconferencing equipment.

The described protocol has been defined during the last years by CCITT Study Group XV with cooperation of national working groups. In one of these working groups the author represented the Daimler Benz Research Institute. To get experience in the field of videophone teleconferencing, this institute has developed and produced functional models of ISDN videophones [MAY89].

2   Goal

The quality of transmitted pictures in a videophone session is strongly dependent on the bitrate which is available for the transmission of video information. Since the pictures become better with higher bitrates, the protocol has to be able to cope with bitrates up to the range of some Mbit/s. On the other hand, it must be assumed that in the near future only 1 or 2 * 64 kbit/s connections (by means of narrowband ISDN) will be available for most of the endusers. For that reason, a protocol designed for both using 64 kbit/s connections, and for working on higher bitrates, too, is desirable.

The goal of the working groups was to define a protocol for audiovisual services (e.g. videophone), which is universal usable, i.e. not specialized for specific networks. Existing recommendations, such as G.704, X.30/I.461, etc. should be taken into account, existing hard- and software should be usable, and the realization of the protocol should be possible on simple microprocessors.

3   Result

The result of the activities is a series of recommendations. To sum up, some of the more important properties are listed below.

In contrast to e.g. the ISDN protocol (D channel), this protocol uses inband signalling, i.e. user data and protocol data are transmitted via one common channel. The protocol makes it possible to multiplex data (merge data from different sources and transmit them over one channel), as well as to split data (send data from one source over more than one channel). It is suitable for application both in networks synchronized centrally and decentrally, and it is designed to work in point-to-point and in point-to-multipoint connections.

The "basic set" of recommendations for this protocol is represented in the following list.

- H.200 - "Framework For Recommendations For Audiovisual Services". This is a survey of 45 recommendations, draft recommendations and titles for planned recommendations, which are defined or have to be defined for audivisual services [CCITT90f].

- H.320 - "Narrowband Visual Telephone Systems And Terminal Equipment". This recommendation describes the technical requirements of terminal equipment to be used for the narrowband videophone service. Narrowband means: The transmission rate is up to 1920 kbit/s [CCITT90a].

- H.221 - "Frame Structure For A 64 To 1920 kbit/s Channel In Audiovisual Teleservices". This recommendation defines the frame structure (syntax and semantics) used by the protocol [CCITT90b].

- H.242 - "System For Establishing Communication Between Audiovisual Terminals Using Digital Channels Up To 2 Mbit/s". This recommendation defines the procedural aspects of the protocol [CCITT90c].

- H.230 - "Frame-Synchronous Control And Indication Signals For Audiovisual Systems". This recommendation describes the exchange of control and indication information between communication partners using the frame structure defined in H.221 [CCITT90d].

- H.261 - "Video Codec For Audiovisual Services At p * 64 kbit/s". This recommendation describes the coding and decoding algorithm for moving video. The factor p is in the scope of 1 to 30 [CCITT90e].

- G.725 - "System Aspects for the Use of the 7 kHz Audio Codec within 64 kbit/s" [CCITT88]. Some definitions in G.725 form a subset of H.221 and H.242.

For the time being, some of these recommendations are being taken over by ETSI (European Telecommunication Standards Institute) and are adapted to European conditions.

4   Videophone Terminal Equipment

The recommendation H.320 "Narrowband Visual Telephone Systems And Terminal Equipment" describes the technical requirements of terminal equipment for the videophone service with a data transmission rate of up to 1920 kbit/s. This recommendation defines the adequate terminal equipment and divides it into several logical blocks. These blocks are defined in their functional behaviour, and relations to appropiate CCITT recommendations (e.g. H.221, H.242, H.230, H.261 and the I.400 series) are given.

Other recommendations which describe terminal equipment suitable for communication with the protocol described in this article, are e.g. G.725 (System Aspects for the Use of the 7 kHz Audio Codec within 64 kbit/s) and H.261 (Video Codec For Audiovisual Services At p * 64 kbit/s).

5   Syntax of Protocol Data Units

Recommendation H.221 "Frame Structure For A 64 To 1920 kbit/s Channel In Audiovisual Teleservices" describes the syntax of protocol data units. The structure defined in this recommendation is not the same as it is used in common protocols like HDLC. Instead, the octet structure of the transmission channels is used for forming eight subchannels (see fig. 1). Each bit of an octet is part of a different subchannel: All bits #1 together build the first subchannel, ... all bits #7 together build the seventh subchannel. The eighth subchannel is called "service channel". It has a more extensive functionality than the other seven subchannels, and it has a special internal structure. Using an ISDN B channel with a transmission rate of 64 kbit/s, each subchannel provides a transmission rate of 8 kbit/s.

Besides this "vertical" division, there is a "horizontal" division: 80 octets (i.e. 80 bits per subchannel) form a "frame", 16 succeeding frames form a "multiframe". A multiframe consists of eight "submultiframes" (SMF), each consisting of two consecutive frames.

The beginning and the end of this frame structure within the octet stream (the "frame alignment" and the "multiframe alignment") are introduced by the structure of the service channel. The alignment is necessary for the correct interpretation of received data. The information concerning the frame limits are coded within the service channel. For that reason, the bits containing this information are called "Frame Alignment Signal" (FAS, fig. 2). The FAS is constituted by the first eight bits of a frame within the service channel.

For it is also necessary to know, whether or not the communication partner has yet found the frame- and multiframe alignment, each partner uses one bit (the 'A'-bit) to signal his state of synchronization on the frame limits.

Another functionality provided by the FAS is the synchronization of multiple channels. Since a connection can be formed by up to six (ISDN B) channels, some more synchronization steps are necessary: The delay between the channels has to be handled, and the channels have to be treated in the correct order. For the first, a multiframe counter is used. By means of this counter, a relative delay of up to +/- 1.28 seconds between two 64 kbit/s channels can be egalized. For the latter, a channel numbering has been introduced.

Following the FAS, the next bits of the service channel form the "Bit-rate Allocation Signal" (BAS). The BAS is used for the transmission of information concerning the capabilities of a terminal, and it is used for signalling the mode in which the local transmitter is working.

Furthermore, the service channel contains the "Encryption Control Signal" (ECS). These bits may, in future, transmit information regarding encryption. If no information about encryption has to be transmitted, these bits contain user data. User data are also contained in bits 25 - 80.

6   Procedures of the Protocol

Recommendation H.242 "System For Establishing Communication Between Audiovisual Terminals Using Digital Channels Up To 2 Mbit/s" describes an inband signalling protocol. This protocol is used to establish, maintain, and disconnect a connection and to react on errors during these phases.

Since a connection can consist of up to six channels, there has been made a distinction between the channel which has been established first ("Initial Channel") and all other channels ("Additional Channel(s)").

Each channel has to send at least synchronization information by means of the FAS, and user data. The initial channel has some more work to do: Additional tasks before and during a communication are the exchange of capability information, sending of command codes and management of additional channels.

While a connection is established, data for control and indication as described in H.320 ("Frame-Synchronous Control And Indication Signals For Audiovisual Systems") can be exchanged.

6.1   Procedures

There are three basic sequences defined for the protocol which are used as building blocks for more complex procedures. All of them make use of the BAS. These basic procedures are:

- The capability exchange sequence for forcing the communication partners into a defined state, and for informing the communication partner about the own capabilities (fig. 3).

- The mode switching sequence for switching the receiver of the communication partner to a mode conforming to the own transmitter's mode (fig. 4).

- The frame reinstatement sequence for changing the transmission mode from an "unframed" format (i.e. without the FAS, BAS and ECS) back to a framed format (fig. 5).

When a communication between two or more partners starts, the partners have to find a common mode for operation. This common mode should make use of as much of the terminal's capabilities as possible, since only in this case all of their facilities can be used. For this, the mode initialization procedure has been defined.

If, during a connection, one of the communication partners wants to change his transmission mode (e.g. from voice to picture), he can do so by using the procedure for dynamic mode switching. It enables him to make use of different capabilities of the connected terminals.

Prior to disconnecting or to using special network services (e.g. call transfer), it is necessary for the connected terminals to work in a mode where 3.1 kHz PCM[1] audio signals can be decoded. For this case, the mode 0 forcing procedure has been defined.

In case of errors due to mode mismatch, the mode mismatch recovering procedure forces a mode reinitialization.

Also, a set of possbile error conditions and the reactions on them has been included into recommendation H.242.

---

[1] pulse code modulation

6.2 Relation to Network Signalling

The establishment and disconnection of channels cannot be achieved by means of the described inband signalling protocol. Instead, network signalling protocols have to be used. In case of an ISDN, this is the D channel protocol.

Prior to disconnecting a channel, the terminals are forced to work in the simplest mode ("mode 0") for being able to e.g. decode audio signals coming from a PABX[2].

7 Literature

[CCITT88] CCITT SGXVIII; Recommendation G.725: System Aspects for the Use of the 7 kHz Audio Codec within 64 kbit/s; Melbourne: July 1988

[CCITT90a] CCITT SGXV; Recommendation H.320: Narrowband Visual Telephone Systems And Terminal Equipment; COM XV-R 37-E, August 1990

[CCITT90b] CCITT SGXV; Recommendation H.221: Frame Structure For A 64 To 1920 kbit/s Channel In Audiovisual Teleservices; COM XV-R 37-E, August 1990

[CCITT90c] CCITT SGXV; Recommendation H.242: System For Establishing Communication Between Audiovisual Terminals Using Digital Channels Up To 2 Mbit/s; COM XV-R 37-E, August 1990

[CCITT90d] CCITT SGXV; Recommendation H.230: Frame Synchronous Control And Indication Signals For Audiovisual Systems; COM XV-R 37-E, August 1990

[CCITT90e] CCITT SGXV; Recommendation H.261: Video Codec For Audiovisual Services At p * 64 kbit/s; COM XV-R 37-E, August 1990

[CCITT90f] CCITT SGXV; Framework For Recommendations For Audiovisual Services; Geneva, 16 – 27 July 1990

[MAY89] May, Franz: Algorithmen und Realisierung eines ISDN-Bildtelefon-Codecs; ntz Bd. 42 (1989) Heft3, S. 130-133

---

[2] private automatic branch exchange

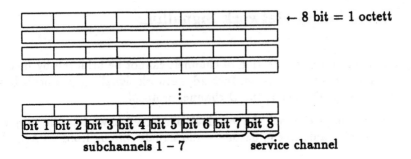

bit 1 | bit 2 | bit 3 | bit 4 | bit 5 | bit 6 | bit 7 | bit 8

subchannels 1 – 7   service channel

Fig. 1: H.221 subchannels

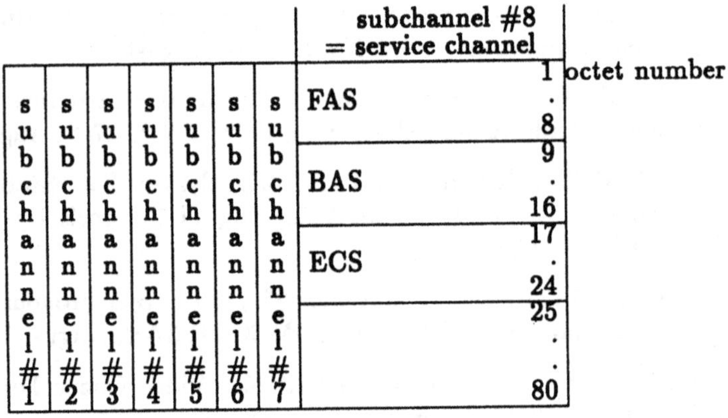

							subchannel #8 = service channel	octet number
s u b c h a n n e l # 1	s u b c h a n n e l # 2	s u b c h a n n e l # 3	s u b c h a n n e l # 4	s u b c h a n n e l # 5	s u b c h a n n e l # 6	s u b c h a n n e l # 7	FAS	1 · 8
							BAS	9 · 16
							ECS	17 · 24
								25 · · 80

Fig. 2: H.221 frame structure

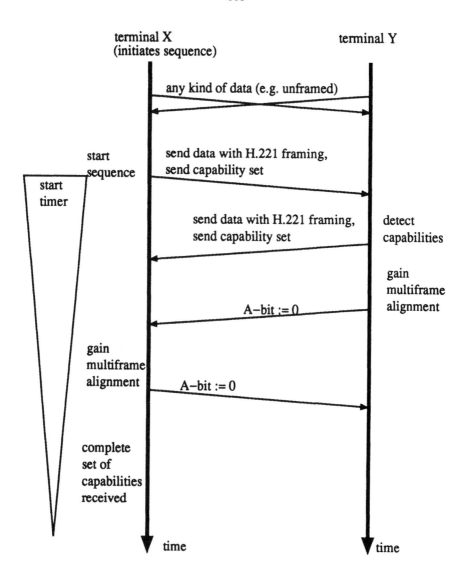

Fig. 3: Successful capability exchange sequence

366

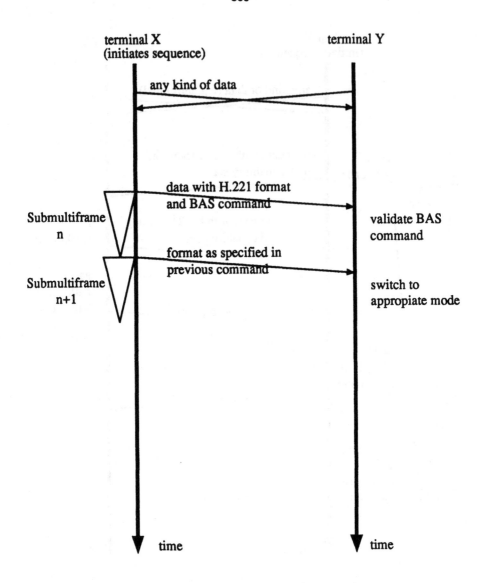

Fig. 4: Mode switching sequence

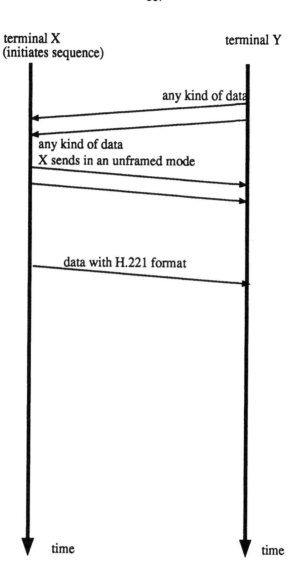

Fig. 5: Frame reinstatement sequence

FDDI-Based Digital Audio Interfaces

U. Zölzer, N. Kalff

Telecommunications Group

Technical University of Hamburg-Harburg

Eissendorfer Str. 40, D-2100 Hamburg 90, FRG

Tel. 49 40 7718 2166, FAX 49 40 7718 2281

Abstract

Digital audio processing is an emerging field of different standards for consumer and professional applications. Besides three main sampling frequencies (professional 48 kHz, consumer 44.1 kHz, broadcasting 32 kHz) two transmission formats have been established, namely a two-channel serial transmission format (AES/EBU-Format) and a multichannel audio digital interface (MADI). The interface techniques between these adopted standards will be discussed considering synchronization problems and the hardware complexity.

1 Introduction

Interfacing different single digital audio systems can be accomplished via the AES/EBU serial transmission link which is a unidirectional two-channel connection running at a data rate up to 3.072 Mbits/sec [1]. A consumer interface format (IEC) is derived from the AES/EBU two-channel format. The MADI interface format allows the transmission of 56 audio channels with a data rate of 100 Mbits/sec [2]. This high-speed link is a unidirectional transmission based on a network specification called the Fiber Data Distributed Interface (FDDI).

The distribution of different AES/EBU two-channel signals and multichannel signals is specially needed in broadcasting and studio applications. A typical configuration of a studio complex is shown in Fig. 1.

The main systems are mixed analog and digital I/O-Systems (ADIO) which contain AD/DA-converters, AES/EBU input and output signals and two-channel systems connected via sampling frequency converters (SFC), digital distribution systems (DD), audio signal processing systems (ASP) and human interfaces. The signal processing systems ADIO, DD and ASP are connected by two-way multichannel audio digital interfaces. Each of these systems is synchronized to an inhouse master sampling clock avoiding sampling frequency conversions. For multichannel sources which can not be synchronized to a common reference clock sampling frequency conversion systems are needed.

Among the switching techniques synchronous data routing can be utilized by time-division multiplexing and asynchronous data routing by traditional crosspoint systems.

In the first part of this paper we will discuss the MADI format, FDDI solutions and the hardware complexity. In the second part we will focus on synchronization problems, sampling frequency conversions and their hardware complexity.

2 FDDI and MADI

- FDDI (Fiber Data Distributed Interface)

FDDI is a high-speed serial interface using fiberoptic cables. Being about 10 times faster than traditional links it serves as point to point connection between two computers or as a fast backbone for a local area network in a ring topology. A single FDDI-based link can carry up to 100 MBits/sec of user data. The actual transmission rate is 125 MBits/sec. The redundancy is added by the transmitter to introduce two additional features. The first is error detection, allowing to recognize bit errors and mark illegal bytes. Second, the user may use up to 15 commands to control the receiver hardware. The conversion from the users data bytes being transmitted with up to 12.5 MBytes/sec into the coded bitstream with 125 MBits/sec and back is performed by AMD's [3] two TAXIchips (Transparent Asynchronous Transmitter/Receiver Interface), the Am7968 transmitter and the Am7969 receiver (see Fig. 2).

Data transmission is initiated by using the transmitters strobe pin. The receiver announces a received data byte with the data strobe pin. During intervals without actual data the transmitter sends a special pattern, the Sync command, to keep the the receivers phase-locked loop locked onto the incoming bit boundary. The chipset within two 28-pin PLCC devices can drive conventional coaxial cables as well as fiberoptic cables.

- MADI (Multichannel Audio digital interface)

MADI is a serial interface format which carries up to 56 channels of digital audio data with a sampling rate between 32 kHz and 48 kHz ± 12.5 %. Using fiberoptic cables a transmission distance up to 2 km is possible, with coaxial cables about 50 m. The channel format is based on the two-channel AES/EBU format, where each channel is represented by 32 bits labelled 0 to 31 (Fig. 3). 24 bits are used for the representation of the audio data. The MADI frame period consists out of the 56 channels which are transmitted sequentially starting with channel 0, and ending with channel 56. The description of bits 4-31 follows those used by the AES/EBU format. The first channel within a sampling period is marked by the first preambel bit. The second preamble bit defines the status of the associated channel to be active or not, the third preamble bit is used for marking a stereo signal and the fourth preamble bit marks a 192-frame block, which is used at the receiver to decode the status (C bit) and user data (U bit).

MADI is based on FDDI and so utilizes the AMD TAXIchips. MADI transports only the pure data and does not carry any information about the sampling frequency.

- Hardware Complexity

Routing the 56 channels from a time-division multiplexing parallel bus system to the transmitter can be accomplished by writing data continuously into a dual-ported ram

(Fig. 4). The composite side is connected to a routing controller which reads the 56 choosen channels. Then the channel data is handed to a multiplexer converting the 32 bit AES/EBU data into four bytes which are transferred to the Am7968. At the receiver a demultiplexer collects the four bytes from the Am7969 and converts it back into the AES/EBU format. Since there are only about 50 ns to pick up the data, an additional buffer is necessary allowing the routing controller to write the data into selected memory locations of a dual-ported ram. The other side of the dual-ported ram is controlled by a time-division multiplexing parallel bus system.

3   Synchronization

The synchronization of systems with different sampling rates is needed when two or more digital systems are brought together in digital distribution systems or digital processing systems (refer to Fig. 1). The first synchronization problem appears when systems are operating at nominally the same sampling rate, but only one system is synchronized to a master reference clock. The second synchronization situation is characterized by systems which are operating with different sampling rates (32 kHz, 44.1 kHz, 48 kHz). If there exists a master system with a master reference clock all signals coming from different systems have to be synchronized by sampling frequency conversion systems to the reference system.

Synchronization for nominally the same sampling rates can be achieved by variable delay techniques in combination with fractional sample-delay algorithms. The more common situation of different plesiochronous sample rates needs sample frequency conversions. The all digital approach of a sampling frequency conversion approximates the digital-to-analog conversion of the source signal and then the analog-to-digital conversion with the new sampling rate [4]. First the sampling rate is increased by a factor of $L = 2^{w-1}$, where w denotes the number of bits for the representation of the audio sample, and then a resampling with the new output sampling rate is performed.

Such systems make use of multirate signal processing techniques to reduce the hardware and as well the software complexity. The algorithm for a sampling frequency conversion can be performed by a digital signal processor system as shown in Fig. 5. A sampling frequency conversion between 44.1 kHz and 48 kHz in both directions for a two-channel AES/EBU signal can be achieved with two digital signal processors and special interface circuits. The performance and the accuracy of such a system is strongly influenced by jitter which might exist on the input and output sampling clock. A more severe problem arises in the case of an asynchronous MADI source, where each channel has to be synchronized by a sampling frequency converter (see Fig. 1). The hardware complexity of a MADI sampling frequency converter can be considered as times 56 of a one channel conversion system (Fig. 6). Therefore the sampling rate of the source must be recovered from the MADI data. Although MADI does not carry information about the sampling rate, it is possible to receive MADI data from a source not being connected to the reference clock of the receiving system. The first bit in a MADI word, the Channel Zero Sync bit, is high for the first channel of a sample period. This periodical pulse at the beginning of the sample period can be used to recover the source sampling rate. This input sampling clock is used for the input bus shown in Fig. 6 and for the sampling frequency converters. The reference sampling clock is connected to the sampling frequency converters, the output bus and the MADI transmitter.

4 Conclusions

FDDI-based MADI transmission links and the hardware complexity of digital audio interfaces have been discussed in the context of all digital distribution systems and audio processing systems. A common master reference clock should be used for synchronization of different digital audio systems avoiding sampling frequency conversion of MADI sources. The outlined interface techniques show the flexibility and effiency of digital audio distribution systems which use MADI transmission links especially in broadcasting and studio environments.

References

[1] Audio Engineering Society, "AES Recommended Practice for Digital Audio Engineering - Serial Transmission Format for Linearly Represented Digital Audio Data," J. Audio Eng. Soc., Vol. 33, pp. 975-984, 1985

[2] Audio Engineering Society, "AES Recommended Practice for Digital Audio Engineering - Serial Multichannel Digital Audio Interface (MADI)," J. Audio Eng. Soc., Vol. 39, pp. 368-377, 1991

[3] Am7968/Am7969 Transparent Asynchronous Xmitter-receiver Interface Data Sheet, Advanced Micro Devices, Sunnyvale, CA, May 1987

[4] R.E. Crochiere, L.R. Rabiner, *Multirate Digital Signal Processing*, Prentice-Hall, Englewood Cliffs, 1983

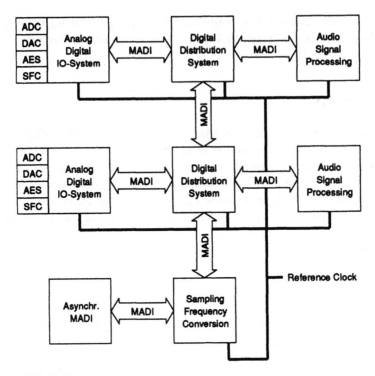

Figure 1: FDDI-based digital audio distribution system

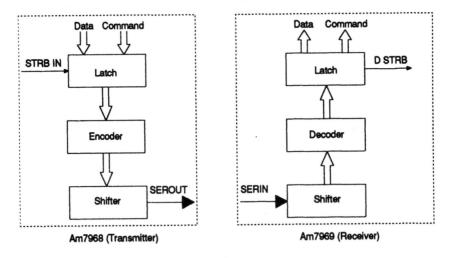

Figure 2: TAXI-Chips

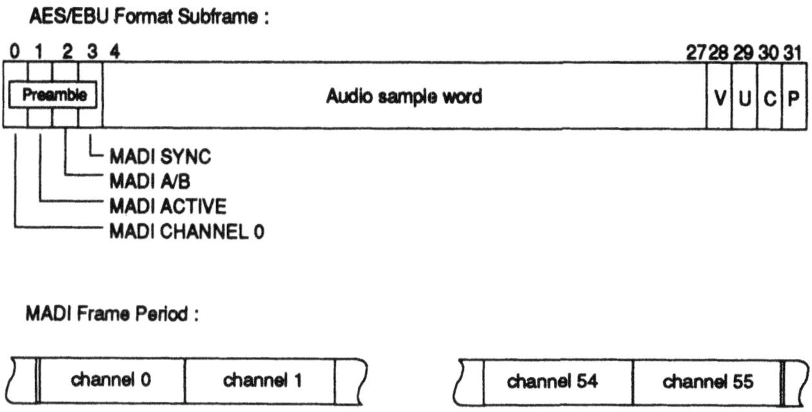

AES/EBU Format Subframe :

0 1 2 3 4		27 28 29 30 31
Preamble	Audio sample word	V U C P

└ MADI SYNC
── MADI A/B
── MADI ACTIVE
── MADI CHANNEL 0

MADI Frame Period :

channel 0	channel 1		channel 54	channel 55

Figure 3: MADI channel-/sampleframe

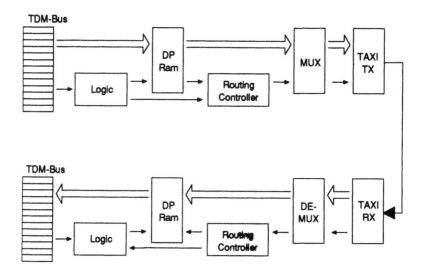

Figure 4: TDM → MADI

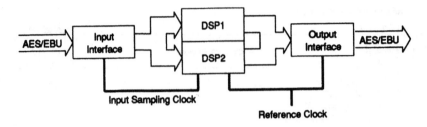

Figure 5: Hardware for two-channel sampling frequency conversion

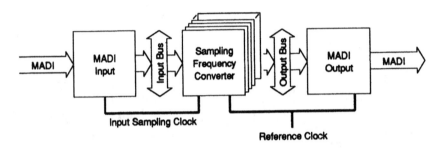

Figure 6: Hardware for MADI sampling frequency conversion

Digital Video Transmission
and the FDDI Token Ring Protocol

B. Cousin
ENSERB - LaBRI*
351 Cours de la Libération
F-33405 Talence cedex
FRANCE

fax : 33 - 56.37.20.23
tel : 33 - 56.84.65.29
e-mail : cb@tirukali.greco-prog.fr

Abstract
This paper proves that the temporal constraints for digital video transmission can be met by the synchronous transmission mode of the FDDI token ring protocol. First, we introduce the FDDI protocol. Second, we establish the temporal constraints for real time image transmission. Then we propose a technique based on an optimal use of emitter and receiver image buffers, and an optimal constituting of transmission units. Finally, we verify that the temporal constraints are fulfilled by the FDDI protocol and the proposed technique. This technique requires only an optimal allocation of the exact bandwidth needed by the image transmission. Moreover, image sample blocking enables larger TTRT to be used, and thus, reduces the overhead induced by the token rotation. We also prove that the proposed technique produces a minimal and constant delay equal to 2*TTRT plus the physical response time of the network, in spite of the aperiodic delivery of the image samples due to the access method of the FDDI protocol.

1. Introduction

Our research group is interested in digital video transmission. Given the very large bandwidth required by digital video, only high speed networks can be considered to achieve the image transmission throughput. We have chosen to focus our study on the FDDI token ring protocol because it is the earliest representative of standardized high speed local area networks, and because its component sets are widely available now, enabling low cost and easy experimentation [FDDI 87, FDDI 88, FDDI 89, AMD 89].

FDDI uses fundamentally a **technique** of asynchronous transmission (that is to say, the delay in transmission is variable), but this protocol stipulates two **modes** of transmission : the asynchronous mode and the synchronous mode. The synchronous mode guarantees a station a pre-allocated bandwidth and the right to transmit with an average periodicity equal to a value negotiated among all the stations. This periodicity is referred to as the Target Token Rotation Time (TTRT). Furthermore, the protocol guarantees a maximum rotation time of the token that cannot surpass 2*TTRT. The synchronous mode is used for those applications whose bandwidth and response time limits are predictable, permitting them to be preallocated. The asynchronous mode is used for those applications whose bandwidth requirements are less predictable (bursty or potentially unlimited) or whose response time requirements are less critical. Asynchronous bandwidth is instantaneously allocated from the pool of remaining ring bandwidths that are unallocated or unused.

---

* Unité de recherche n° 1304 associée au CNRS

At first glance, it seems easy to transmit digital voice or video by means of the synchronous mode of the FDDI protocol. But let us recall that, the digital image traffic is relatively unpredictable, first because a compression process can be required to enable high definition images to be transmitted on a 100 Mbit/s network, and second, in order to have a low frequency of access to the medium to reduce the overhead induced by the token rotation, it is necessary to block together and send simultaneously several images or several parts of different images. Moreover FDDI protocol does not propose isochronous transmission : the token ring access method can lead to variations in the transmission delay of images due to variation of the network load.

An upward-compatible enhancement, called FDDI-II [FDDI 9b, Ross 90, Teener 90], adds a circuit switched (isochronous) service to the existing packet service of the basic FDDI. FDDI-II employs a cycle structure to control the multiplexing of packet data and circuit switched data in the same ring. This structure repeats on the ring every 125 microseconds. The allocation of isochronous channels is allowed with a variable granularity. Supported channels sizes include 8, 16, 32 and 64 Kbit/s plus any multiple of 64 Kbit/s up to 6.144 Mbit/s, providing a maximum of 98.304 Mbit/s isochronous payload. In addition, the aggregate of any or all of the isochronous channels may be used as one circuit service, satisfying the needs of heavy load applications. The remaining and unallocated bandwidth is dedicated to the packet service.

We have chosen to use the synchronous mode of transmission of the FDDI protocol and not the isochronous mode of FDDI-II. In first place because FDDI-II network is not yet available, and in second place, because we do not suppose that the sample traffic occurs in precise amounts on a time basis. This is due to the compression process which is required to enable high definition image to be transmitted on a 100 Mbit/s network. So our sample traffic is relatively unpredictable, and the use of circuit switched service would need overallocation of isochronous channels.

At the emitter, to enable the synchronous allocation of the right mean bandwidth required by the video transmission to be fulfilled, our intention is to carefully undertake of the constituting of each synchronous transmission unit. If the token is late, the transmission unit carries all the image samples produced from the instant of the previous token arriving at the instant of the expected token arriving (TTRT after the arriving of the previous token). If the token is early, the transmission unit carries all the remaining samples, because the control of the quantity of data transmitted at each capture of the token does not have to be managed at the level of the FDDI sender, the normal throughput of the emitter of images naturally assuming this role. The buffering induced by this technique can be usefully exploited to block several samples in one unit of transmission enabling the transmission to be improved, first because the overhead induced by the structure of the transmission unit (starting and ending delimiters, addresses, frame control, and so on) is spread over a large number of samples, second because the token rotation overhead is reduced with its frequency. This way of doing the constituting, enables the emitting delay to be minimized and needs only an optimal allocation of the exact bandwidth required by the image transmission.

At the receiver, in order to allow the transmission of periodic information with asynchronous techniques, it is necessary to supply sufficient buffer space in the memory at the level of the receiver to accommodate the inevitable variations in transmission delay. These memory buffers imply a systematic delay inhospitable to interactive applications, but these memory buffers can be usefully exploited to detect and then to correct loss, corruption, and duplication in all or part of the images of the video, and to synchronize the flow of data to correct the inherent drift problem due to the presence of distinct clocks in the sender and in the receiver.

In short, the problems of data corruption, of synchronization, and of adaptation of the receiver to the flow of data, all oblige both the sending and the receiving stations to buffer part of the images. Buffering at the level of the sender allows us to send large frames onto the network and thus enables the transmission to be optimized. The sender buffer is also needed to adapt the flow of the image samples to the token delay. Buffering at the level of the receiver allows us to manage the aperiodic delivery of the image samples due to the previous phenomena. In fact, the accumulation of these different bufferings is quite acceptable physiologically; a delay of several images occasions a delay of less than a tenth of a second, a negligible delay in human terms. We recall that this delay is applicable to all the images, and thus only the starting of the video is affected by it. Nevertheless, this buffering does demand a great deal of memory.

2. FDDI

In a former work, we have developed a model of the FDDI protocol [Cousin 91]. Accordingly to the standard statements, we have formally proved that the time between successive token arrivals at any given station has an upper limit of twice the Target Token Rotation Time (TTRT) negotiated by the stations during the initialization phase of the protocol minus the token delay measured at the given station during the previous rotation. This limit is tantamount to proving that the token rotates quickly enough to satisfy the standard statement : "The protocol guarantees an average response time (TRT : Token Rotation Timer) not greater than the TTRT, and a maximum TRT not greater than twice TTRT". In [Johnson 87] Johnson proves a similar but weaker result to assure that the token rotates quickly enough to prevent initiation of recovery unless there is failure of a physical resource or unless the network management entity within a station initiates the recovery process. A formal proof of the two properties can be found in [Sevcik 87], but the study is applied on a lightly modified and improved FDDI protocol.

We will let N denote the number of stations in the network, assuming one network connection per station. The stations around the ring will be numbered from 0 to N-1 in a clockwise manner.

The protocol FDDI uses an access method called Token Ring [Johnson 86, Ross 89]. FDDI stations are connected in a ring where a token passes round. Any station that wants to transmit over the ring has to capture the token. At the end of its transmission, that station must release the token. Every station, therefore, obtains the right to use the medium turn by turn.

We denote $Tj(I)[k]$ the moment where the token is received at the station k during its Ith rotation. So, $Tj(I+1)[k]-Tj(I)[k]$ is the duration of the Ith rotation of the token, duration measured at the station k.

During the initialization phase of the FDDI protocol, all the stations connected to the ring negotiate the value of TTRT. The TTRT chosen is the smallest. TTRT has to lie between the two values, TTRTmin and TTRTmax ($TTRTmax \geq TTRT \geq TTRTmin$). The value TTRTmin corresponds to the minimum time for the management and rotation of the token. A TTRT value less than TTRTmin would not even allow the token to reach all of the stations, and thus such a time is unacceptable. A TTRT value greater than TTRTmax is conceivable without major problems except that it creates a partition of the medium that is somewhat prejudicial to equal access because a station holding the token could very well keep it a very long time. Furthermore, a TTRT value greater than TTRTmax slows the detection of errors and in addition the reconfiguration of the ring.

Let us note that the smaller the value of TTRT, then the more important becomes the amount of time dedicated to the management of access to the medium. In fact, the number of rotations of the token per unit of time is inversely proportional to TTRT. Thus the token consumes a great part of the bandwidth. We have every interest in sustaining as great as possible a TTRT within the limits of foreseeable applications [Dykeman 88].

To provide the required service towards the token rotation duration the standard specifies a timer in each station called TRT (Token Rotation Timer). It is used to control scheduling during normal operation and to detect and recover from serious error situations. Whenever the TRT expires, it is reinitialized to the TTRT value and the variable "Late_Ct" is set. If the token arrives at the station before the expiration of the timer (early), it is reinitialized to the value TTRT and the variable "Late_Ct" is reset.

Although in the standard the FDDI timing control is assured by both the value of the TRT and the Late_Ct variable, in order to simplify the timing model we use only one parameter. We denote by **TRT(I)[k]** the value of the timer at the station k during the Ith token rotation. It can be recursively defined by the token rotation time Tj :
(TRT definition) :

$\forall k, \text{ TRT}(0)[k] = Tj(1)[k] - Tj(0)[k], \text{ and } \forall I \neq 0,$
if $\text{TRT}(I-1)[k] \leq \text{TTRT}$ then $\text{TRT}(I)[k] = Tj(I+1)[k] - Tj(I)[k]$,
else $\text{TRT}(I)[k] = (\text{TRT}(I-1)[k] - \text{TTRT}) + Tj(I+1)[k] - Tj(I)[k].$

At the first rotation the TRT value is initialized. Afterwards if the token is early, that is to say if the previous rotation respects the negotiated periodicity TTRT, the next TRT is equal to the token rotation duration. Otherwise, if the token is late, the next TRT is equal to the sum of the token rotation duration and the token delay. This sum enables the delay to be taken into account from one rotation to the next, and thus, enforces the periodicity.

Each station has another timer called by the standard : THT (Token Holding Timer). It contains the maximum duration during which the stations can transmit in asynchronous mode. It is set at each early token reception with the token gain. We denote by **THT(I)[k]** the value of the timer at the station k during the Ith token rotation :
(THT definition) :

$\forall I, \text{ if } \text{TRT}(I)[k] \leq \text{TTRT} \text{ then } \text{THT}(I)[k] = \text{TTRT} - \text{TRT}(I)[k] \text{ else } \text{THT}(I)[k] = 0.$

Similarly, $\delta_j(I)[k]$ denotes the delay of a late token :
(δ_j definition) :

$\forall I, \text{ if } \text{TRT}(I)[k] > \text{TTRT} \text{ then } \delta_j(I)[k] = \text{TRT}(I)[k] - \text{TTRT} \text{ else } \delta_j(I)[k] = 0.$

We denote Ts(I)[k] the synchronous transmission duration during the Ith token rotation. Using the previous notations, the maximum token rotation duration can be described by :
(Token rotation duration property) :

$\forall I, \forall k, \text{ TRT}(I)[k] \leq 2 * \text{TTRT} .$

So now, we have to show that the synchronous transmission mode of the FDDI protocol enables real time image transmission to be carried out.

3. Images

3.1 Temporal constraints

Both transmission of digital images and digital voice have temporal constraints that we do not ordinarily encounter in conventional data transfer. These temporal constraints associate samples. A sample is that portion of a signal that is digitalized. For example, a sample could be a group of bits, one byte of coded sound, or a line of an image.

The set of the samples makes a sequence {ei}. So we can associate to each sample its running number in the sequence. We indicate the moment of production of the sample by the emitter with the notation **Te**. Likewise, we use the notation **Tv** (visualization) to indicate the moment when the sample can be displayed on the visual equipment. Te and Tv are strictly increasing functions.

The preservation of the quality of the video during transmission requires that two constraints must be satisfied. First constraint : the delay after the emission of the video must be humanly tolerable, virtually instantaneous. We refer to the time that one must wait to see the first image of a video as Tmax. This time is critical if the user intervenes in the unfolding of the video; that is, if the video is in any sense interactive. Second constraint: the images should appear on the screen of the receiver at the same speed relative to one another as they are produced by the emitter. If these two constraints are satisfied, then the video is received with temporal integrity. Two relations suffice to express these constraints:

Tolerable delay constraint $\quad : \forall i, \ Tv(i) < Tmax + Te(i).$

Temporal integrity constraint $\quad : \forall i, \forall j, \ Tv(j) - Tv(i) = Te(j) - Te(i).$

These temporal constraints exist only if the video should be visualized on its arrival at the receiver (in real time, no less!). These constraints do not exist if the video is broadcast in deferred time (for example, if it is pre-recorded for later broadcast), and if it is thus consequently stored on its arrival at the receiver. In such a case, the transmission of the video can simply be treated as the transmission of a large file.

3.2 Transmission

In fact, in as much as they are located on distinct sites, the receiver of images is completely independent of the sender of images, and it is thus difficult to respect the two previous constraints. Several phenomena intervene in the variation of transmission delay : the access method, the image compression and decompression processes, the blocking of samples, the clock drift, etc.

Conventionally, the clock of the receiver of images is slaved to the clock of the emitter by means of a synchronization included in the signal. Since the conventional methods of transmission use **isochronous** (circuit switched) technique, the intervals of time between samples are preserved during their transmission. The synchronization of the receiver with the sender of the images is therefore easily achieved. It suffices to slave the receiver's clock to the flow of the received images. Only a constant delay is added (the propagation delay).

The most current techniques of transmission now use **asynchronous** transmission technique. With this technique, the delay in the transmission of the samples varies: it depends on the access method to the medium, on the resolution of collisions, on the load of the network, etc. Consequently the time separating two samples at their reception may differ from the time separating them at their production. We can no longer count on slaving the clock of the receiver directly on the flow of received images.

The technique of asynchronous transmission permits a better use of support than does isochronous transmission because sporadic flows can be compensated for. Asynchronous transmission works well with dynamic allocation methods of the bandwidth between different links as a function of the load. Nevertheless, in order to be efficient, the overhead introduced by this dynamic management must be compensated for by a better allocation of the traffic. In contrast, isochronous techniques can use a method of static allocation that requires little or no management overhead.

We denote **Tr** the moment of reception of a sample. We denote by **Tt** the response delay of a sample over the network. These moments are described by the relation :
(Tr definition) :

$\forall i, \ Tr(i) = Te(i) + Tt(i).$

3.3 Blocking

The blocking of several samples in one unit of transmission enables the transmission to be improved. The overhead induced by the structure of the transmission unit (starting and ending delimiters, addresses, frame control, and so on) is spread over a large number of samples. As all the samples blocked in the same transmission unit are sent and received at the same moment, this technique produces variation in the transmission delay. Consequently the time separating two samples at their delivery may differ from the time separating them at their production.

The response delay Tt of local area networks like FDDI consists of the access delay **Ta**, the transmission delay **Td**, and the propagation delay **Tp**. The propagation delay depends on the propagation speed and the length of the media. The propagation delay can be regarded as constant. The transmission delay depends on the data rate and the length of the transmission unit. The access delay depends on both the load and the access method used by the protocol. Access delays fluctuate in most LAN. They are related by the relation :
(Tt definition) :

$$\forall i, \; Tt(i) = Ta(i) + Td(i) + Tp.$$

As the samples, the transmission units makes a strictly increasing sequence {Si}. So we can associate to each unit is number in the sequence. We denote **I**(i) the number of the transmission unit associated to the sample of number i. We denote **deb**(I) the number of the first sample of the unit I. And we denote **fin**(I) the number of the last sample of the unit I.

If two consecutive samples do not belong to the same transmission unit then they belong to two distinct but consecutive units.
(Consecutive units definition) :

$$\forall i, \; \text{if } I(i+1) \neq I(i) \text{ then } I(i+1) = I(i)+1.$$

Previously, we noted that all the samples blocked in the same transmission unit are sent and received at the same moment, because the samples blocked in a same transmission unit are available at the receiver when the transmission unit is entirely received.
(Unit receiving moment definition) :

$$\forall i, \; \forall j, \; \text{if } I(i) = I(j) \text{ then } Tr(i) = Tr(j).$$

3.4 The usable synchronous bandwidth

Let S (and R respectively) the station number of the sender (the receiver) of samples. If we use the FDDI protocol to send the images, a station begins to send when it receives the token. So Tj(I)[S] is also the moment where the transmission units associated with the Ith token rotation are sent. As FDDI protocol uses token ring as the access method, the moments Te, Tj, and Ta are related by :
(Access method definition)

$$\forall i, \; Tj(I(i))[S] = Te(i) + Ta(i).$$

To use the synchronous transmission mode of the FDDI protocol to send a video, first, we need to know the average synchronous throughput Ds[k] required by each station k to transmit the images in real time, to ensure that we always maintain the following relation :
(Synchronous bandwidth definition) :

$$\sum_{k \in [n, n+N[} Ds[k] \leq D.$$

That is to say, the sum of throughput sent should be lower than the effective throughput D of the network. This avoids overallocation of the medium. The effective throughput is obtained by starting from the nominal throughput minus the throughput used to manage the network, essentially the packaging of the frames and the management of the token. Network management has the responsibility for maintaining this statement. Every station requesting to transmit in synchronous mode calls the network management for a reservation of the average throughput required ([FDDI 88] and [FDDI 87]).

Secondly, once the TTRT is fixed, once we know the average synchronous throughput $Ds[k]$ required by each station k to transmit the images in real time, to achieve correct protocol operation we have to maintain the previously established relation :
(Synchronous transmission duration definition) :

$$\forall I, \forall n, \quad \sum_{k \in [n,n+N[} Ts(I)[k] \leq TTRT.$$

To maintain this relation for any rotation, the evident solution is to limit the duration of the synchronous transmission to $Ts[k]$.
(Maximum synchronous transmission definition) :

$$\forall k, Ts[k] = Ds[k].TTRT+D.$$

Unfortunately, the load on the network can make the moment at which the token arrives at a station vary greatly, remembering that a station must capture the token before it can transmit. This moment is remembered by the token rotation timer (TRT) local to each station. It may be early or late with respect to the negotiated TTRT period. Logically, in order to maintain the inequality of the synchronous bandwidth relation, each station k should have the right to transmit at most $Ds[k].TRT$ bits. This quantity is extremely difficult to manage because the TRT varies as a function of the load with each rotation of the token. ($0<TRT<2*TTRT$). Moreover, the implementation of FDDI does not permit us to get the value of TRT in time. We risk, then, to exceed the duration $Ts(I)[k]$ attributed to each station k, and thus to violate the inequality of the synchronous transmission duration relation, if we do not adapt the length of a frame to the rotation time of the token.

However, if the token is early (TRT<TTRT), this indicates that the network is underloaded, and thus it is permissible to transmit $Ds[k].TTRT$ bits, but impossible. It is impossible because the image emitter has not yet produced enough samples in such a short time. Our proposition is to send all the produced samples when an early token arrives. In that case, we know that the synchronous bandwidth relation and the synchronous transmission duration relation are obviously enforced by the regular throughput of the image emitter.

Inversely, if the token is late (TTRT<TRT), then the FDDI operations ensure that the delay cannot surpass 2*TTRT, even if all of the stations transmit the entirety of their throughput synchronously. But if we want to respect the synchronous transmission duration relation, the stations are allowed to send at most $Ds[k].TTRT$ bits at each token rotation. The remaining $(TRT - TTRT).Ds[k]$ bits are not sent with the first arrived token, but the FDDI operation guarantees that the delay will not be cumulative, so the remaining bits will be sent with the next units. In fact, the protocol is self-regulating because the overload induced by the token diminishes if the time between two passes of the token grows. Furthermore, if one of the stations does not use the entire throughput allocated to a synchronous transmission, then the unused time will be recovered, first of all, to ensure other synchronous transmissions, in recovering the delay, and in re-establishing the negotiated frequency of rotation of the token; secondly and ultimately to authorize asynchronous transmissions.

So, we propose building each synchronous transmission unit in such a way (Figure 1). If we denote $J(I)$ the set of sample numbers carried by the transmission unit associated with the Ith token rotation, then the relation of correct constituting of the transmission unit property is defined by :

(Correct constituting of transmission unit definition) :

$\forall I, \forall i \in J(I), Tj(I)[S] - TRT(I-1)[S] \leq Te(i) < Tj(I)[S] - \delta j(I-1)[S].$

In conclusion, we propose to use the synchronous mode of the FDDI protocol to transmit images of a video. The mean transmission rate $Ds[k]$ necessary for the transmission of the video should be known, and the application requires the network manager to make an appropriate reservation for the duration of the video to guard against congestion of the media. The negotiation procedure for the TTRT could then be started, if required. The smaller the value of the required TTRT, the smaller the delay in transmission. However, we have already raised the idea that the efficiency of the FDDI protocol will be accordingly weakened. The calculations that we have undertaken indicate that the ideal value lies in the neighborhood of twenty milliseconds [Cousin 90]. Independently of the fact that the negotiated value of the TTRT should lie between TTRTmin and TTRTmax to ensure the proper global functioning of the network, our application can accommodate a large range of values for TTRT. If the token is late, the application should be able to transmit at most $Ls[k] = Ds[k].TTRT$ bits at each rotation of the token. If the token is early, the control of the quantity of data transmitted at each capture of the token does not have to be managed at the level of the FDDI sender, the normal throughput of the emitter of images naturally assuming this role.

3.5 Validation

First of all, we have to prove that the definition of the correct constituting of transmission units enables the maximum synchronous transmission duration requirement, and thus the synchronous bandwidth definition, to be fulfilled.

(Maximum synchronous transmission duration property) :

$\forall k, \forall I, Ts(I)[k] < Ts[k].$

Proof :

Assuming that $Ls(I)[k]$ is the number of bits of the transmission unit associated with the Ith token rotation, by definition of $Ts(I)[k]$:

$Ts(I)[k] = Ls(I)[k] + D.$

The number of bits of a transmission unit has an upper limit of the duration between the first and the last sample of the transmission unit multiplied by the effective throughput of the sample.

$\Rightarrow Ts(I)[k] \leq (Te(fin(I))-Te(deb(I))) . Ds[k] + D.$

Assuming the correct unit constituting definition :

$\Rightarrow Ts(I)[k] \leq ((Tj(I)[S] - \delta j(I-1)[S]) - (Tj(I)[S] - TRT(I-1)[S])) . Ds[k] + D.$

$\Rightarrow Ts(I)[k] \leq (TRT(I-1)[S] - \delta j(I-1)[S]) . Ds[k] + D.$

Two cases appear :

1. Either $TRT(I-1)[S]<TTRT$ then $\delta_j(I-1)[S]=0$:

$\Rightarrow Ts(I)[k] \leq TRT (I-1)[S] . Ds[k] + D.$

Which can have an upper limit, according to the assumption :

$\Rightarrow Ts(I)[k] \leq TTRT . Ds[k] + D.$

According to the definition of maximum synchronous duration :

$\Rightarrow Ts(I)[k] \leq Ts[k]. (\lozenge)$

2. Or $TRT(I-1)[S]>TTRT$ then by definition $\delta_j(I-1)[S]=TRT(I-1)[S] - TTRT$:

$\Rightarrow Ts(I)[k] \leq TTRT . Ds[k]+ D.$

By definition of $Ts[k]$:

$\Rightarrow Ts(I)[k] \leq Ts[k]. (\lozenge)$

The temporal integrity constraint can be achieved, first, if the samples are buffered between the receiver and the image visualization equipment. The buffer has to be large enough to contain all the samples produced during 2*TTRT duration. Secondly, we can proved that all the samples are received in time at the receiver (i.e. before being displayed).
(Correct timing visualization property) :

$\forall i, \ Tv(i) > Tr(i).$

To prove this property, we need to prove the property of correct reception. If the visualization moment of the first sample is delayed by twice the Target Token Rotation Time then the reception moment of the samples is limited by the visualization moment of the first and the last sample of the same transmission unit.
(Correct reception property) :

$\forall i, Tv(0)=Tr(0)+2*TTRT \implies Tv(fin(I(i)))<Tr(i)+2*TTRT-\delta j(I(i)-1)[S]\leq Tv(deb(I(i)+1))$
Proof :
According to the correct constituting relation :

$\forall I, \forall i \in J(I), Tj(I)[S] - TRT(I-1)[S] \leq Te(i) < Tj(I)[S] - \delta j(I-1)[S].$

For i = fin(I) : (1)$\forall I, Tj(I)[S] - TRT(I-1)[S] \leq Te(fin(I)) < Tj(I)[S] - \delta j(I-1)[S],$

and for i = deb(I') : (2) $\forall I', Tj(I')[S]-TRT(I'-1)[S]\leq Te(deb(I'))<Tj(I')[S]-\delta j(I'-1)[S].$

According to the TRT definition : $\forall I, TRT(I)[S] = Tj(I+1)[S] - Tj(I)[S] + \delta j(I-1)[S],$

which can be rewritten : $\forall I, Tj(I+1)[S] - TRT(I)[S]=Tj(I)[S] - \delta j(I-1)[S].$
Let I'=I+1, then the relations (1) et (2) can be rewritten :
(3) $\forall I, Te(fin(I))<Tj(I)[S]- \delta j(I-1)[S]\leq Te(deb(I+1)).$
According to the Te definition :

$\forall i, Tv(i) = Te(i) + Tv(0) - Te(0).$
The relation (3) can be rewritten :

$\implies \forall I, Tv(fin(I)) < Tj(I) - \delta j(I-1)[S] + Tv(0) - Te(0) \leq Tv(deb(I+1)).$
From the assumption about the visualization moment of the first sample :
$Tv(0) = Tr(0)+2*TTRT.$

$\implies \forall I, Tv(fin(I)) < Tj(I) - \delta j(I-1)[S] + Tr(0) + 2*TTRT - Te(0) \leq Tv(deb(I+1)).$
According to the Tr definition, the Tt definition and the access method definition :

$\forall i, Tr(i) = Tj(I(i))[S] + Td(i) + Tp.$

$\implies \forall i, Tv(fin(I(i)))<Tj(I(i))[S]-\delta j(I(i)-1)[S]+Tj(I(0))[S]+Td(0)+Tp+2*TTRT-$
$$Te(0)\leq Tv(deb(I(i)+1)).$$

Assuming that the transmission delay is constant for a fixed data rate : $\forall i, Td(i)=Td.$

$\implies \forall i, Tv(fin(I(i))) < Tr(I(i))-\delta j(I(i)-1)[S]+Tj(I(0))[S]+2*TTRT-Te(0) \leq Tv(deb(I(i)+1))$
From the assumption of the sending moment of the first sample : $Te(0) = Tj(I(0))[S],$

$\implies \forall i, Tv(fin(I(i))) < Tr(I(i)) - \delta j(I(i)-1)[S] + 2*TTRT \leq Tv(deb(I(i)+1)). (\Diamond)$

Then, we prove the correct timing visualization property :

$\forall i, Tr(i) < Tv(i).$
Proof :
Recurrent demonstration :
1. For i=0, the relation is obvious because, according to the visualization moment of the first sample assumption $(Tv(0) = Tr(0) + 2*TTRT)$ then : $Tr(0) < Tv(0). (\Diamond)$

2. Assuming that the recurrent assumption is true for $i \in [0,n]$, two cases appear :
2.1 Either the samples n and n+1 belong to the same transmission unit : "I(n)=I(n+1)".
Then, according to the Tv definition :
$Tv(n+1) = Tv(n) + Tv(n+1) - Tv(n).$

According to the second temporal constraint :

$Tv(n+1) = Tv(n) + Te(n+1) - Te(n)$.

According to the strictly increasing function Te :

$Tv(n+1) > Tv(n)$.

According to the recurrent assumption :

$Tv(n+1) > Tr(n)$.

According to the initial assumption : if $I(n)=I(n+1)$ then $Tr(n)=Tr(n+1)$.

$Tv(n+1) > Tr(n+1)$. $(\Diamond)$

2.2 Either the samples n and n+1 do not belong to the same transmission unit : "$I(n) \neq I(n+1)$".

Then, we know that "$I(n+1)=I(n)+1$", because the samples and the transmission units are numbered in an strict increasing manner.

According to the Tr definition, the Tt definition and the access method definition :

$Tr(n+1) = Tj(I(n+1))[S] + Td(n+1) + Tp$. Which can be rewritten :

$Tr(n+1) = Tj(I(n))[S] + Tj(I(n+1))[S] - Tj(I(n))[S] + Td(n+1) + Tp$.

According to the following assumption : $\forall i, Td(i)=Td$.

$Tr(n+1) = Tr(n) + Tj(I(n+1))[S] - Tj(I(n))[S]$.

From the token rotation duration property, the TRT definition and the δj definition :

$Tr(n+1) \leq Tr(I(n)) + 2*TTRT - \delta j(I(n)-1)[S]$.

According to the correct reception property :

$\forall i, Tv(fin(I(i))) < Tr(I(i)) + 2*TTRT - \delta j(I(i)-1)[S] \leq Tv(deb(I(i)+1))$. Then :

$Tr(n+1) < Tv(deb(I(n)+1))$.

According to the definition of the deb and fin functions : if $I(n) \neq I(n+1)$ then $fin(I(n))=n$ and $deb(I(n)+1) = n+1$.

We deduce that : $Tr(n+1) < Tv(n+1)$. $(\Diamond)$

The Tolerable delay constraint introduces a delay Tmax, which can be deduced from the previous definitions and relations :

(Tmax lower limit property) :

$Tmax > Tt(0) + 2*TTRT$.

Proof :

According to the Tolerable delay constraint :

$\forall i, Tmax > Tv(i) - Te(i)$.

Which can be rewritten :

$\forall i, Tmax > Tv(0) + (Tv(i) - Tv(0) - Te(i) + Te(0)) - Te(0)$.

According to the temporal integrity constraint :

$Tmax > Tv(0) - Te(0)$.

According to the visualization moment of the first sample :

$Tmax > Tr(0) + 2*TTRT - Te(0)$.

According to the Tt definition :

$Tmax > Tt(0) + 2*TTRT$. $(\Diamond)$

Accordingly, at the price of a delay due to the buffering equal to 2*TTRT plus the physical response time of the network, we prove that it is useless to request a rotation time equal to half of the delay required by the application, as that would let it over-determine the maximum rotation time guaranteed to be more than 2*TTRT.

4. Conclusion

In view of this study, we can observe the image transmission has to resolve two problems, namely the preservation of the synchronization between images, and the minimization of the buffer length.

Yet the ring topology in the FDDI protocol both necessitates and allows controlled access to the medium, and favors the management of a method of access favorable to the transmission of images by the creation of two modes of transmission : the synchronous and the asynchronous modes. The synchronous mode of transmission guarantees a station an average throughput and the right to transmit with a periodicity, on the average, equal to a value --the TTRT -- negotiated among all the stations. Moreover, this mode guarantees that the maximum rotation time cannot exceed 2*TTRT.

Nevertheless the service provided by the FDDI synchronous transmission mode is not sufficient to satisfy the temporal constraints required by the image transmission. To succeed in satisfying these constraints, we propose to carefully undertake the constituting of each transmission unit. If the token is late, the transmission unit carries all the image samples produced from the instant of the previous token arriving at the instant of the expected token arriving. If the token is early, the transmission unit carries all the remaining samples. To deal with the asynchronous transmission technique and the aperiodic delivery of the samples, we propose to buffer samples at the level of the sender and at the level of the receiver.

First, we prove that this proposed way of constituting and buffering enables the temporal constraints to be satisfied. Second, we prove that the overall delay of any sample is equal to 2.TTRT plus the physical response time of the network. We prove that it is possible to allocate only the exact mean throughput to achieved image transmission in real time. This allocation optimizes the use of the medium bandwidth. Moreover, sample blocking enables our application to be adapted to a large range of TTRT, enabling an efficient image transmission process.

We observe that in order to allow the transmission of periodic information with asynchronous techniques, it is necessary to transmit temporal information explicitly; it is further necessary to supply sufficient buffer space in memory at the level of the sender to deal with the token delay and at the level of the receiver to accommodate the inevitable variations in transmission delay. These memory buffers imply a systematic delay inhospitable to interactive applications. In short, the great throughput required by the transmission of images obliges one to use a great quantity of rapid access memory. Yet the memory buffers necessary at the level of the receiver in order to allow the use of an asynchronous technique of transmission can be usefully exploited to detect and then to correct loss, corruption, and duplication in all or part of the images of the video.

A prototype is under development which will put into practice what we prove. This prototype consists of three subsystems : the video subsystem, the communication subsystem, and the computer interface. The communication subsystem uses the AMD SuperNet components for the FDDI network. Each subsystem is linked to the others by a very high speed bus. This architecture enables both the local and distant image distribution not to be restricted by the bus throughput of the computer. Specific components enabling fast image processing (real time compression, image analysis,...) will be included in the prototype in the near future. This prototype will enable us to study the network management, principally the synchronous allocation bandwidth process which is not largely developed in the current FDDI network proposed.

References

[AMD 89] AMD, "The SuperNet family for FDDI", Advanced Micro Devices technical manual, July 1989.
[Cochennec 85] J.Y.Cochennec, P.Adam, T.Houdoin, "Asynchronous time-division Networks : Terminal Synchronization for Video and Sound Signals", GLOBECOM'85, 1985.

[Cousin 90] B.Cousin, L.Kamoun, "ATM, HSLAN and Images", 3rd IEEE COMSOC International MultiMedia Workshop (Multimedia'90), Bordeaux - November 1990.

[Cousin 91] B.Cousin, R.Castanet,"Evaluation du protocole FDDI pour la diffusion d'images en temps réel", colloque francophone sur l'ingénierie des protocoles (CFIP'91). Pau - France, 16-20 Septembre 1991.

[Dykeman 88] D.Dykeman,W.Bux, "Analysis and Tuning of the FDDI Media Access Control Protocol", IEEE Journal on Selected Areas in Communications, vol SAC6 n°6, p997-1010, july 1988.

[FDDI 87] "FDDI Token Ring Media Access Control (MAC)", ANSI X3.139, 1987.

[FDDI 88] "FDDI Token Ring Physical Layer Protocol (PHY)", ANSI X3.148, 1988.

[FDDI 89] "FDDI Token Ring Physical Layer Medium Dependent (PMD)", Draft Proposed ANSI X3.166, March 1989.

[FDDI 9a] "FDDI Token Ring Station Management (SMT)", Draft Proposed ANSI X3T9.5, May 1990.

[FDDI 9b] "FDDI Token Ring Hybrid Ring Control (HRC)", Draft Proposed ANSI X3.186, 11 May 1990.

[Iyer 85] V.Iyer, S.P.Joshi "FDDI's 100 Mbit/s protocol improves on 802.5 spec's 4 Mbit/s limit", EDN Electronics Data Network, p151-160, May 1985.

[Johnson 86] M.J.Johnson, "Reliability Mechanisms of the FDDI High Bandwidth Token Ring Protocol", Computer Networks and ISDN Systems n°11, North Holland, 1986.

[Johnson 87] M.J.Johnson, "Proof that Timing Requirements of the FDDI Token Ring Protocol are Satisfied", IEEE transactions on communications vol COM35 n°6, June 1987.

[Lam 78] S.S.Lam, "A New Measure for Characterizing Data Traffic", IEEE transaction on communications, vol COM 24 n°1, January 1978.

[Ross 89] J.E.Ross, "An Overview of FDDI : The Fiber Distributed Data Interface", IEEE Journal on Selected Areas in Communications, vol SAC7 n°7, p1043-1051, September 1989.

[Ross 90] J.E.Ross, J.R.Hamstra, R.L.Fink, "A LAN among MAN's", Computer communication review vol 20 n°3, July 1990.

[Sevcik 87] K.Sevcik, M.Johnson, "Cycle Time Properties of the FDDI Token Ring Protocol", IEEE transactions on software engineering vol SE13 n°3, March 1987.

[Teener 90] M.Teener, R.Gvozdanovic, "FDDI-II Operation and Architectures", 14th conference on Local Computer Networks,1990.

[Ulm 82] J.M.Ulm, "A Timed Token Ring Local Area Network and its Performance Characteristics", 7th conference on local computer networks, p50-56, Minneapolis-USA, February 1982.

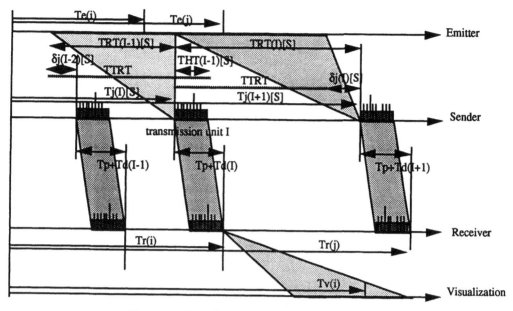

Figure 1 - Constituting of the transmission units

Session XI: Open Session

Chair: Radu-Pospescu Zeletin, Technical University of Berlin and GMD-FOKUS

The open session of the workshop had no talks scheduled. It served as the first part of a workshop conclusion, trying to identify the main multimedia research issues for the future. The session was based on short ad-hoc presentations and an open discussion on the different issues raised in Heidelberg.

The first issue discussed were new error recovery techniques in end-to-end multimedia communication. The main characteristics of end-to-end data communication for text and graphics is that high reliability is required. A large variety of techniques have been developed for this purpose in the past. Most of these techniques are based on hand-shaking protocols.

For audio and video data, one often sees the statement that they do not require error correction methods. This is not really always true, e.g., compressed video streams are quite sensitive to errors. Therefore, the question was raised if there exist mechanisms which provide reliability, but are based on unidirectional communication, thereby avoiding the delay problems inherent to handshaking.

Ernst Biersack from Bellcore presented results from experiments based on forward error correction (FEC) algorithms in an ATM environment. He integrates his mechanisms in the transport protocol (but in his own words, he "collapses everything between the application and the ATM interface into the transport layer"). Using FEC on a cell basis, a very low end-to-end error probability was achieved in the experiments presented. The induced redundancy in the data to be transported, however, affects the number of connections which can be supported.

FEC is a promising research topic due to the better integration of different types of information (data, voice, and video) with different characteristics and reliability requirements in one end-to-end stream. Ernst indicated that his first experiments showed that FEC works best for mixed scenarios of bursty and continuous traffic. It is also a promising approach for reliable multicast. However, FEC does not come for free − there are cases where the amount of redundancy is too high and others where it is too low. With selective retransmission, bandwidth is used to only the amount actually needed.

The second main topics of the session were flow control and synchronization. The discussion focused on which flow control mechanisms are needed for multimedia and how they can be combined with synchronization functions.

It was agreed that flow control on a rate basis is the best solution. Domenico Ferrari summarized this with: "Window flow control is out, rate control is in." For audio and video retrieval, some workshop participants mentioned that much, however, can also be achieved by output buffering and backpressure mechanisms. The induced delay may prevent the use of these mechanisms for conversational services.

During the workshop, different speakers have used the term "synchronization" with different meanings. During the open session we tried to clarify and accommodate different views and definitions. Ralf Guido Herrtwich presented a taxonomy of the different meanings and definitions of "synchronization" used during the workshop. On one hand, the term is used in upper layers of multimedia communication for expressing the required relation between events in a multimedia context. This synchronization is also called "synthetic synchronization" and is part of the multimedia scripting or document editing process. On the other hand, the term is used in the lower layers to express either the inter-stream or intra-stream synchronization. Intra-stream synchronization is also often called "jitter control" or "pacing." Ralf said that while inter-stream and intra-stream synchronization are different problems they may have a common solution (and presented the analogy to operating systems where semaphores are one common solution to both mutual exclusion and condition synchronization).

Low-level synchronization (inter- and intra-stream) was the subject of the following discussion. It was noted that several sources of timing exist in the system and that mechanisms have to be provided for an application to specify in relation to which time source synchronization shall be performed. Time sources may both be actual clocks and regular data sources. Referring to Ralf's previous statement of a common solution for low-level synchronization, Domenico Ferrari mentioned that with jitter control for single streams inter-stream synchronization comes for free.

Francois Horn from CNET concentrated on high-level synchronization. His definition of synchronization is that "any identified relation between events in a multimedia system constitutes a synchronization point." He identified temporal synchronization as a subset of a more general synchronization problem. To express synchronization one needs a language to describe these relations. He advocated for ESTERELLE as a language to be used in this context because of the precise formal specification that comes with it.

The last item addressed during the open session was the hardware architecture of future multimedia workstations. It was the general opinion that the multimedia workstation will be based on an internal switch as opposite to bus systems today. This is motivated mainly by the required processing concurrency in the multimedia environment. Andy Hopper made the remark that regardless of how the architecture really looks like, the most important thing is to avoid to change gear when communicating locally and remotely.

The open session provided the best forum during the workshop to elaborate on a wide variety of topics of common interest. Due to the intensive discussions in the audience it is recommended that future workshops should allocate more time for such type of events.

A Performance Study of Forward Error Correction in ATM Networks

Ernst W. Biersack
Bellcore
445 South Street
Morristown, NJ 07962-1910
E-mail: erbi@bellcore.com

November 1991

Abstract

The asynchronous transfer mode technique (ATM) has been standardized as the transport system for future broadband communication networks. ATM allows the multiplexing and transmission of data from different services over the same network. If the cell loss rate in the network is higher than the loss rate requested by the service, the protocols in the endsystems must make up for the difference in loss rate. Traditionally, a retransmission-based scheme (i.e. ARQ) is used to recover from loss of data. In high bandwidth delay-product networks the latency introduced by retransmission-based error recovery schemes may become unacceptably high. *Forward error correction* (FEC) schemes do not have this drawback. When FEC is used, the transmitter sends with the original data some redundant data that can be used by the receiver to reconstruct lost pieces of the original data without requesting their retransmission. FEC trades off an increase in the bandwidth required for the capability to recover from partial loss.

We use simulation to study an FEC scheme that can recover a fixed number of lost cells within a block of consecutive cells. The results obtained for different traffic scenarios demonstrate that FEC can reduce the loss rate by multiple orders of magnitude. FEC is shown to be most effective when used by video sources in a heterogeneous traffic scenario consisting of video and burst sources.

1 Introduction

Asynchronous Transfer Mode (ATM) is the internationally agreed upon transfer mode for broadband networks [ATM 89]. ATM provides high-bandwidth low-latency multiplexing and switching [MINZ 89]. The basic unit of multiplexing and switching in ATM is called a *cell*. A cell has a fixed length of 53 bytes: 48 bytes of data (*payload*) and 5 bytes of control information such as *virtual path identifier* (VPI), *virtual channel identifier* (VCI), and a *cyclic redundancy checksum* (CRC) based header error control. An ATM-type network will experience three types of errors: bit errors due to noise, switching errors due to undetected corruption of the cell header, and cell losses due to congestion. Losses due to congestion are expected to be far more common than the other two types of errors.

One of the reasons for adopting ATM is for the integration of services. ATM will be used by different applications that require services with widely varying *quality of service* (QOS) requirements. One QOS requirement is reliability of the transmission. If the degree of reliability provided by the network is lower than the reliability requested by an application, the endsystems must make up for the difference.

The two basic mechanisms available to improve reliability are *automatic repeat request* (ARQ) and *forward error correction* (FEC). ARQ is a *closed-loop* technique based on retransmission of data that were not correctly received by the receiver. ARQ requires the transmitter and receiver to exchange state information about the status of individual messages. Each retransmission of a messages adds at least one round-trip time of latency. Therefore, ARQ may not be applicable for transmitting data from applications with low latency constraints. Constrained latency services are necessary for human interaction, process control, remote sensing, etc. As the name implies, data for applications using this service is worthless if it does not arrive within a certain time. Services such as face-to-face audio or video require that the end-to-end latency be less than 100 milliseconds if the system is to provide acceptable performance. A connection across the continental USA could easily have a retransmission time of more than 100 milliseconds. Another disadvantage of ARQ based schemes is the complexity required to keep track of a potentially very large number of outstanding messages.

FEC is an alternative to ARQ that avoids its shortcomings and is well suited for operation in high bandwidth-delay product networks. FEC involves the transmission of redundant information along with the original data so that if some of the original data is lost, it can be reconstructed using the redundant information. In data communications, the use of FEC is attractive for supporting services that cannot rely on retransmission, such as real-time services over a high latency network. The amount of redundant information is typically small, so that FEC remains efficient. If ARQ is not feasible because of the additional latency, and the network itself does not have any other means of providing different streams with different QOS, FEC makes the operation of the network more cost-effective by allowing it to operate at a higher utilization. Without FEC, the network must be operated at a utilization where the loss rate never exceeds the most stringent loss rate required by an application. In this case, all applications would receive this low loss rate, independent of their actual need. FEC can also be used for an endpoint-based support of QOS. If a networks does not support different degrees of reliability, FEC is a viable mechanism to achieve the requested reliability.

The rest of the paper is organized as follows. The next section describes in detail the operation of FEC. Section 3 describes our experiment and the assumptions made. Section 4 presents the simulation results illustrating the effectiveness of FEC.

2 Operation of Forward Error Correction

Coding theory distinguishes two types of data corruption: an *error* is defined as a bit with an unknown value in an unknown location; whereas an *erasure* is a bit with an unknown value in a known location. If the FEC decoder is able to take advantage of erasure information, replacing an error with an erasure approximately doubles the error correcting power of the code. In ATM, one byte of the five byte header is a CRC for header error control that can correct one-bit errors and detect two-bit errors. For bit error rates smaller than 10^{-9}, which are typical for fiber optics transmission systems, the cell loss rate due to random bit errors is negligible. Therefore, congestion losses are the dominant form of error on ATM networks and the network can be modeled as a well-behaved erasure channel. Congestion losses are typically not random losses but then occur in bursts and often effect consecutive cells of a single stream. For FEC to be effective, it is important to recover the loss of multiple cells in a group of consecutive cells. Not only do congestion errors appear as erasures but occur in multiples of the cell size and align on cell boundaries. In order to be able to determine which cells are missing, cells must contain a sequence number. The FEC system used in this paper takes k cells as input and produces $k + h$ cells as output. At the receiver, any k of which is enough to recover the original information as long as none of the received cells is corrupted by bit errors (erasures only). A group of k cells from which the h redundant cells are generated is referred to as a *block*. The redundant cells are referred to as *over-code*. Advances in implementation technology make FEC possible even at speeds of several hundred Mb/s. At Bellcore, a Reed-Solomon-based code has been developed that has the desired performance and can be implemented on a single chip operating at 400 Mb/s with $h = 16$ or 1 Gb/s with $h = 4$ (and any k) [MCAU 90]. The FEC encoder and decoder are almost identical, which simplifies implementation. Figure 1 illustrates the operation of the FEC system for $k = 3$ and $h = 2$ cells[1]. The FEC encoder at the transmitter produces 2 redundant cells (fec-1, fec-2) for every block of 3 cells. In the example, one data cell (data-2) and one redundant cell (fec-2) get lost. The FEC decoder at the receiver uses fec-1 and the two correctly received data cells to reconstruct the missing data cell (data-2). The sum of the traffic generated by all sources is referred to as the (normalized) *load* λ. λ assumes values between 0 and 1, where $\lambda = 1$ means that cells are generated at the same rate as the multiplexer can handle them. Since FEC increases the total amount of traffic in the network by the number of redundant cells generated by the FEC encoder, the cell loss rate will increase whenever FEC is used. For FEC to be effective, it must recover enough lost cells to reduce the cell loss rate **after decoding** to a level lower than the cell loss rate when no FEC is used. FEC is applied by individual sources. A source that applies FEC is referred to as an *FEC source*, a source that does not apply FEC as a *non-FEC source*. If λ is the aggregated load generated by the sources, the load in the network after FEC is applied is referred to as the *effective load* λ_{eff}. We have $\lambda_{eff} = \lambda \times (1 + \frac{h}{k})$ if all

---

[1]Typically, the ratio k/h is much larger than in this example.

sources apply FEC.

A cell is a very small unit of data. To avoid the complexity of dealing with individual cells and to amortize the higher-layer headers over a larger unit of data, functions such as error detection and error correction are done over blocks of cells. We assume a block consists of k cells and is the unit of error detection and FEC. A block that has a least one cell missing/lost is said to be *corrupted*. When FEC is applied and h redundant cells are generated, a block is considered *lost* if more than h of its $k + h$ cells are lost. Our main measure for the performance of FEC will be the reduction of the block loss rate. When no FEC is applied the block loss rate $P_B(\lambda)$ is equal to *Pr($i > 0$ cells are lost in a block of k cells | load λ)*. When redundant cells are generated, the block loss rate **before** cell loss recovery at the receiver will rise to $P_B(\lambda_{\text{eff}})$. The block loss rate is a more meaningful measure of the network performance than the cell loss rate. It is, for instance, useful in estimating the expected losses in an ATM system that performs segmentation and reassembly of cells into blocks. The block loss rate **after** cell loss recovery was performed is referred to as $P_{Brec}(\lambda_{\text{eff}})$ and is equal to *Pr($i > h$ cells are lost in a group of $k + h$ cells | load λ_{eff})*. The gain G due to FEC is defined as $G \stackrel{\text{def}}{=} \log_{10}(\frac{P_B(\lambda)}{P_{Brec}(\lambda_{eff})})$. G measures the reduction of the block loss rate in terms of orders of magnitude. ($G = j$, $j \in 0, 1, 2, \ldots$, means that FEC reduces the block loss rate by j orders of magnitude.) For FEC to be effective, we require $G > 0$. When FEC is used only by a subset of the sources, the non-FEC sources will observe an increase in their block loss rate and their penalty D is defined as $D \stackrel{\text{def}}{=} \log_{10}(\frac{P_B(\lambda_{eff})}{P_B(\lambda)})$.

The following example illustrates the tradeoff between the increased cell loss rate and the recovery of lost cells due to FEC. Let us assume that the increase in load due to FEC causes the block loss rate to increase by a factor of ten, i.e. $P_B(\lambda_{\text{eff}}) = 10 * P_B(\lambda)$. For FEC to be effective and achieve $G > 0$ it must reconstruct enough cells to recover more than 90% of all corrupted blocks. If 99% of the corrupted blocks can be recovered, we have $P_{Brec}(\lambda_{\text{eff}}) = \frac{1}{100} * P_B(\lambda)$ and therefore $G = \log_{10}(10) = 1$, i.e. a block loss rate reduction by one order of magnitude over $P_B(\lambda)$.

3 Experimental Setup

To study the effect of FEC, we have built a simulator that models a multiplexer with N input ports and one output port (see figure2). The multiplexer is output-buffered with a single shared buffer of finite capacity B at the output port. Each cell time the multiplexer checks all inputs for a newly arriving cell and puts these cells in the output buffer, if there is any space, and drops them otherwise. To make the service to the different input ports fair, the multiplexer starts each cell time with a different input port: if at time T there is at least one cell arriving and the first port to be checked is port i, then at time $T + 1$ the first input port to be checked will be port $(i + 1) mod\ N$. Each input port has one source connected to it. Every source generates on the average the same amount of traffic. An ATM network will carry traffic from different types of applications with different statistics. In our simulation model we distinguish between two different types of sources: (1) *burst sources* representing applications such as bulk data transfer or transactions and (2) *video sources* representing variable bit rate (VBR) video sources such as entertainment video or video conferencing. A burst source is characterized by the *interarrival time between bursts, burst size,* and *cell separation* of cells within the burst. The interarrival time is geometrically distributed, the number of cells per burst is a constant, and the spacing between cells is fixed. A spacing of x means that during the transmission of a burst, a cell is transmitted every x cell times. The video source we use has been derived from entertainment video and therefore reflects the complex nature of this type of application. M. W. Garrett at Bellcore has generated the video data set by encoding a two hour-long action movie using an intra-frame 8×8 Discrete Cosine Transform coding scheme with run-length and Huffman encoding [GARR 91]. The data set contains the number of bytes per frame produced by the encoder. The duration of one frame is $\frac{1}{24}$ of a second, the total number of frames is 171,000. The statistics of this data — in terms of bandwidth — are: Maximum: 15.06 Mbit/sec, Mean: 5.34 Mbit/sec, Minimum: 1.79 Mbit/sec, and Maximum/Mean (burstiness): 2.82. In the simulator, the data for each frame is broken up into cells of 48 bytes and the transmission of the cells is spaced out equally over the duration of the frame. The different video sources are *unsynchronized*, i.e. they start at different points in the movie. Unsynchronized video sources exactly represent a situation where different people watch the same movie at different times (video on demand) and simulate the multiplexing of different video sources. The cells of a video source are spaced equidistantly within one frame-time. When FEC is applied to a video source, the spacing between cells is adjusted such that the original cells together with the redundant cells are spaced equidistantly within one frame-time. Since the video sources have a fixed bandwidth, the capacity of the multiplexer is altered to yield different values of λ. For a particular scenario, all sources generate the same amount of traffic.

Table 1 lists all the parameters of the simulation.

Parameter	Value
Load λ	$0.7 - 0.95$
Total number of sources N	32
Number of video sources v_srcs	24, 32
Number of burst sources b_srcs	8
Number of FEC sources f_srcs	0, 4, 8, 12, 16, 24, 32
Burst interarrival time	geometric distribution
Burst size	50 (cells)
Block length k	50 (cells)
Cell separation for video sources	equidistant over frametime
Cell separation sep for burst sources	10
Percentage $\frac{h}{k}$ of redundant cells per block	0, 10, 20, 30
Size of switch output buffer B	100 (cells)

Table 1: Simulation Parameters

4   Simulation study of FEC

The effectiveness of FEC depends very much on the cell loss behavior of the network, because FEC can only recover a limited number of lost cells per block. For FEC to work well, it is necessary that most corrupted blocks have less cells missing than the FEC-decoder can recover.

Since the traffic mix in future ATM networks is unknown, we use various homogeneous and heterogeneous traffic patterns to investigate the effect of FEC on the block loss rate. For each scenario we present various measures that illustrate certain key parameters. Let CL be a discrete random variable that represents the percentage of cells lost in block and can assume any value from $0, 1, \ldots, 100$. $F_{CL}(x) \stackrel{\text{def}}{=} Pr(CL \leq x \mid CL > 0)$ is the cumulative distribution function of the percentage of cells lost per block, provided at least one cell is lost. Therefore, $F_{CL}(x) = p$ means, that with probability p the percentage of lost cells in a corrupted block is less or equal than x. The larger the value of $F_{CL}(x)$ for a given x, the more corrupted blocks can be recovered with an over-code of x-percent. Another measure used in this experiment is the block loss rate $P_B(\lambda)$.

4.1   All video sources

Our first traffic scenario, referred to as **V1**, has 32 video sources that are all unsynchronized, a buffer size of 100, and a block size of 50. For different loads the percentiles of CL and the block loss rate $P_B(\lambda)$ are given in table 2. For a plot of $F_{CL}(x)$ see figure 3.

Load λ	Block Loss Rate $P_B(\lambda)$	90.0	99.0	99.9
		Percentiles of CL		
0.900	2.8e-06	4	6	6
0.925	2.6e-03	6	8	14
0.950	2.4e-02	6	12	24

Table 2: Scenario V1, all video sources.

Up to a load of $\lambda = 0.85$ no loss is observed. Increasing the load from 0.90 to 0.95 results in a sharp increase of the block loss rate by 10^4. The percentiles of CL show that most corrupted blocks lose only a small percentage of cells and a few redundant cells are sufficient to recover most corrupted blocks. At $\lambda = 0.925$, 90% of all corrupted blocks lose less or equal than 6% of the cells in a block (90.0 percentile), 99% of all corrupted blocks lose less or equal than 8% of the cells in a block (99.0 percentile), and 99.9% of all corrupted blocks lose less or equal than 14% of the cells in a block (99.9 percentile). However, when FEC is applied, the additional load due to the redundant cells will significantly increase the block loss rate of the non-FEC sources, imposing a high penalty on them. This is confirmed by a simulation with $\lambda = 0.90$ where 4 of the 32 video sources apply FEC with a 20% over-code (ten redundant cells per block of fifty cells). While the block loss rate $P_{Brec}(\lambda_{eff})$ for the FEC sources is reduced to zero the block loss rate $P_B(\lambda_{eff})$ for the non-FEC sources increases by a factor of 10^3, i.e. $D = 3$. When FEC is applied by more than 4 sources, FEC is not effective at all since $P_{Brec}(\lambda_{eff})$ is higher than $P_B(\lambda)$, i.e. $G < 0$. In general, the usefulness of FEC in a scenario with all video sources is limited. FEC can not be applied whenever the over-code necessary to achieve the required gain G causes the penalty D for the non-FEC sources to be higher than acceptable. Instead, the load must be kept low enough, to meet the most stringent loss requirement of any source.

4.2 Mixed Traffic Scenario

Future ATM networks will most likely carry data from different applications. Therefore, a more realistic traffic scenario is heterogeneous with 24 video sources and 8 burst sources. The parameters for the burst sources are burst length 50 and cell separation 10. The video sources are unsynchronized. This scenario is referred to as **VB24-8**. For VB24-8, the first losses are observed at $\lambda = 0.7$ as compared to $\lambda = 0.9$ where the first losses occurred for V1. For $\lambda \leq 0.90$, the cell losses perceived by the video sources are due to the interferences with the burst sources.

Figure 4 plots $F_{CL}(x)$ and table 3 gives the block loss rates and the percentiles of CL for the video sources. Not shown are the values for the burst sources. The block loss rates for the burst sources are the same as for the video sources. Due to the higher burstiness of the burst sources their percentiles of CL are significantly higher than the ones for the video sources. For VB24-8, independent of the load, 90% of all corrupted video blocks lose less or equal to 10% of their cells, 99% of the blocks never lose more than 20% of their cells. Therefore, for the same over-code, the gain due to FEC will be much higher when FEC is applied to the video sources in VB24-8 as compared to when FEC is applied to the burst sources in B1.

Load	Block Loss Rate	90.00	99.00	99.90	99.99
λ	$P_B(\lambda)$	Percentiles of CL			
0.70	5.1e-05	4	8	10	12
0.75	4.0e-04	4	8	14	18
0.80	2.7e-03	6	10	16	24
0.85	1.3e-02	6	14	20	30
0.90	5.0e-02	8	18	26	36
0.95	1.4e-01	10	20	32	44

Table 3: Scenario VB24-8, Percentiles of CL for the video sources.

We performed additional simulations with FEC applied by some or all of the video sources. Scenario **VB-FEC-1** is derived from VB24-8 and has $\lambda = 0.7$. The over-code of the FEC-sources is 10%, the number of video sources applying FEC is 4, 8, 12, 16, or 24. Figure 5 shows how FEC effects the block loss rates $P_B(\lambda_{eff})$ and $P_{Brec}(\lambda_{eff})$. The block loss rate is given for each of the 32 sources individually. The station numbers 0–23 are allocated to the video sources, numbers 24–31 to the burst sources. When j stations apply FEC, their station numbers range from 0 to $j - 1$. For VB-FEC-1 an over-code of 10% reduces the block loss rate $P_{Brec}(\lambda_{eff})$ for the eight FEC sources to $5 * 10^{-7}$, yielding a gain $G \geq 2$. Figure

6 plots the gain and the penalty for VB-FEC-1. Both, the gain and the penalty are linear functions of the number of sources that apply FEC. As the number of FEC sources increases, the penalty D increases. However, the penalty never exceeds $D = log_{10}(\frac{2*10^{-4}}{5*10^{-3}}) = 0.875$, which is noticeably smaller than the gain.

Scenario **VB-FEC-2** is derived from VB24-8 and has $\lambda = 0.8$. Four video sources use FEC and the amount of over-code is 10, 20, or 30 percent. Table 4 shows the effectiveness of FEC. When applied by four video sources, FEC reduces their block loss rate by about one order of magnitude per 10% over-code and has little impact on the block loss rate of the non-FEC sources.

Over-code	Gain for Video FEC	Penalty for Video non-FEC	Penalty for Burst
10 %	1.9	0.12	0.11
20 %	2.9	0.26	0.24
30 %	3.8	0.39	0.47

Table 4: Gain and Penalty for scenario VB-FEC-2.

We have seen that among all the scenarios investigated, FEC is most effective for the heterogeneous traffic scenario. Even for high loads, FEC can reduce the block loss rate by multiple orders of magnitude. The gain achieved for the FEC sources by far exceeds the penalty for the non-FEC.

5   Related work

Loss recovery using FEC has been studied previously [OHTA 91, SHAC 90, ZHAN 91]. Our work differentiates itself in several ways.

(1) Our traffic model for the video sources is derived from an actual source while previous models used to evaluate the performance of FEC assume that the inter-arrival times are exponentially or hyper-exponentially distributed.

(2) The code used to generate the redundant cells is more powerful. The redundant cells are generated via a modified Reed-Solomon code that is able to recover any h cells lost out of $k + h$ cells.

(3) Our main measure of performance is the block loss rate, while other studies focus on the cell loss rate.

6   Conclusion

We have investigated the performance of an FEC scheme that can recover up to h lost cells out of $k + h$ cells. The model consisted of N sources generating traffic for a multiplexer with a finite output buffer. The arrival for the burst sources follows a geometric distribution, the one for the video sources is derived from a real movie after it was encoded. We performed simulations with different traffic mixes to obtain the gain and penalty due to FEC. FEC was most effective for a scenario consisting of mixed burst and video sources. When applied by the video sources, FEC can reduce the block loss rate by many orders of magnitude, while the non-FEC sources are only marginally effected. For a homogeneous scenario, FEC is less effective. In the case of all video sources FEC is only effective if applied by a few sources and the penalty for the non-FEC sources is severe.

In summary, FEC can be very effective in reducing the block loss rate. The gain depends on the traffic scenario and varies with the load, the amount of over-code, and the number of FEC sources. If possible, FEC should be used only by a subset of the sources. For a fixed over-code, as the burstiness or the load increase, the gain FEC decreases.

Acknowledgment

I would like to thank M. W. Garrett for providing me with the data for the video source and C. J. Cotton for writing a first version of the simulator. The comments of C. J. Cotton, M. W. Garrett and A. McAuley on earlier versions of this paper are very much appreciated.

References

[ATM 89]   Comité Consultatif International de Télégraphique et Téléphonique, "Broadband Aspects of ISDN", *Recommendation I.121*, 1989.

[GARR 91] M. W. Garrett and M. Vetterli, "Congestion Control Strategies for Packet Video", *Proc. 4th Int. Workshop on Packet Video*, Kyoto, Japan, August 1991.

[MCAU 90] A. J. McAuley, "Reliable Broadband Communications Using a Burst Erasure Correcting Code", *Proc. ACM SIGCOMM 90*, pp. 287–306, Philadelphia, PA, September 1990.

[MINZ 89] S. E. Minzer, "Broadband ISDN and Asynchronous Transfer Mode (ATM)", *IEEE Communications*, 27(9):17–24, September 1989.

[OHTA 91] H. Ohta and T. Kitami, "A Technique to Detect and Compensate Consecutive Cell Loss in ATM Networks", *Proc. IEEE INFOCOM '91*, pp. 781–790, Bal Harbour, FL, April 1991.

[SHAC 90] N. Shacham and P. McKenny, "Packet Recovery in High-Speed Networks using Coding", *Proc. INFOCOM 90*, pp. 124–131, San Francisco, CA, June 1990.

[ZHAN 91] L. Zhang and K. W. Sarkies, "Modelling of a Virtual Path and its Applications for Forward Error Recovery Coding Schemes in ATM Networks", *Proc. SICON '91*, Singapore, September 1991.

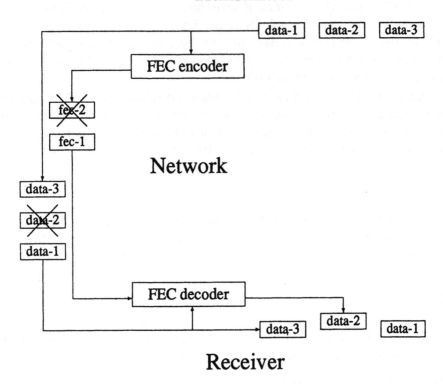

Figure 1: Operation of FEC.

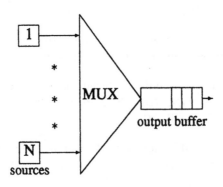

Figure 2: Model of the output-buffered multiplexer.

401

Cumulative Distribution Function of Percentage of Cells Lost per Block

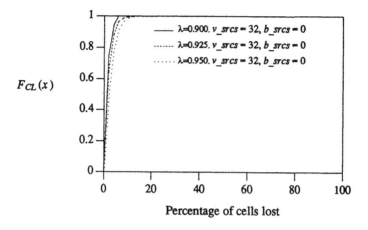

Figure 3: $F_{CL}(x)$ for scenario V1.

Cumulative Distribution Function of Percentage of Cells Lost per Block

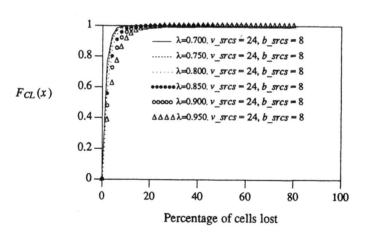

Figure 4: $F_{CL}(x)$ for the video sources in scenario VB24-8.

Block loss

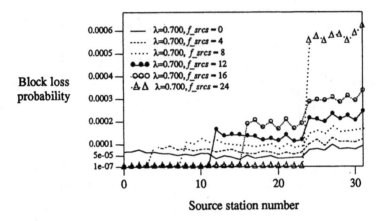

Figure 5: $F_{CL}(x)$ for scenario VB-FEC-1.

Gain or Penalty

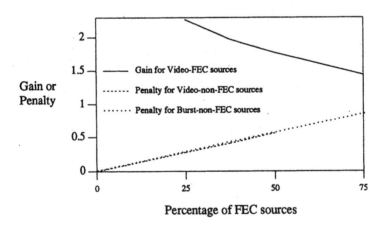

Figure 6: Gain and Penalty for scenario VB-FEC-1.

Session XII: Workshop Wrap-Up

Chair: Ralf Guido Herrtwich, IBM European Networking Center

While Session XI already reviewed many of the areas covered during the workshop, the final session further summarized what was accomplished during the two days in Heidelberg. Doug Shepherd of Lancaster University presented an excellent (and humorous) workshop conclusion from a very personal perspective.

Doug started by coming back to the last topic of the open session. He said that one of the most surprising results of the workshop for him was that the community more or less agreed on the need for new workstation technology to better support multimedia. While no talk really dealt with this issue, the discussion repeatedly came back to the topic of bus-based vs. switch-based systems, where some even want to integrate an ATM switch in the workstation to avoid message transformation from local to remote communication. The notion of "desktop networking" was used to describe this scenario.

More than one year after the Berkeley Workshop, Doug now sees agreement that operating systems need to be modified to handle multimedia. Doug was pleased not to see anyone voting for UNIX to support multimedia, but he sees that many people will modify UNIX for multimedia processing and that then the marketing people will step in and tell everybody "that they always said that UNIX would be suitable for multimedia." Real-time mechanisms need to find their way into a general-purpose operating system − they can no longer just be used in a dedicated environment. Abstractions that were developed in connection with modern, distributed operating systems such as Mach, Chorus, and Amoeba seem to be particularly useful for multimedia. Threads are one example: They allow to program applications that do not work with the traditional event-feedback loop common to handling discrete media. While a lot of abstractions were presented during the workshop for high-level programming with multimedia, it is yet a research issue how to map these abstractions to the operating system primitives.

During the workshop, many different proposals for multimedia communication protocols were made. Doug stressed the importance of standardization which was obviously not an attractive topic for most of the workshop participants. Yet, standards for multimedia communication will evolve anyway so it is the researchers' task to influence this process. Doug urged the participants to work together towards a common protocol for audio and video transport − though he was not sure whether a single protocol would suffice or whether different protocols for different purposes would be needed. In any case, he wants to have one protocol per problem, not "many protocols all doing the same, which is the current situation." Many schemes (e.g., for jitter control) are variations of methods already presented in the literature − yet, reference to them was not always made during the workshop. At the end, Doug asked for more recognition of work by others and more cooperation among people in the community.

Lecture Notes in Computer Science

For information about Vols. 1–529
please contact your bookseller or Springer-Verlag